MW ML 120
ML CTC
96000262 WCFS

# AIR POLLUTION
## *and*
# LUNG DISEASE
## *in*
# ADULTS

**Edited by**
**Philip Witorsch, M.D.**
**Samuel V. Spagnolo, M.D.**

**CRC Press**
**Boca Raton   Ann Arbor   London   Tokyo**

**Library of Congress Cataloging-in-Publication Data**

Air pollution and lung disease in adults / edited by Philip Witorsch and
    Samuel V. Spagnolo.
       p.   cm.
    Includes bibliographical references and index.
    ISBN 0-8493-0181-5
    1. Lungs—Diseases—Environmental aspects.  2. Air—Pollution—
Health aspects.  3. Lungs—Dust Diseases.  I. Witorsch, Philip.  II. Spagnolo, Samuel V.
RC756.A34  1994
616.2′4—dc20

                                                        94-5086
                                                            CIP

© 1994 by CRC Press, Inc.

No claim to original U.S. Government works
International Standard Book Number 0-8493-0181-5
Library of Congress Card Number 94-5086
Printed in the United States of America  1  2  3  4  5  6  7  8  9  0
Printed on acid-free paper

# PREFACE

The purpose of this book is to provide a general approach to problems involving air pollutants and respiratory disorders in adults. The book is directed not only at physicians but also at toxicologists, environmental scientists, other biological scientists, industrial hygienists, other health professionals, and even interested nonscientists, such as attorneys, regulators, and administrators. Consistent with our multidisciplinary target audience, we have attempted to include diversity among our authors, whose areas of expertise and interest include pulmonary medicine, toxicology, allergy-immunology, pathology, environmental engineering, and industrial hygiene. Also consistent with this broad target audience, we have attempted to be more general than specific and have emphasized general approaches applicable to many problems, with only a few of the chapters addressing specific conditions. The specific entities addressed have been selected to illustrate the general approaches and because of their particular importance or timeliness. We have made no attempt to be comprehensive in our coverage of the area. There are many excellent comprehensive works available, and one more would add little to the medical literature. It is our hope that our readers will come away with a broad understanding of approaches to problems of air pollution and lung disease, rather than specific detailed information that will likely be out of date in a relatively short time.

Thus, we begin with a chapter on exposure and environmental characterization (written by an environmental engineer), followed by a general discussion of direct and indirect injury to the respiratory tract (authored by a respiratory toxicologist). There then follows a discussion of carcinogenesis and lung cancer by an oncologist and a chapter addressing the pathology of environmental lung disease by a pulmonary pathologist. The next chapter deals with the general area of causality assessment and causal inference and toxicology, written by two toxicologists and a pulmonary-environmental-occupational physician, followed by a general discussion of the approach to clinical evaluation of the individual patient, diagnosis, and differential diagnosis. The next chapter remains general, as two pulmonary internists address the role of spirometry and cardiopulmonary exercise testing in impairment evaluation. The remaining three chapters are the only ones that deal with specific entities that we considered to be important or timely enough to address or that served well to illustrate the general principles and approaches. These three chapters, respectively, address immunologic mechanisms, including asthma, hypersensitivity pneumonitis, and related disorders (co-authored by two allergist-immunologists); pneumoconioses, chronic and interstitial pulmonary fibrosis, and bronchiolitis (co-authored by three pulmonary internists); and three conditions with an uncertain relationship to air pollution, i.e., sick building syndrome, multiple chemical sensitivities, and chronic fatigue syndrome.

Whether we have accomplished our goals of reaching a broad-based audience and of emphasizing general principles as opposed to specifics will be determined by you, the reader. We are grateful to the contributing authors for their most scholarly endeavors and to the publishers for their infinite patience with us. It is our hope that this book will prove of some use to all who have to address the problem of air pollution and lung disease in adults from another perspective.

Philip Witorsch, M.D.<br>
Samuel V. Spagnolo, M.D.

## THE EDITORS

*Philip Witorsch, M.D.*
Clinical Professor of Medicine
Adjunct Professor of Physiology
The George Washington University
    School of Medicine
        -- and --
Adjunct Professor of Pharmacology
Georgetown University School of
    Medicine
Washington, DC

*Samuel V. Spagnolo, M.D.*
Professor of Medicine
The George Washington University
    School of Medicine
Washington, DC

## THE CONTRIBUTORS

*Nancy J. Balter, Ph.D.,*
Associate Professor of Pharmacology
Georgetown University School of
    Medicine
Washington, DC

*Richard C. Bernstein, M.D.*
Fellow in Pulmonary Diseases
The George Washington University
    School of Medicine
Washington, DC

*Salvatore R. DiNardi, Ph.D.*
Associate Professor of Public Health
University of Massachusetts
Amherst, MA

*Jordan N. Fink, M.D.*
Professor of Medicine
Chief, Allergy/Immunology Section
Medical College of Wisconsin
Milwaukee, WI

*Donald E. Gardner, Ph.D.,*
Vice President and Chief Scientist
ManTech Environmental Technology, Inc.
Research Triangle Park, NC

*David W. Kamp, M.D.*
Associate Professor of Medicine
Chief, Pulmonary Section VALMC
Director, MICU VALMC
Northwestern University School of
    Medicine
Chicago, IL

*Talmadge E. King, Jr., M.D.*
Professor of Medicine
University of Colorado School of
    Medicine
        -- and --
Executive Vice President for Clinical
    Affairs
National Jewish Center for Immunology
    and Respiratory Medicine
Denver, CO

*Steven H. Krasnow, M.D.*
Associate Professor of Medicine
The George Washington University
    School of Medicine
        -- and --
Chief, Oncology Section
Veterans Affairs Medical Center
Washington, DC

Gail M. McNutt, M.D.
Attending Physician
Marshfield Clinic Lakeland Center
Minocqua, WI

Ralph J. Panos, M.D.
Associate Professor of Medicine
Director, Interstitial Lung Disease
Northwestern University School of
    Medicine
Chicago, IL

Arnold M. Schwartz, M.D., Ph.D.
Professor of Pathology
The George Washington University
    School of Medicine
Washington, DC

Sorell L. Schwartz, Ph.D.
Professor of Pharmacology
Director, Toxicology and Applied
    Pharmacokinetics
Georgetown University School of
    Medicine
Washington, DC

Samuel V. Spagnolo, M.D.
Professor of Medicine
The George Washington University
    School of Medicine
Washington, DC

Philip Witorsch, M.D.
Clinical Professor of Medicine
Adjunct Professor of Physiology
The George Washington University
    School of Medicine
    -- and --
Adjunct Professor of Pharmacology
Georgetown University School of
    Medicine
Washington, DC

# AIR POLLUTION
## *and*
# LUNG DISEASE
## *in*
# ADULTS

# TABLE OF CONTENTS

# Chapter 1

# Exposure and Environmental Characterization

*Salvatore R. DiNardi, Ph.D.*

## CONTENTS

## INTRODUCTION

The U.S. Environmental Protection Agency (EPA), under the direction of the Clean Air Act, has established National Ambient Air Quality Standards (40 CFR. part 50, 4–12) for six significant air pollutants: total suspended particles, sulfur dioxide, carbon monoxide, lead, ozone, and nitrogen dioxide. Several of these pollutants are also found in indoor air. The existing standards now apply to ambient or outdoor air, but indoor environments possess their own unique family of pollutants derived from various indoor activities, construction materials, and ambient air. In addition, energy conserving ventilation systems and "tight" building (low infiltration) construction methods have contributed to elevated indoor pollutant concentrations. Moreover, studies of the patterns of human activity in developed countries have revealed that people on average spend more than 23 hours per day in indoor environments (home, work, transit, etc.).[1,2] The indoor air environment in residential and office settings can present a potent route of exposure to air pollutants. These contaminants include carbon monoxide, formaldehyde, nitrogen oxide, carbon dioxide, volatile organic compounds, asbestos, radon, bioaerosols, and environmental tobacco smoke. Many of the sources involved are common to residences and offices.

This chapter briefly characterizes air pollutants found in both ambient and indoor air. It addresses measurement of pollution and methods for modeling the distribution of pollutants over broad areas as well as in enclosed areas. However, it does not deal at length with the special situations of occupational exposure in specific industrial settings.

## CHARACTERIZING AIR POLLUTANTS

Air pollutants can be classified as chemical, physical, or microbiological. Among the chemical pollutants are gases, vapors, and fumes. Physical pollutants are particles suspended in the air, characterized as both total and respirable particulate matter, and may include dusts of both organic and inorganic origin. Microbiological pollutants include viruses, bacteria, molds, spores, fungi, and protozoa.

The physical state of air pollutants dictates how they can be sampled and analyzed. These methods are developed based on the chemical and physical properties of the contaminants. Polarity, dipole movement, acidity, and alkalinity are the chemical properties used to remove contaminants from air. Particle size and aerosol behavior are physical properties used to remove solid contaminants from air.

**GAS** is any material in the gaseous state at 25° C and 760 mm Hg. Normally a formless fluid, gas expands to fill the spaces or enclosure. Gases can be changed to the liquid or solid state by the combined effect of increased pressure and decreased temperature. Examples of concern are welding gases, exhaust gases from internal combustion machines, and waste gases from refining or sewage, such as hydrogen sulfide, waste anesthesia gases, ammonia, hydrogen, helium, argon, and nitrogen. Size ranges are usually <0.0005 micrometers ($\mu$m).

**VAPOR** is a gaseous phase of a substance that is ordinarily liquid or solid at 25° C and 760 mm Hg. Evaporation is the process by which a liquid changes to a vapor state and mixes with the surrounding atmosphere. Solvents with low boiling points will volatilize readily. Examples of substances that emit vapors are trichloroethylene, methylene chloride, and mercury. Size ranges are usually <0.005 $\mu$m.

**DUST** denotes solid particles capable of temporary suspension in air or other gases. Produced from larger masses, the application of physical forces (e.g., handling, crushing, grinding, rapid impact) is usually implied. Typical dusts are rock, ore, metal, coal, wood, and grain as well as animal dander, insect parts, and fungal spores found in residential settings. Size ranges are between 0.1 to 30.0 $\mu$m. Dust particles occur in sizes up to 300 to 400 $\mu$m, but those above 20 to 30 $\mu$m usually do not remain airborne.

**FUME** denotes solid particles generated by condensation from the gaseous state, generally after volatilization (evaporation) from melted substances and often accompanied by a chemical reaction, such as oxidation. Examples are zinc oxide from welding on galvanized metal, lead oxide from soldering, and copper oxide from smelting, as well as fumes from cleaning agents and office products. Size ranges are between 0.001 and 1.0 $\mu$m.

**MIST** is a dispersion of suspended liquid particles, many large enough to be individually visible to the unaided eye. Particles are generated by condensation from the gaseous to the liquid state or by breaking up into a liquid in a dispersed state, such as splashing, spraying, foaming, or atomization. Mist forms when a finely divided liquid is suspended in the atmosphere. Examples are the oil mist produced during cutting and grinding operations, acid mist from electroplating, acid or alkali mist from pickling operations, and spray mist from painting operations. Size ranges are between 0.01 to 10.0 μm.

**SMOG** is a combination of smoke and fog. The term is usually applied to extensive atmospheric contamination by aerosols arising from a combination of natural and anthropogenic sources.

**SMOKE** is carbon or soot particles <1.0 μm in size. These small, gas phase particles created by incomplete combustion consist predominantly of carbon and other combustible materials. Smoke generally contains droplets and dry particles. Size ranges are usually between 0.01 to 1.0 μm.

## METEOROLOGICAL CONSIDERATIONS IN AMBIENT AIR

Weather patterns play a key role in the movement of outdoor air pollutants from their sources to the individuals who are exposed to them. The behavior of the atmosphere at a given time inevitably plays a role in the pollution levels. If the air moves slowly past a pollution source—a smokestack, for example—then the downwind plume will carry high concentrations of the pollutant. Conversely, if air flows quickly past a source, the pollutant will be dispersed in a larger volume of air. This movement of air is characterized as "ventilation." Ventilation occurs at very different rates in the atmosphere under different meteorological conditions. Of primary concern in looking at ambient air pollution, large scale ventilation is the critical factor in the buildup of air pollution in a given locale.

In a meteorological low-pressure system, air flows toward the center in an attempt to bring about equilibrium. This convergence causes updrafts in the center of the low. Although the winds are generally light very near the low-pressure system's center, winds farther away from the center but still associated with the system tend to be moderate, resulting in overall increased rates of ventilation.[3] Low-pressure systems typically cover fairly small areas and tend to be transient, seldom remaining in one place for very long. They are also often accompanied by clouds, which can cause precipitation that tends to remove some pollutants from the atmosphere through sedimentation. Cloudy skies also minimize the variation in mixing between levels of the atmosphere (atmospheric stability) from day to night. Moderate horizontal wind speeds and upward vertical movement associated with low-pressure systems generally result in good ventilation—that is, a significant volume of air moving past a given location. Typically, air pollution is low in an area where a low-pressure system is present.[3]

High-pressure systems, on the other hand, tend to be characterized by poor ventilation resulting in increased pollution buildup. Winds flow outward from the high-pressure center, pulling down air from higher in the atmosphere, where long-distance pollution mixes in, to compensate for the horizontally expelled volume. This sinking air causes a subsidence inversion. This subsidence usually results in clear skies, allowing a maximum flow of radiated heat, inward during the day and outward at night. As a result, the atmosphere tends to be unstable by day but stable at night, often causing radiation inversions (the trapping of warm air at the surface by colder air aloft). High-pressure systems tend to extend over large areas, and although they are transient, they move slowly and carry only light winds. Ventilation in the vicinity of a high-pressure system is typically much less than ventilation around a low-pressure system.[3]

In many cases, ventilation becomes extremely poor in the weak pressure gradient near the center of a high-pressure system. If the high has a warm core, very little air moves near the center, causing stagnation. Winds are very light, and skies are free of clouds, contributing to the formation of surface base radiation inversions at night.[3]

## HVAC ENGINEERING CONSIDERATIONS IN INDOOR AIR

Adequate dilution ventilation within a structure is of concern in handling potential problems of indoor air pollution. Ventilation is the primary method to maintain adequate indoor air quality in the home as well as in the nontraditional workplace (schools, offices, shopping malls, etc.). The assumption is that indoor air contains the typical contaminants—e.g., particles, odor, low levels of volatile organic compounds (VOCs), formaldehyde, bioaerosols and the accumulation of carbon dioxide—that result from the collection of people, plants, pets, human activities, furnishings, etc., within a space. After eliminating or controlling sources, the usual method of maintaining indoor air quality is to dilute indoor air pollutants with outdoor air. Of course, the outdoor air quality must be acceptable, that is, it must meet or exceed the EPA standards for ambient air quality.

Carbon dioxide ($CO_2$), while generally not a significant air pollutant in its own right, can serve as a surrogate for measuring general indoor air quality. As people breathe, they exhale $CO_2$. The accumulation of $CO_2$ is not sufficient to cause an adverse health effect. More specifically, it does not approach the Occupational Safety and Health Administration (OSHA) permissible exposure level (PEL). But indoor $CO_2$ concentration can be measured easily.

The concentration of indoor $CO_2$ is an indicator of the amount of outdoor air (dilution air) introduced to the space. The standard promulgated by the American Society of Heating, Refrigeration, and Air-Conditioning Engineers, ASHRAE 62-1989, recommends that $CO_2$ levels be maintained below 1000 ppm. In the standard, there is another requirement that the outdoor air be delivered to the occupied space. Many complaints about indoor air quality are reported, however, when the $CO_2$ concentration is between 750 and 1000 ppm. This may have to do with the quality of the mixing that the air undergoes on its way through the space. These $CO_2$ levels often result in complaints of "stuffiness." Indoor air that is maintained at <1000 ppm $CO_2$ may contain other pollutants. These indoor pollutants and other indoor air quality problems might exist even when indoor $CO_2$ levels are below 1000 ppm.

The outdoor air that is introduced to the indoor environment mixes with the indoor air contaminants and thus improves the overall indoor air quality. As the indoor $CO_2$ concentration is diluted, all other indoor air pollutants (particles, gases, and vapors) are also reduced.

Figure 1 is a schematic representation of a typical ventilation system with partial recirculation. It is sometimes difficult or impossible to measure the outdoor air volume flowrates needed to dilute indoor $CO_2$ in an actual operating air handler. ASHRAE[66] recommends a means to estimate outdoor air volumes (dilution ventilation) that depends on measurements of outdoor, indoor, and return air temperature. There are two techniques based on $CO_2$ measurements using a nondispersive infrared analyzer to estimate the quantity of outdoor air moving into an indoor space. One method depends on calculating the recirculation fraction (RF),[67] which is the ratio of the quantity of air flowing through the mixed air chamber or recirculated to the supply air volume flowing into the room. The other estimates the outdoor air volume moving into an indoor space using the method described by the EPA.[68] This approach estimates the fraction (or percent) of outdoor air that enters a space. The outdoor air fraction (OAF) is the ratio of the flowrate of outside air to the flowrate of the supply

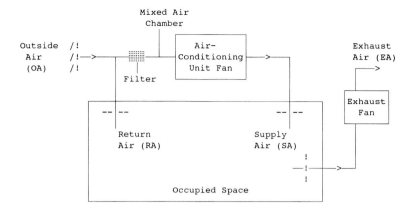

C_i = carbon dioxide concentration in the "i"th location.
Q_i = air volume flowrate in the "i"th location.

| CARBON DIOXIDE CONCENTRATION | AIR VOLUME FLOWRATE |
|---|---|
| $C_o$ = Outside air (OA). | $Q_o$ = Outside air (OA) into the space. |
| $C_m$ = Mixed air chamber (MA). | $Q_m$ = Recirculation air volume flowrate. |
| $C_s$ = Supply air (SA). | $Q_s$ = Supply air into the room. |
| $C_r$ = Return air (RA). | $Q_r$ = Return air from the room. |
| $C_e$ = Exhaust air (EA). | $Q_e$ = Exhaust air out of building. |
| $C_{room}$ = Concentration in the center of room. | |

**Figure 1** A typical heating ventilating air conditioning system with partial air recirculation.

air into a space. The outdoor air fraction (OAF or OA%) method varies slightly from the recirculation fraction (RF) method.

## MEASURING CONTAMINANTS

Carbon monoxide (CO) measurement is based on nondispersive infrared spectrophotometry (NDIR).[3] The measuring device contains a source of infrared radiation, a sealed reference cell, and a detection chamber that is flushed with ambient air. Since CO absorbs infrared radiation, the infrared radiation reaching the detector cell is proportional to the CO present in the sample. Water vapor interferes with this measurement technique, so CO detectors contain a moisture trap.

Ozone is measured through chemiluminescence.[4] When ozone and ethylene react chemically, they release light at levels directly proportional to the concentration of ozone.

Nitrogen oxides are also measured by chemiluminescence, but by a two-part measurement of nitric oxide (NO) in reaction with ozone and nitrogen dioxide ($NO_2$) in reaction with ozone.[4]

There are several methods to measure sulfur dioxide ($SO_2$) concentrations in air, but the EPA reference method[5] specifies a technique that involves collecting $SO_2$ by bubbling the air sample through an aqueous solution of potassium tetrachloromercury (TCM). The absorbed $SO_2$ forms a stable complex with TCM. Dye and formaldehyde are added to form red-purple pararosaniline methylsulfonic acid. Optical absorption at 548 nm is linearly proportional to the concentration of $SO_2$.

Particulate pollutants are measured by removing particles from air through filtration or impaction and weighing either the filter or impaction surfaces before and after removal. Multistage impactors, where each stage removes particles of a different size range, are used to characterize the distribution of particle size. Electrical mobility of particles and their light-scattering properties may also be used to determine particle size distribution. Chemical constituents are determined from the deposited particles through various chemical and spectrophotometric techniques used in conventional chemical qualitative analysis. Measurement of radioactive particles (radon and its progeny) is usually accomplished through filtration or electrostatic precipitation followed by radioactivity measurement with a scintillation counter. Radon and radon progeny in groundwater are usually measured through liquid scintillation counting.

A number of volatile organic compounds are generally measured through gas chromatography combined with mass spectroscopy; if the concentrations are sufficiently high, infrared spectroscopy can be used.

## MODELING METHODS

Modeling is an exercise that enables the simulation of a system or process using mathematical tools. The mathematical tools are an equation or set of equations developed to predict changes in a system as various parameters in that system change. Modeling is used to calculate the concentration of power plant effluents downwind and in the surrounding area. These are known as atmospheric dispersion models; they are large (many mathematical terms), and some require high-speed personal computers with large memories to handle dozens of input variables. There are mathematical techniques (pharmacokinetic models) used to predict the fate of drugs in patients and solvent exposures among workers. Mathematical models are used to predict the environmental fate of chlorofluorocarbons (CFCs) in the ozonosphere and the impact of greenhouse gases on global warming.

### OUTDOOR AIR MODELING

The modeling of air pollution behavior on a broad scale is a complex process, and a wide variety of models are available. They are generally applied in locating new industrial facilities to gauge the environmental impact of air pollution emissions. The models are usually distinguished by type of source, pollutant, transformation and removal mechanisms, distance of transport, characteristics of the terrain, building shape, and time. In the simplest form, a model requires information on source or sources (including emission rate) and meteorological data such as wind velocity, direction, and turbulence. The model then predicts, through mathematical equations, how the pollutant is transported and dispersed, and its chemical and physical transformations (including removal).[6]

### INDOOR AIR MODELING

Calculating contaminant concentration permits indoor air quality professionals to estimate possible exposures so that appropriate air monitoring can be developed. A carbon dioxide model is useful in diagnosing building-related occupant complaints (sometimes called "tight building syndrome").

Modeling the concentration of gases, vapors, and $CO_2$ in indoor air requires a cursory understanding of the similarities and differences between the two types of ventilation used to control contaminants in the indoor environment. Ventilation is the control of the working environment by removing contaminants at the source using local exhaust ventilation or dilution of the contaminant with uncontaminated air. When considering ventilation as a solution to control contaminants, the designer must clearly understand that differences exist between local exhaust and dilution ventilation. In a traditional workplace, the designer is obligated to choose a system that protects the worker's breathing zone. A frequent error made by ventilation designers is to use dilution ventilation as the technique of choice to control hazardous exposures.[69]

**Dilution ventilation** improves air quality in a space by diluting the contaminants in that space with uncontaminated air, thus reducing the contaminant concentration below a given level. **Local exhaust ventilation** removes a contaminant at the source, before it reaches the workers' breathing zone.

Dilution ventilation serves many useful purposes in nontraditional workplaces, (e.g., commercial and public buildings, offices, and schools). It is used primarily to control nuisances, odors, trace quantities of volatile organic compounds, tobacco combustion products, and $CO_2$, after source control, elimination, substitution, and isolation are implemented. ASHRAE recommends ventilation rates to control or avoid the accumulation of indoor air pollutants and carbon dioxide, in Standard 62-1989.[70] The assumption is that if carbon dioxide is kept within the limits proscribed by ASHRAE, the dilution ventilation is also controlling many other possible indoor air pollutants.

Dilution ventilation is almost never used to protect the workers' breathing zone. ASHRAE Standard 62-1989 has limited usefulness in traditional workplaces (plants, machine shops, assembly lines, foundries, etc). Dilution ventilation is not a substitute for correctly designed local industrial exhaust ventilation in the traditional workplace. Dilution ventilation may not assure worker protection and typically handles larger air volumes and may be more costly to operate then local exhaust ventilation systems. The control of occupational exposures using dilution ventilation appears to be simple and straightforward to the uninformed. The volume of air needed to dilute contaminants to a predetermined level may be determined by mathematical modeling using well-established ventilation equations.

Dilution ventilation, however, often requires huge volumes of air if a pollution source is particularly strong or unusually toxic. Local exhaust ventilation is almost always the preferable choice for maintaining air quality in industrial settings because it removes the potential for exposure by removing the pollutant before it enters workers' breathing space. In many cases it will be supplemented by dilution ventilation.

## OUTDOOR AIR POLLUTANTS: PROPERTIES, SOURCES, EFFECTS

Pollutants in ambient air have both natural and anthropogenic sources, but far and away the largest point sources are natural processes, such as volcanic eruptions. The greatest anthropogenic contributor to air pollution is undoubtedly combustion processes. The EPA has established National Ambient Air Quality Standards (NAAQS) for several pollutants recognized as posing health risks.[7]

### SUSPENDED PARTICLES

Suspended particles in ambient air derive from a variety of sources, including natural processes. Volcanic eruptions and wind driven soil erosion both supply substantial particle loads to the atmosphere as dusts, fumes, and mists. But combustion processes, especially the burning of fossil fuels, supply a substantial quantity of partially burned hydrocarbons to the atmosphere. Automobile pollution is frequently cited as a major source of suspended particles, but other major sources include fly ash from industrial combustion processes, such as smelting and asphalt drying, as well as from burning fossil fuels to generate electricity and heat buildings.

The constituents of suspended particles are extremely complex; they include carbon particles as well as metal oxides and metal particles. Health effects vary a great deal for individuals, but exposure to high levels of particulate matter is associated with susceptibility to respiratory infection. The NAAQS for Total Suspended Particles is 75 $\mu g/m^3$ (annual average) and 260 $\mu g/m^3$ (24-h average). Beginning at approximately 260 $\mu g/m^3$, health effects include mild aggravation of respiratory symptoms in susceptible persons and irritation in the general population. At 375 $\mu g/m^3$, persons with heart or lung disease show decreased tolerance of exercise. At levels above 875 $\mu g/m^3$, premature death of ill and elderly persons can be expected, and the general population will experience symptoms that affect normal activity.[7]

## SULFUR DIOXIDE

The primary sources of sulfur dioxide ($SO_2$) in the atmosphere are fossil fuel combustion and volcanic eruptions, with human activity accounting for approximately 90% of the $SO_2$ pollution. $SO_2$ is a particularly noxious air pollutant, in that it forms an irritating acid on contact with moist mucous membranes.

The NAAQS for $SO_2$ exposure is 0.03 ppm annually or 0.14 ppm in a 24-h period, although these standards may be lowered. A level of 0.07 ppm in the presence of large quantities of particulate matter appears to aggravate chronic respiratory disease in children. The odor threshold for sulfur dioxide is 0.5 ppm, a level that produces increased airway resistance in exercising asthmatics. A level of 8 ppm will produce throat irritation in healthy adults, whereas a 10-min exposure to 10 ppm may produce bronchospasm.[7]

## CARBON MONOXIDE

Carbon monoxide (CO) is added to the atmosphere primarily by incomplete combustion, the leading source being automobile exhaust. Other fossil fuel combustion processes also contribute to CO pollution, as do natural events such as forest fires. NAAQS primary standard is an annual 1-h maximum of 35 ppm, not to be exceeded more than once per year, and an annual 8-h maximum of 9 ppm, also not to be exceeded more than once per year.[7]

CO outcompetes oxygen for binding sites on hemoglobin and thereby interferes with the body's oxygen transport system. Health effects of CO at low levels are controversial. Outside of industrial exposure, CO levels in ambient air are less of a health problem than those found in indoor air. (See next section for a discussion of health effects of CO as an indoor air pollutant.)

## NITROGEN OXIDES

U.S. NAAQS are set for nitrogen dioxide ($NO_2$), since nitric oxide (NO) tends to oxidize rapidly to that state. NAAQS sets exposure at a maximum level of 0.05 ppm.[7] The primary source is fossil fuel combustion. While nitrogen dioxide is a major constituent of smog, unhealthy concentrations are found more frequently in indoor air than in ambient air. (See next section for discussion of health effects of nitrogen oxides as an indoor air pollutant.)

## OZONE

Ozone ($O_3$), the highly reactive form of oxygen, is much more prevalent as an ambient air pollutant than as an indoor air pollutant, since the exposed surfaces of a building provide many points of reaction. Surfaces serve as a sink to reduce $O_3$ concentrations in outdoor air that enters a building. The NAAQS for $O_3$ is an annual maximum, not to be exceeded once per year, of a 1-h exposure at 0.12 ppm.[7]

A major constituent of urban smog, $O_3$ is formed by the photochemical reaction of hydrocarbons and nitrogen oxides in the presence of sunlight. Elevated levels of $O_3$ produce short-term irritation of the respiratory system in healthy adults and can cause long-term lung damage, especially in children.

## Lead

Lead is a special case among the suspended particle pollution because even low concentrations are deleterious to growing nerve tissue and produce anemia. Thus, the movement toward nonleaded gasoline was among the first actions of the EPA. The NAAQS is 1.5 µg $Pb/m^3$ air, based on the assumption that this level contributes only 0.5 µg Pb/dl of blood.[7] Environmental lead is a major concern among public health officials. The U.S. Congress passed the Residential Lead Based Paint Hazard Reduction Act of 1992 (also called Title X).[73] Among other aspects of lead, Title X is concerned with lead paint hazards, lead-contaminated dust, and lead-contaminated soil. Lead-contaminated media can be a substantial local ambient air pollution problem as well as an indoor air pollutant.

## INDOOR AIR POLLUTANTS: PROPERTIES, SOURCES, EFFECTS

### FORMALDEHYDE

Formaldehyde (HCHO) is a colorless, volatile gas with a pungent odor and high water solubility. Formaldehyde in the home or office can be derived from a number of sources, such as some particle board and plywood products, wall coverings, some carpet backings, adhesives, cigarette smoke, various synthetic resins, paper products, and combustion processes. However, public awareness of the potential health hazards of formaldehyde arose from the use of urea formaldehyde foam insulation (UFFI) in as many as 200,000 homes in the U.S.[8] In the office environment, furniture, furnishings, and carbonless copy paper are additional potential sources.

The off-gassing of formaldehyde from sources inside the building leads to elevated concentration in the indoor environment. Measurements of the concentration of formaldehyde inside homes insulated with UFFI were significantly elevated over homes without UFFI (0.1 to 0.8 ppm as compared to 0.03 to 0.07).[9] The extensive use of particle board in the construction of mobile homes tends to increase the concentration of formaldehyde when compared to conventional homes. One study found that 50% of mobile homes examined had a formaldehyde level >0.47 ppm.[10] Office buildings have also been shown to have elevated formaldehyde concentrations in the range of 0.1 to 0.3 ppm.[11]

There are no reported acute health effects associated with formaldehyde concentration <0.05 ppm.[12] However, the high degree of water solubility leads to irritation of the upper respiratory system and eyes at 0.05 to 0.1 ppm. Lower respiratory effects are not present until concentrations exceed 5.0 ppm.[12] All of these levels are subject to a large degree of interindividual variability. Many questions remain concerning the chronic health effects of formaldehyde exposure. Recent studies with laboratory rats have demonstrated a positive association between nasal cancer and formaldehyde exposure.[13,14] Formaldehyde has been investigated as a risk factor for other cancers, such as lung, buccal-pharyngeal, brain, prostate, skin, kidney, and digestive system. Formaldehyde has been associated with chronic decrease of lung function and asthma at concentrations that might be found in a home insulated with UFFI.[12] Formaldehyde exposure in the home and in the office has been linked with a prevalence of neuro-psychologic symptoms, such as headache, memory lapse, fatigue, and insomnia.[12]

## OXIDES OF NITROGEN

Oxides of nitrogen ($NO_x$) are gaseous compounds formed from atmospheric nitrogen and oxygen during combustion at high temperatures. Although NO is the primary oxide of nitrogen-based pollutant that results from combustion, it rapidly oxidizes to $NO_2$. Other possible byproducts that may be present in trace amounts include nitrous and nitric acid ($HNO_2$ and $HNO_3$), nitrite anions, and nitrate salts. $NO_2$ is the most significant indoor pollutant of the nitrogen oxide family.

The indoor environment may be contaminated with $NO_2$ through the use of un-vented gas cooking appliances, wood stoves, kerosene space heaters, and other heaters that are fueled with natural gas and propane. About 50% of the homes in the U.S. use gas ranges for the preparation of meals.[15] Sources from outside the home may also contribute to the indoor levels.

Indoor concentrations of $NO_2$ often exceed the level outside the building. The ambient $NO_2$ concentration ranges from 3.5 to 13.5 ppb. The unvented daily use of a gas cooking range adds 25 ppb to the background concentration of $NO_2$ in the home,[16] yielding indoor concentrations in the 25 to 75 ppb range.[15] In the kitchen or the room where a gas heater is located, the values may peak at 200 to 400 ppb.[17] The levels of $NO_2$ in homes using electric cooking and heating devices are often lower than the ambient concentration.[16,18,19]

$NO_2$ can damage the lungs via its oxidant properties, and it also renders the lungs more susceptible to respiratory infection.[19,21] $NO_2$ has been shown to have deleterious effects on the lung's natural defense mechanisms, mucociliary clearance, alveolar macrophage, and the immune system.[18,20] Some studies have demonstrated that levels comparable to those measured in homes with gas appliances may increase the reactivity of the airways of some asthmatics.[19] The chronic effects of long-term, low-level exposure is currently being studied, but early results indicate that such exposure leads to a general deterioration of lung function.[20]

## CARBON MONOXIDE

CO is an odorless, colorless gas arising from incomplete combustion of carbonaceous fuels. The widespread use of carbon fuels for home heating and cooking makes CO a prevalent indoor contaminant. In the home, emissions from gas appliances and cigarette smoke and emissions from internal combustion engines operating near air inlets all contribute to indoor exposures. Gas ovens used for heating purposes (a common occurrence in low-income households) and kerosene space heaters are additional sources.

The indoor concentration of CO (typically between 0.5 and 5 ppm) usually exceeds the corresponding outdoor levels. The concentrations in homes while a gas range is in use range from 2 to 6 ppm, and occasional peaks have been measured at 12 ppm.[24] The use of gas ovens for heat may increase the interior CO level by 25 to 50 ppm.[25] Heavy cigarette smoking has the potential to contribute up to 10 ppm to the indoor environment.[26]

The ability of CO to outcompete oxygen for binding sites on hemoglobin accounts for the compound's severe interference with the oxygen transport system. Tissues with the highest oxygen demands, such as muscle, brain, and heart, are the first to suffer from CO poisoning (higher level exposures). The health effects of CO at low levels remain controversial, but there is some indication that such exposures may increase the risk of cardiac arrhythmias and myocardial infarctions.[27] Those who suffer from diseases that diminish oxygen transport (coronary artery or chronic obstructive pulmonary disease) are particularly vulnerable to the effects of low-level CO exposure.

## CARBON DIOXIDE

The HVAC Engineering Considerations section above characterized $CO_2$ as not normally considered an indoor pollutant. It is often employed as a surrogate for monitoring general indoor air quality and ventilation efficiency. This colorless, odorless gas is one of the major products of the complete combustion of carbon-based fuels. The primary source of $CO_2$ is the people and animals who occupy the interior space. Animal metabolism is essentially the combustion of glucose to produce $CO_2$ and water vapor, which are exhaled. Other sources include unvented kerosene and gas space heaters, tobacco smoke, and the outside air (although at much lower levels than found indoors).

The ambient concentration of $CO_2$ is generally 360 ppm. Indoor environments with an inadequate supply of fresh air may measure in the range of 1000 ppm with a normal amount of occupants, and crowded areas may exceed 3000 ppm. During the flight of the space shuttle Endeavor, $CO_2$ concentrations onboard rose to about 4000 ppm.

The adverse health effects associated with excess $CO_2$ arise from the displacement of oxygen from the air at high concentrations of $CO_2$ >5000 ppm. Symptoms of $CO_2$ exposure include headaches, dizziness, drowsiness, and shortness of breath. The concentration of $CO_2$ necessary to invoke these effects varies a great deal among individuals; most people, however, do not notice effects until the concentration of $CO_2$ exceeds 800 to 1000 ppm.

## VOLATILE ORGANIC COMPOUNDS

Volatile organic compounds (VOC) comprise a diverse group of organic chemical molecules that have the ability to volatilize significantly at room temperature. VOC include aliphatic hydrocarbons, halogenated hydrocarbons, and aromatic compounds. Hundreds of VOC have been detected in indoor environments.[29] Prevalent VOC include benzene, toluene, styrene, trichloroethylene, dichlorobenzene, methylene chloride, and chloroform. Due to the number of possible VOC contaminants, the list of sources of VOC in residential and office indoor environments is seemingly endless. The more significant sources include paints, adhesives, cleansers, building materials, cigarette smoke, printed material, photocopiers, plasticizers, and combustion processes.

VOC concentrations are usually two to five times higher indoors than outdoors. Indoor concentrations vary from <1 ppb to >5 ppm, while ambient levels remain below 3.5 ppm. Some VOC concentrations remain relatively constant over time, and others display wide fluctuations.[29] Typical indoor concentrations of selective VOC are as follows: benzene (<7 ppb), toluene (<15 ppb), trichloroethylene (3.7 ppb), m,p xylene (3.4 ppb), m,p dichlorobenzene (0.7 ppb), chloroform (0.4 ppb), and styrene (0.5 ppb).[12] Each of these VOC can be elevated through many of the sources previously mentioned.

Each VOC is associated with its own set of potential adverse health effects. Most VOC are respiratory and eye irritants. Furthermore, some of the most commonly detected semivolatile organic compounds (SVOC), such as polycyclic aromatic hydrocarbons, are either established or suspected carcinogens. The role of VOC in building-related illnesses (BRI) has been examined.[30] High concentrations of VOC have been detected in buildings where BRI symptoms have been reported, which include mucous membrane and eye irritation, cough, fatigue, and headache.[31] However, no study has been able to implicate a particular VOC as an isolated cause of BRI. VOC are generally present in complex mixtures, and the health effects associated with such combinations are difficult to predict.

## ENVIRONMENTAL TOBACCO SMOKE

Environmental tobacco smoke (ETS) is derived from both the mainstream smoke (15%) that is inhaled by the smoker and the sidestream smoke (85%) that is released directly from the burning tip of the cigarette. Because of the lower temperature of the periphery of the burning cigarette, combustion producing sidestream smoke is less complete than the combustion yielding mainstream smoke. Therefore, sidestream smoke may have a higher concentration of many toxins, and mainstream smoke may have more $CO_2$ and water. Furthermore, the concentration of CO is 2.5 times higher in sidestream smoke, and the smaller particles generated by sidestream smoke are more likely to be deposited in the alveoli.[32] ETS is ubiquitous, and exposure at some level will occur in almost all indoor environments, whether or not smoke is generated in the vicinity.[34]

More than 3800 chemicals have been identified in tobacco smoke, of which 50 have been classified as carcinogens in either humans or animals.[34] Smoking in indoor environments has been shown to increase the concentrations of respirable particles, nicotine, polycyclic aromatic hydrocarbons, CO, acrolein, $NO_2$, and others. Smoking one pack a day contributes about 20 $\mu g/m^3$ to indoor particle concentrations.[35] Homes with smokers had cadmium, arsenic, and iron levels that exceeded outdoor concentration.[35] Smoking has also been shown to generate 50 mg of CO per cigarette.[26]

Many of the identified substances that comprise tobacco smoke are eye and respiratory irritants. The 1986 report to the Surgeon General concluded that exposure to passive smoke was associated with increased frequency of lower respiratory illness and chronic respiratory symptoms in children.[36,37] Some studies also suggest that parental smoking may be associated with increased risk of asthma[38] and middle ear disease.[39] The chronic effects of prolonged low-level exposure to ETS are controversial; however, some studies have reported increased lung cancer risks for non-smoking women who live with smokers,[40,41] while others have not substantiated that link. The theory that passive smoke can lead to cancer draws biological plausibility from the fact that no threshold dose for respiratory carcinogenesis has been observed.[12] One controversial risk assessment claims that 5000 lung cancer deaths annually in the U.S. can be attributed to ETS exposure.[42] Other recent reports have suggested that ETS possibly increases the risk for cancers in other organs than the lung and may also be associated with the development of cardiovascular disease.[12] The EPA now considers ETS a human carcinogen.[71] Some controversy on the EPA risk assessment continues to be voiced by the tobacco industry.[72]

## RADON

Radon is a naturally occurring, chemically inert radioactive substance that is formed in the decay of uranium, which is a ubiquitous component of soil and rock. Because it is an inert gas, radon has the ability to diffuse out of the ground and enter the atmosphere. Therefore, the predominant source of indoor radon is the subslab soil.[43,44] Radon diffuses through the building foundation into the basement and then distributes throughout the indoor space. Other sources of radon include ground water (when it is enclosed indoors, as in showers), stone-derived building materials, and natural gas.

Typical indoor radon concentrations for the U.S. range from 0.01 to 4 pCi/L.[46,47] It was estimated in 1984 that 1 to 3% of homes (more than 1 million) in this country have radon levels that exceed 8 pCi/L.[46,47] Some homes in uranium-rich regions have radon levels that exceed 20 pCi/L. Determination of tissue dose is not a factor of concentration alone but also depends on such factors as characteristics of the carrier aerosol, degree of radon progeny equilibrium, pattern of respiration, and location of particle deposition in the respiratory system. The decay of radon produces a series of short-lived isotopes known as radon progeny,[47,49] two of which, Po-214 and Po-218, emit alpha particles. When attached to small particles, radon progeny may be inhaled into the respiratory tract, where alpha decay is theorized to produce the cellular damage that leads to malignancies. This association between radon progeny exposure and lung cancer has been proven for highly exposed uranium miners.[49,50] The incidence of lung cancer has been found to correlate with exposure to radon in the home.[43,51,52] Of the estimated 135,000 lung cancer cases annually in the U.S., recent risk assessment models predict that 10,000 of them may be due to radon progeny exposure.[55,56,57] If these estimates are accurate, radon alone is a greater hazard than all other air contaminants combined.[56]

## ASBESTOS

The term ''asbestos'' refers to a variety of naturally occurring mineral silicates. This flexible fibrous material is incombustible, has tensile strength, and has many desirable thermal and electric insulating properties. Asbestos may be used in roofing and flooring materials, textiles, papers, filters, cement, panels, pipes, coating material, and thermal and acoustic insulation. Although many materials contain asbestos, the amount of fibers naturally released is small in comparison to the exposure that can potentially result from disruption of the material.[57] Material that has been applied through a spraying process may be more susceptible to fiber release.

A report issued by the EPA in 1988 reported that 733,000 public and commercial buildings contained asbestos. The concentration of airborne fibers is reported to range from 0 to over 100 fibers per cubic centimeter (f/cc), depending on the condition of the asbestos-containing material and the proximity of activity.[58] However, a study of 49 government buildings in 1989 concluded that while airborne asbestos levels in buildings containing asbestos are slightly elevated over the values of buildings that do not contain any asbestos, all asbestos levels were low enough as not to warrant concern.[57]

Asbestos exposure in the occupational setting has been demonstrated to cause four distinct diseases:[59] asbestosis (a nonmalignant scarring of the lung tissue), broncho-genic carcinoma (a malignancy of the lining of the air passages), mesothelioma (a diffused malignancy of the lining of the chest cavity), and possibly cancers of the stomach, colon, and rectum. Results from indoor asbestos concentration studies indicate that residents of a building constructed with asbestos-containing materials will not develop severe adverse health effects, such as asbestosis or cancer, for the exposures are too low.[59] This prediction does not include situations of remodeling or renovation where the asbestos-containing material is likely to be disturbed.

It is notable that of the asbestos used historically in the U.S., approximately 90% is chrysolite, which is considered to be less harmful to the body than some other forms of asbestos. Chrysolite is serpentine and collects in bunches; hence, it is easily cleared by the lung. Approximately 3% of the asbestos employed in the U.S. is amosite and crocidolite. These fibers have barbed surfaces and are not easily cleared by the lung; exposure to them appears to be associated with mesothelioma.[60]

## BIOLOGICAL AGENTS (BIOAEROSOLS)

Many biological agents qualify as potential indoor air contaminants because they can induce adverse health effects upon exposure. The most commonly found components of bioaerosols include viruses, bacteria, actinomycetes, fungal spores, algae, amebae, arthropod fragments and excrement, and animal and human dander.[61,62] The critical requirement for microorganism proliferation is moisture; not surprisingly, standing water in humidifiers and condensate drip pans in air conditioning systems are potential sources of bioaerosols. Other potential sources include bathrooms, damp basements, evaporator pans of refrigerators, and shower heads.[62] Most bacteria found in the indoor environment can be derived from humans, whereas spores, molds, and fungi generally arrive from the outdoor air.

The concentration of indoor air microorganism contaminants has not been well characterized because sampling procedures have not been standardized.[12] The American Conference of Governmental Industrial Hygienists (ACGIH) does recommend an air sampling and analysis protocol for indoor bioaerosols.[74] However, the numbers of colony forming units (CFU) in different indoor environments and in the ambient air have differed so greatly that a qualitative assessment is impossible. It is possible for the concentration of biological contaminants in the indoor air to be slightly below ambient values or to exceed outdoor values by a factor of 10 or more.[63] In general, if the CFU of air in a residential or office setting exceeds that in ambient air by more than a factor of 10, there is likely to be a microorganism amplification problem caused by the building.

Bioaerosols induce illness mainly through infection of the respiratory tract and through the induction of an immune response. The most well-known adverse health effect due to a biological agent is legionnaires' disease. The disease is spread by the aerosolization of the bacteria *Legionella Pneumophilla,* as in the 1976 Philadelphia epidemic.[64] House dust mites, animal proteins, and fungal spores may produce an asthmatic attack through immediate hypersensitivity.[29] Aspergillus[65] and other infectious agents have demonstrated airborne transmission routes.

## REFERENCES

1. **Szalai, A.,** Ed., *The Use of Time: Daily Activities of Urban and Suburban Populations in Twelve Countries,* Mounton, The Hague, 1972.
2. **Chapin, F. S., Jr.,** *Human Activity Patterns in the City,* Wiley-Interscience, New York, 1974.
3. **Dailey, W. V., and Fertig, G. H.,** *Anal. Instrum.,* 77, 79, 1978.
4. **Stevens, R. K., and Hodgeson, J. A.,** *Anal. Chem.,* 45, 443A, 1973.
5. U.S. Environmental Protection Agency, Fed. Regist. 40, CFR 50, 6-1, July 1979.
6. **Stern, A. C., Boubel, R. W., Turner, D. B., and Fox, D. L.,** *Fundamentals of Air Pollution,* Orlando, Academic Press, 1984.
7. U.S. Environmental Protection Agency, Guidelines for Public Reporting of Daily Air Quality Pollutant Standards Index, 450/2-76-013, Washington, D.C., 1976.
8. Committee on Indoor Pollutants, National Research Council, Indoor Pollutants, National Academy Press, Washington D.C., 1981.
9. **Golfish, T.,** Formaldehyde and building related illness, *J. Environ. Health,* 44, 116, 1981.
10. **Dally, K., Hanrahan, L., Woodburg, M., and Kanarek, M.,** Formaldehyde exposure in nonoccupational environments, *Arch. Environ. Health,* 36, 277, 1981.
11. **Breysse, P. A.,** The office environment—how dangerous? in *Indoor Air, Vol. 3: Sensory and Hyperreactive Reactions to Sick Buildings,* Swedish Council for Building Research, Stockholm, 1984, 315.
12. **Samet, J. M., Marbury, M. C., and Spengler, J. D.,** Health effects and sources of indoor air pollution, Part II, *Am. Rev. Respir. Dis.,* 137, 221, 1988.

13. **Olsen, J. H., Jensen, S. P., Hink, M., Faurbo, K., Breum, N. O., and Jensen, O. M.,** Occupational formaldehyde exposure and increased nasal cancer risk in man, *Int. J. Cancer,* 34, 639, 1984.

14. **Hayes, R. B., Raatgever, J. W., Debruyn, A., and Gerin, M.,** Cancer of the nasal cavity and paranasal sinus, and formaldehyde exposure, *Int. J. Cancer,* 37, 487, 1986.

15. Bureau of the Census, U.S. Department of Commerce, 1980 Census Housing, Vol 1, U.S. Government Printing Office, Washington, D.C., 1983, T.82 and T.153. HC80-1-B1.

16. **Spengler, J. D., Duffy, C. P., Letz, R., Tibbitts, T. W., and Ferris, B. G.,** Nitrogen dioxide inside and outside 137 homes and implications for ambient air quality standards and health effects research, *Environ. Sci. Technol.,* 17, 164, 1983.

17. **Spengler, J. D., and Sexton, K.,** Indoor air pollution: a public health perspective, *Science,* 221, 9, 1983.

18. **Yocom, J.,** Indoor/outdoor air quality relationships: a critical review, *J. Air Pollut. Control Assoc.,* 32, 500, 1982.

19. **Quakenboss, J. J., Kanarek, M. S., Spengler, J. D., and Letz, R.,** Personal monitoring for nitrogen dioxide exposure: methodological consideration for a community study, *Environ. Int.,* 8, 249, 1982.

20. Subcommittee on Nitrogen Oxides, Committee on Medical and Biological Effect of Environmental Pollutants, National Research Council, Nitrogen Oxides, National Academy of Science, Washington, D.C., 1976.

21. **Jakab, G. J.,** Nitrogen dioxide induced susceptibility to acute respiratory illness: a perspective, *Bull. N.Y. Acad. Med.,* 56, 847, 1980.

22. **Orehek, J., Massari, J. P., Gayard, P., Grimaud, C., and Charpin, J.,** Effects of short term, low level nitrogen dioxide exposure on bronchial sensitivity of asthmatic patients, *J. Clin. Invest.,* 57, 301, 1976.

23. **Detel, R., Tasshkin, D. P., Sayre, J. W., Rokaw, S. N., Massey, F. J., Coulson, A. H., and Wegman, D. H.,** The UCLA population studies of CORD: X. A cohort study of changes in respiratory function associated with chronic exposure to $SO_x$, $NO_x$, and hydrocarbons, *Am. J. Public Health,* 81, 350, 1991.

24. **Moschandreas, D. J., Stark, J. W. C., McFadden, J. E., and Morse, S. S.,** Indoor Air Pollution with Residential Environments, U.S. Environmental Protection Agency, 1978, 600/7-78-223.

25. **Sterling, T. D., Dimich, H., and Kobayashi, D.,** Use of gas ranges for cooking and heating in urban dwellings, *J. Air Pollut. Control Assoc.,* 32, 162, 1981.

26. **Brookman, E. T., and Birenzvigi, A.,** Exposure to Pollutants from Domestic Combustion Sources: Preliminary Assessment, NTIS Report pb81-111536, TRC Environmental Consultants, Inc., 1981.

27. **Dolan, M. C.,** Carbon monoxide poisoning, *Can. Med. Assoc. J.,* 133, 392, 1985.

28. **Wallace, L. A., Pallizeri, E. D., and Gordon, S. M.,** Organic chemicals in indoor air, *Indoor Air and Human Health,* Lewis Publishers, Chelsea, MI, 1985, 361.

29. **Gammage, R. B., White, D. A., and Gupta, K. C.,** Residential measurements of high volatility organics and their sources, in *Indoor Air,* Swedish Council for Building Research, Stockholm, 1984, 157.

30. **Bach, B., Molhave, L., and Pederson, O.F.,** Human reaction during controlled exposure to low concentration of organic gasses and vapors known as normal indoor air pollutants, in *Indoor Air,* Swedish Council for Building Research, Stockholm, 1984, 397.

31. **Sterling, T., and Sterling, E.,** Impact of different ventilation levels and fluorescent lights on building illness, *Can. J. Public Health,* 74, 385, 1983.

32. U.S. Department of Health and Human Services, The Health Consequences of Smoking. A Report of the Surgeon General, Washington, D.C., 1982, (79-22071).
33. **Cummings, M. K., Markello, S. J., Mahoney, M., Bhargava, A. K., McElroy, P. D., and Marshall, J. R.,** Measurement of current exposure to environmental tobacco smoke, *Arch. Environ. Health,* 45, 74, 1990.
34. **Hoffman, D., Haley, N. J., Brunneman, K. D.,** Cigarette sidestream smoke, presented at the U.S.-Japan Meeting on Lung Cancer, Honolulu, March 21–23, 1983.
35. **Moschandreas, J. D., Zabranski, J., and Peltas, D. J.,** Comparison of Indoor and Outdoor Air Quality. Electric Power Research Inst., Menlo Park, CA, 1981.
36. Office on Smoking and Health, U.S. Department of Health and Human Services, The Health Consequences of Smoking. A Report of the Surgeon General, Washington, D.C., 1984, (84-50205).
37. Office on Smoking and Health, U.S. Department of Health and Human Services, The Health Consequences of Involuntary Smoking. A Report to the Surgeon General, Washington, D.C., 1986.
38. **Burchfield, C. M.,** Passive Smoking, Respiratory Symptoms, Lung Function and Inhalation of Smoke, Ph.D. Dissertation, University of Michigan, Ann Arbor, 1984.
39. **Pukander, J., Loutonen, J., Tinomen, M., and Kramer, P.,** Risk factors affecting the occurrence of acute otitis media among urban children, *Acta Otolaryngol,* 100, 260, 1985.
40. **Hirayama, T.,** Non-smoking wives of heavy smokers have a higher risk of lung cancer, *Br. Med. J.,* 282, 183, 1981.
41. **Trichopoulos, D., Kalandid, A., Sparros, L., MacMahon, D.,** Lung cancer and passive smoking, *Int. J. Cancer,* 27, 1, 1981.
42. **Repace, J. L., and Lowery, A. H.,** A quantitative estimate of non-smoker's lung cancer risk from passive smoking, *Environ. Int.,* 11, 3, 1985.
43. National Council on Radiation Protection and Measurement, Exposure to the Uranium Series with Emphasis on Radon and its Daughters, NCRP Report Number 77, Bethesda, MD, 1984.
44. **Nero, A. V.,** Airborne radio nucleotides in buildings: a review, *Health Phys.,* 45, 303, 1983.
45. **Nero, A. V.,** Indoor concentrations of radon222 and its daughters, *Indoor Air and Human Health,* Lewis Publishers, Chelsea, MI, 1985, 43.
46. **Nero, A. V., Svhecho, M. B., Nazaroff, W. W., and Revon, K. L.,** Distribution of Airborne Radon22 Concentrations in U.S. Homes, U.C. Berkeley, 1984, report LBL-18274.
47. **Evans, R. D.,** Engineer's guide to the elementary behavior of radon daughters, *Health Phys.,* 17, 229, 1969.
48. **Holaday, D. A., Rushing, D. E., Coleman, R. D., Woolrich, P. F., Kusnetz, H. L., and Bale, W. F.,** Control of Radon and Daughters in Uranium Mines and Calculations on Biological Effects, U.S. Government Printing Office, Washington, D.C., 1957, no 494.
49. **Lundin, F. D., Wagoner, J. K., and Archer, V. E.,** Radon Daughter Exposure and Respiratory Cancer, Public Health Service, Washington, D.C., 1971.
50. Committee on the Biological Effects of Ionizing Radiation, National Research Council. The Effects on Population of Exposure to Low Levels of Ionizing Radiation, National Academy of Sciences, Washington, D.C., 1980.
51. **Hess, C. T., Weiffenbach, C. V., and Norton, S. A.,** Environmental radon and cancer correlations in Maine, *Health Phys.,* 45, 339, 1983.
52. **Bean, J. A., Isacson, P., Hahne, R. M., and Kohler, J.,** Drinking water and cancer incidence in Iowa, *Am. J. Epidemiol.* 116, 924, 1982.

53. **Cohen, B. L.,** Health effects of radon and insulation in buildings, *Health Phys.,* 39, 937, 1980.

54. **Steinhausler, F., Hofmann, W., Pohl, E., and Pohl-Rushing, J.,** Radiation exposure of the respiratory tract and associated cancer risk due to radon daughters, *Health Phys.,* 45, 331, 1983.

55. National Council on Radiation Protection and Measurement, Evaluation of Occupational and Environmental Exposure to Radon and Radon Daughters, NCRP Report Number 78, Bethesda, MD, 1984.

56. **Burton, J. D.,** Risks of radon exposure significant, but controls are relatively simple, *Occup. Health Safety,* 60, 46, 1991.

57. **Chesson, J., Hatfield, J., Schultz, B., Dutrow, E., and Blake, J.,** Airborne asbestos in public buildings, *Environ. Res.* 51, 100, 1990.

58. Committee on Non-occupational Health Risks of Asbestiform Fibers, National Research Council, Asbestiform Fibers: Non-occupational Health Risks, National Academy Press, Washington, D.C., 1984.

59. **North, C.,** Occupational respiratory disease, *Safety Health,* Sept., 40, 1989.

60. **Mossman, B. T., Bignon, J., Corn, M., Seaton, A., and Gee, J. B. L.,** Asbestos: scientific developments and implications for public policy, *Science,* 247, 294, 1990.

61. **Ager, B. P., and Tickner, J. A.,** The control of microbiological hazards associated with air conditioning and ventilation, *Ann. Occup. Hyg.,* 27, 341, 1983.

62. **Burge, H. A.,** Indoor sources of airborne microbes, in *Indoor Air and Human Health,* Lewis Publishers, Chelsea, MI, 1985, 139.

63. **Reynolds, S. J., Streifel, A. J., and McJilton, C. E.,** Elevated airborne concentrations of fungi in residential and office environments, *Am. Ind. Hyg. Assoc. J.,* 51, 601, 1990.

64. **Fraser, D. W., Tsai, T. R., and Orenstein, W.,** Legionnaires' disease: description of an epidemic, *N. Engl. J. Med.,* 297, 1189, 1977.

65. **Stone, H. H., Cuzzell, J. Z., Kolb, L. D., Moskowitz, M. S., and McGowen, J. E.,** Apergillus infection of the burn wound, *J. Trauma,* 19, 765, 1979.

66. *ASHRAE Handbook—1991 HVAC Applications Volume,* American Society of Heating, Refrigeration and Air Conditioning Engineers Inc., Atlanta, Chap. 34, p. 34.2.

67. **Göthe, C. J., Bjurström, R., and Ancker, K.,** A simple method of estimating air recirculation in ventilation system, *Am. Ind. Hyg. Assoc. J.,* 49, 66, 1988.

68. USEPA/NIOSH Building Air Quality, A Guide for Building Owners and Facility Managers, S/N 055-000-00390-4, Pittsburgh, 1991.

69. American Conference of Governmental Industrial Hygienists, Industrial Ventilation, A Manual of Recommended Practice, 21st ed., Cincinnati.

70. ASHRAE Standard 62-1989, Ventilation for Acceptable Indoor Air Quality, American Society of Heating, Refrigeration, and Air Conditioning Engineers Inc., Atlanta, 1989.

71. USEPA Assessment of Lung Cancer in Adults and Respiratory Disorders in Children, EPA 600/6-90/006A, Superintendent of Documents, Washington, D.C., 1991.

72. **Janofsky, M.,** Tobacco groups sue to void rule on danger in secondhand smoke, *The New York Times,* June 23, 1993.

73. American Industrial Hygiene Association, White Paper: Environmental lead, *Am. Ind. Hyg. Assoc. J.,* 54, A180, 1993.

74. American Conference of Governmental Industrial Hygienists, Guideline for the Assessment of Bioaerosols in the Indoor Environment, Cincinnati, 1989.

# Chapter 2

# Direct and Indirect Injury to the Respiratory Tract

*Donald E. Gardner, Ph.D.*

## CONTENTS

## INTRODUCTION

Human life has always entailed exposure to airborne chemicals. Nearly five centuries ago, Leonardo da Vinci provided evidence that inspired air could carry harmful substances into the body causing serious damage to the lung. Exposure to airborne contaminants is not only limited to the working environment nor to the polluted urban air but is also present in homes. The EPA has ranked the American home as fourth in its list of serious health hazards. Indoor airborne chemicals include benzene, formaldehyde, radon, nitrogen dioxide, ozone, and numerous others. The health effects associated with airborne contaminants are not always limited to the respiratory tract since this route may also be the portal of entry for substances that can then be translocated from the respiratory tract to other systemic sites. To produce an effect that is beyond the pulmonary system, it is necessary that the chemical, its metabolite(s), or its reactive product(s) be transported and made available to some specific target site.

If one compares the potential exposure via inhalation to other exposure routes, the seriousness and importance of inhalation toxicology becomes evident. At rest, the average adult breathes about 15 kg of air each day. When one compares this amount with the daily intake of food (1.5 kg) and water (2.0 kg), the potential exposure through inhalation becomes significant. The lung also has nearly 4 times the total surface area interfacing with the environment as does the total combined surface area of the gastrointestinal tract and the skin. This makes the lung a prime target organ for a wide range of noxious substances, which can cause not only serious dysfunction of normal respiratory functions but can also adversely alter many of the nonrespiratory functions of the lung. The nonrespiratory functions of this organ play an important role in the pathogenesis of pulmonary disease. Such functions include, for example, the lungs ability to (1) elaborate or inactivate compounds such as prostaglandins, histamine, interleukin-1, kinins, and angiotensin II converted from angiotensin I, which have potent effects on bronchial and vascular smooth muscle, (2) efficiently clear deposited contaminants from the pulmonary system, (3) phagocytize, kill, or neutralize inhaled microorganisms and particles, (4) activate both cellular and humoral immune systems in response to foreign antigenic material, and (5) perform other functions, such as excretion, sense of smell, and trapping of blood emboli.

In spite of vast exposure to many viable microorganisms and nonviable airborne agents, the normal respiratory system is efficient in clearing the lung of unwanted substances and maintaining pulmonary health. When these normal pulmonary defenses are compromised, inhaled toxic substances have the potential for causing structural changes in the lung and altering important physiological and/or changing biochemical functions, all of which can increase an individual's risk of disease.

This chapter will focus on providing the reader with a useful overview of the principles of respiratory toxicology, an understanding of the basic experimental techniques presently being used in assessing and identifying pulmonary responses to airborne chemicals, and specific examples of research findings as they apply to toxicology of the lung. The toxicologist needs to be able to understand, evaluate, and make sound scientific judgement on the effects of airborne chemicals on the respiratory system, to assess potential risks, and to extrapolate these effects to the human population. Table 1 lists various factors that must be considered in the designing of scientific, sound inhalation experiments and in the interpretation of toxicological data. There are several excellent review documents available for the reader who seeks a more in-depth discussion of the various topics discussed in this chapter.[1-5]

## CLASSIFICATION OF AIRBORNE SUBSTANCES

Physical classification of inhaled chemicals is one of the determinants of toxicity. The term "gas" refers to a chemical substance that is in a gaseous state at room temperature and pressure. "Vapor" refers to chemicals that are in a gaseous state but normally are a solid or liquid at room temperature and pressure. One can distinguish a gas from a vapor by its critical temperature. A gas has a critical temperature less than ambient (20° C), and a vapor has a critical temperature >20° C; thus, a vapor can be condensed to a liquid at 20° C with increased pressure.

The term "aerosol" is used to describe suspensions of solid particles or liquid droplets in the air. Aerosols can be defined as "dusts," which are solid particles that are formed by milling or grinding, whereas "fumes" are the results of combustion, sublimation, or condensation of vaporized material. "Mist" and "fog" are aerosols of liquid droplets formed by the condensation of liquid on particles in the air. Gases and vapors may also interact with airborne particles and be adsorbed onto the surface of aerosols and carried into the lung on respiratory-sized particles. Particles that have a length along one axis that is at least three times (aspect ratio ≥3:1) greater than

Table 1    **Factors in Interpreting Toxicological Data**

Types of Human Exposure Data
  • Controlled human exposure
  • Occupational exposure
  • Accidental overexposure
  • General population exposure (prospective and retrospective)
Types of Animal Exposure Data
  • Route of exposure
  • Concentrations
  • Duration
  • Species
  • Biological response
Disparity of Biochemical, Toxicological, or Pathological Effects
  • Marked species variation in response
  • Distinct species differences in metabolic pathways
  • Capacity to produce known pathological changes
Essential Prerequisites for Validation of Data
  • Specification and properties of test materials
  • Analytical methods for determining compounds and metabolites
  • Prior knowledge of pharmacology, toxicology, and metabolism of compounds
Other Factors
  • Likelihood of accumulation within body
  • Assessment of dose–response relationship
  • Ascertaining the development of tolerance/adaptation
  • Reversibility of effects
  • Additive and synergistic responses

along the other two are referred to as "fibers." Asbestos, mineral wool, carbon, glass, and plastic fibers are examples of such particles.

## ANATOMICAL AND STRUCTURAL FACTORS INFLUENCING TOXICITY

### INTRODUCTION

For the scientist interested in assessing and predicting the toxicity of airborne substances, an understanding of the basic biology of the respiratory system is essential. This section will provide a general overview of the structural and related functional characteristics of this organ system. The uniqueness of the pulmonary anatomy plays an important role in determining the total dose of toxicant delivered to the respiratory target tissue and the etiology of a disease at a specific locale. Thus, understanding the anatomy of the respiratory tract is an important factor in determining and predicting the toxicological response.[3-8]

The respiratory system is divided into three different anatomical regions, each of which is discussed separately below. Figure 1 is a schematic showing these three major regions of the respiratory tract. The nasal pharyngeal region is most proximal; it begins at the anterior nares and includes all structures of the respiratory system down to the level of the larynx, which is at the entrance to the trachea. The tracheobronchial region consists of the primary conducting airways of the lung, from the trachea to the beginning of the pulmonary region (terminal bronchioles). The pulmonary region is the beginning of the respiratory portion of the lung, consisting of the terminal bronchiole, respiratory bronchioles, alveolar ducts, alveolar sacs, and alveoli. It is

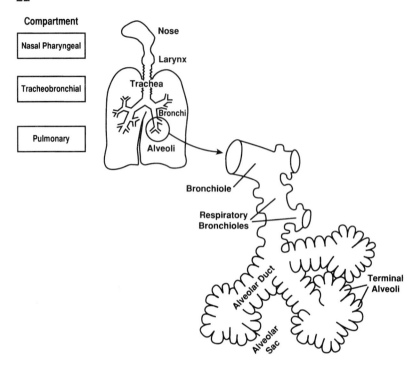

**Figure 1** Schematic illustration of the three compartmental regions of the respiratory tract.

within this pulmonary or alveolar region of the respiratory tract that gas exchange takes place.

## NASAL PHARYNGEAL (NP) REGION

The complicated morphology of the nasal airway is not only related to the role of the nose as an olfactory organ but also includes a number of other physiological functions. This region, as the ''porte d'entree'' to the respiratory tract, plays an important role in warming and humidifying the incoming air during its passage along the conducting airways, and it is considered to be the respiratory tract's first line of defense against inhaled chemicals. It is not possible to go into great detail regarding the nasal structure and physiology in this chapter, but an extensive review on this subject is available.[6] However, understanding certain factors regarding the gross architecture of the nasal passages and its function in removing potentially harmful inhaled substances is necessary in considering the toxicological consequences from inhaled chemicals.

While the nose can provide an early warning of the presence of respiratory irritants, the detection of a noxious odor may not necessarily be related to any direct health effect. Also, since many chemicals can cause olfactory fatigue, this defense system cannot always be considered to be a reliable protective mechanism. However, such responses need to be considered as a potential indicator of exposure. Because the nose can also be a portal of entry of foreign substances that may be antigenic, it has capabilities for responding to such substances with intranasal antibodies, reinforced by cellular immune responses, which can be rapidly activated even before clinical symptoms are manifested. Nasal lesion distribution caused by inhaled toxicants is thought to result from regional differences in (1) susceptibility of tissue types, (2) uptake patterns within the nasal cavity, or (3) a combination of these factors.

## TRACHEOBRONCHIAL (TB) REGION

The tracheal dimensions and the number of divisions of the bronchi vary considerably from species to species, and both the diameter and the length of the major airways change during breathing. Such information can be critical in understanding the deposition of particles and gases in the respiratory tract and the possible health effects associated with such an exposure.

In the TB region, active mechanisms exist for removal of deposited substances. Since particles entering this region will differ in their physical and chemical makeup, they will be deposited differently. The smallest particle will be deposited more distally than the larger particles. These conducting airways are lined with specialized fine organelles called cilia, which are coated with a layer of fluid produced by specialized mucus-secreting goblet cells. The motion of the cilia that continuously beat in a wave-like motion sweeps substances deposited on this mucus blanket to the nasopharynx, where the mucus can be expectorated or swallowed. In a healthy individual, the clearance rates from this region have been estimated to vary from <0.5 h to >24 h. The rate of movement of this mucociliary escalator is slowest in the smallest airways and tends to increase proximally. Hence, the larger particles deposited higher in the TB region would be expected to be cleared faster.

When the diameter of the bronchi decreases in size to approximately 1 mm or less, they are then referred to as bronchioles. As the bronchioles continue to divide, they finally result or terminate in the last division, the terminal bronchiole. Beyond the terminal bronchiole, the airway becomes a very thinly walled tube and is referred to as the respiratory bronchiole, which has numerous small air sacs (alveoli) protruding from its walls.

## PULMONARY REGION

The pulmonary functions of this pulmonary region are gas exchange (i.e., exchanging oxygen for carbon dioxide), clearance, and immunological defense. The total surface area of this region is over 50 m², which approximates that of a tennis court. The alveoli are made up of primarily two types of cells, the Type 1 and Type 2 cells. Gas exchange occurs across the Type 1 cell, which has a smooth surface, is very thin, and covers the greatest surface area. The Type 2 cell is larger, contains numerous microvilli, and produces and secretes surfactant(s), which, by reducing the surface tension, serves to reduce the tendency of the alveoli to collapse. The Type 1 cells seem to be quite susceptible to injury during hyperoxia, exposure to oxidants (e.g., ozone, nitrogen dioxide) and irradiation.

When a gas molecule, such as oxygen or a volatile organic toxicant, is transported from the alveolus to the circulation, it must pass through several barriers. First, such chemicals must penetrate the thin fluid surfactant layer covering the alveolar surface. Other barriers include the basement membrane and the endothelial cell wall of the blood capillary. The continuous exchange of oxygen and carbon dioxide between the blood and the atmosphere is accomplished by the simultaneous processes of pulmonary ventilation, pulmonary perfusion, and alveolocapillary diffusion. This transfer of gas between the capillary blood and the alveolar gas is felt to be a passive process, dependent upon the relative partial pressure of each particular gas in the two spaces.

Inhaled chemicals that are deposited in the pulmonary region of the respiratory system can either (1) exert toxic action at the local site, (2) penetrate and be absorbed into the systemic circulation, (3) be transported out of the region by the phagocytic action of another pulmonary cell, the alveolar macrophage, or (4) be transferred out of the lung through the pulmonary lymphatic channels. The kinetics of such clearance indicate that the retention time in this area of the lung can range from a few hours to many years. Such a long retention time can significantly contribute to the pathogenesis of chronic diseases in this region of the lung.

## UNDERSTANDING HEALTH RISK ASSOCIATED WITH AIRBORNE CHEMICALS

### INTRODUCTION

Occupational and environmental toxicologists are faced with the responsibility of providing sound scientific data that can be used to ensure individuals that they can conduct their daily activities without undergoing any undue risk that might potentiate the development of disease.

Determining and understanding the potential health risk to the respiratory tract from airborne contaminates is actually a three-part problem—a *chemical problem* involving the detection, identification, and determination of the concentration of the toxic compounds in the air; an *exposure problem* involving estimates of exposure and the number of people exposed, especially the susceptible subpopulations; and a *biological problem* involving measurements of health indicators and the assessment of health risk from such exposure. The National Academy of Sciences has addressed these problems and has recommended certain steps in the evaluation and assessment of human health risk.[9] To cause a health risk, a chemical must both be present in the environment at some significant level and be toxic. Figure 2 depicts the major elements of risk assessment and how available research data support this process. Once

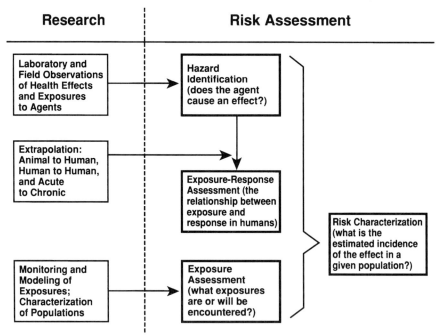

**Figure 2** Elements in the risk assessment requiring the results of scientific research. (From Vainio, H., Hemminki, K., and Wilbourn, J., *Carcinogenesis*, 6, 1653, 1985. With permission.)

either hazard identification or exposure–response assessment are available, predictions can be made that certain effects can occur if certain exposure conditions exist. Exposure assessment addresses the issue of whether effective exposures actually occur and, if so, what populations are exposed. Interrelating the exposure assessment with exposure–response assessment provides the risk characterizations.

## EPIDEMIOLOGICAL, CLINICAL, AND LABORATORY ANIMAL STUDIES

Being confronted with the formidable task of risk assessment or designing studies to support it, the inhalation toxicologist must carefully evaluate and interpret the available data collected from epidemiological, human clinical, and animal studies. The goals of these methods are the same: to gather sufficient data about the potential adverse effects of exposure to a chemical to enable one to predict the kind of adverse effects that may occur and the exposure situations that are likely to cause these effects. However, each method approaches this goal by a different path. Table 2 summarizes the various advantages and limitations of these various approaches. If the data are adequate, a critical assessment of this information can provide the scientific data base necessary to predict possible health consequences from exposure to airborne chemicals, to assess their relevance, and to estimate what level of exposure is likely to be safe. Usually the epidemiological and human clinical data base is quite small compared to the data available from animal studies. The optimal data base for the assessment of toxicity has information from all three approaches, since this minimizes the inherent limitations of each. For example, epidemiological studies cannot prove causative relationships because of the many confounding variables. Nevertheless, epidemiological data reflect the real world of human exposure, which is quite different from a controlled laboratory scenario. Epidemiological studies may show that an association between exposure and effects can qualitatively identify the potential for health risk. However, obtaining quantitative exposure–response information requires quite extensive studies, which are often cost-prohibitive and lengthy to conduct.

While controlled human clinical studies offer the best opportunity to relate cause and effect of a pollutant exposure to humans directly, they too have deficiencies. Because the safety of the volunteer is of paramount importance, the test exposure can have no residual effects and is generally restricted to only short-term exposures. Chronic exposures cannot be performed, and only rather typical medical diagnostic

**Table 2  Assessment of Pulmonary Response to Inhaled Toxicants**

| Factors | Epidemiological Studies | Clinical Studies | Animal Studies |
|---|---|---|---|
| Chemical agents | Mixtures | Less toxic | Extensive |
| Dose response | Difficult | Limited | Wide range |
| Health parameters | Noninvasive | Noninvasive | Multitudinous |
| Number of subjects | Many | Few | Unlimited |
| Length of exposure | Acute, chronic | Short-term | Acute, chronic |
| Cost | Expensive | Expensive | Relatively less |
| Mechanism of action | No | Yes, limited | Yes |
| Sensitive subpopulation | Many | Limited | Limited |
| Use of data for standards | Direct | Direct | Indirect |

procedures, such as pulmonary function and blood clinical chemistries, can be performed.

The strengths of well-designed animal studies are that they can provide complete evaluation of the toxicology of the test substance. The researcher has the choice of a wide range of concentrations, exposure regimens, chemical agents, biological parameters, and test species. Such studies are most useful in providing meaningful dose–response relationships. Many physiological mechanisms are common to animals and humans, leading to the hypothesis that if a chemical causes a particular health effect in several animal species, it is likely to cause a similar effect in humans. However, quantitative extrapolation of the effective airborne toxicant concentration in animals to humans is not yet possible. Relevance to humans may be limited by insufficient knowledge of metabolism and pharmacokinetics of the pollutant in both humans and animals. Metabolism of a chemical may lead to detoxication or may result in metabolic activation leading to a toxic effect. Thus, when using various toxicological databases, several factors must be considered before a scientifically sound and defensible decision can be made regarding the toxicological assessment associated with a known exposure.

## INHALED DOSE VS. CONCENTRATION

The concentration is the amount of a substance in the air of the breathing zone of the exposed species. The dose is defined as the amount of a substance that is actually delivered to the target, preferably the molecular target, but more often the tissue or whole body, depending on the information available. Developing a quantitative understanding of the relationship between exposure, dose, and a specific pulmonary response is a fundamental goal of all toxicologists. A fundamental tenet of toxicology is that it is not the chemical but the dose that determines the health risk. This amount can be determined more easily when using more conventional routes of exposure, such as gavage or injection. When the exposure is by inhalation, determining total dose is more difficult, and a number of factors need to be included.

In most inhalation studies, the actual amount of test chemical deposited in the animal is seldom measured, and exposure is most often expressed by simply stating the concentration and duration of the exposure or by multiplying the two (concentration × time). Many factors are involved in determining the actual dose the animal receives (see next section). In an attempt to better define dose and effect relationships, Haber, in 1924, stated that the response of an animal to a given gas could be related to the product of the concentration and the exposure time necessary to produce some specific physiological effect, i.e., death. This is referred to as Haber's Law, which states that Concentration (C) × Time (T) = K, where K is a constant. Care must be taken in using this formula, since it is valid only for a certain limited number of substances and for only certain combinations of concentrations and brief durations of exposure. Although poorly defined for most chemicals, a more accurate expression is likely to be $C^a \times T^b = K$, where the exponents a and b are estimated from the data. This expression allows for the fact that C and T do not always contribute equally to the observed toxicity.

## DOSIMETRY FACTORS FOR INHALED TOXICANTS

Studies in animals are the only research approach that can provide cause–effect data under controlled and defined conditions with virtually unlimited exposure conditions and methods, enabling a rather complete description of a very broad array of reversible or irreversible toxic effects. However, the concentration of toxicants in the breathing zone of the animal does not by itself define the dose delivered to the respiratory tract. To be able to accurately and quantitatively extrapolate observed effects seen in animals to humans and to define mechanisms of action, significant improvements are needed in the area of respiratory tract dosimetry. Dosimetry refers to determining the dose (the amount of the parent compound or its toxic metabolite) that reaches specific target sites in the animal and in the human as a function of exposure concentration. To achieve this objective requires the incorporation of a number of factors, including species anatomical and ventilatory differences and various physicochemical properties of the pollutant parameters; the study must also be based upon factors that govern transport, retention, and removal of the pollutant. In inhalation toxicology, specific terminology is applied to these processes. The term *deposition* refers specifically to the amount of inhaled particles or gases from the inhaled air that is deposited in regions/cells of the respiratory tract. The term *clearance* refers to the subsequent translocation (movement of material within the lung or to other organs), transformation, and removal of deposited substances from the respiratory tract or from the body. *Retention* refers to the temporal pattern of residual particles or gases that are deposited and stay in the respiratory tract and are not cleared.

## GENERIC FACTORS INFLUENCING DEPOSITION
Anatomy and ventilation strongly influence deposition of both gases and particles. Additional properties specific to gases but different from particles are discussed below. Ventilation is important because (1) the physics of air flow are important to deposition mechanisms and (2) the rate and depth of breathing influence the volume of air and hence the mass of pollutants entering the respiratory tract and the total surface area over which deposition can occur. Another important element is the route of breathing (oral, nasal, or oronasal), since this influences filtering efficiency of inhaled materials and thus impacts the dose delivered to the lower respiratory tract.

Although most adults are nasal breathers at rest, they may resort to chronic or periodic mouth breathing under certain conditions such as exercise, nasal obstruction, or in the presence of chemical irritants. About 15% of the population breathes through the mouth even at rest. As respiratory demands increase, the proportion of air entering via the mouth also increases. For those who are nasal breathers at rest, there is a switch to oronasal breathing when ventilatory demand reaches about 35L per min, which responds to the demand of moderate exercise. Such action can significantly alter the pattern of deposition and thus the toxicological response to the inhaled substance. For instance, nearly 100% of particles having an aerodynamic size of about 10 μm or larger are deposited in the nasopharyngeal region during nasal breathing. This compares to only about 65% deposition of such particles under conditions of oronasal breathing. There is also increased penetration of larger particles deeper into the respiratory tract with oronasal breathing. While nasal breathing offers an effective means of protecting sensitive lower respiratory tract tissues from airborne toxicants, it should be remembered that rodents cannot breathe through the mouth, a factor that must be taken into consideration when extrapolating such animal data to man. Also, there exists a great difference in the complexity of the nasal passages, resulting in differences in nasal airflow patterns between man and the laboratory animal, which in turn may account for species-specific lesion distribution following inhalation exposure to certain highly water-soluble gaseous irritants. While some of these morphological factors may be useful in protecting the deeper regions of the respiratory tract, they also make the nose more vulnerable to the effects of particles, soluble gases, and vapors.

Pollutants can alter physiological responses during exercise by causing pulmonary function changes (e.g., increased airway resistance through constriction), which tend to decrease the volume of air penetrating to the alveoli and which can result in a shift to rapid, shallow breathing. Exercise has been shown to have a pronounced effect on pulmonary tissue uptake of gaseous pollutants but to have little effect on TB tissue. The deposition, clearance, and retention of inhaled particles and reactive gases has been extensively reviewed.[4,17]

## DOSIMETRY FOR PARTICLES

For particles, the overriding factors influencing regional respiratory tract deposition are those based upon aerodynamic properties, which in turn depend upon a variety of physical properties. Particles of the same physical size do not necessarily behave the same aerodynamically. For example, a denser particle will tend to fall faster than a less dense particle of the same size. Therefore, the size of airborne particles is expressed in terms of their "aerodynamic diameter," which is the diameter equivalent of a theoretical spherical particle with a density of 1 that has the same terminal settling velocity (i.e., behaves aerodynamically in the same way) as the particle in question. Using such an adjustment permits a more valid comparison between particles of different physical sizes. These differences between actual and aerodynamic sizes are important in predicting respiratory deposition of inhaled particles. Aerosols of two entirely different chemicals, such as a 1 μm nonspherical particle of uranium dioxide and 3 μm water droplet, will have the same settling velocity, allowing for differences in both shape and density. Fibers are illustrative of the importance of aerodynamic diameter. They tend to line up in the air stream with their long axes parallel to the direction of flow. Such particles have aerodynamic properties that are most strongly influenced along their diameter (width) rather than their length. Thus, long fibers can be deposited more deeply in the lung than spherical particles of the same volume. The health hazards of inhalation of fibers suggest that the specific disease risk may be associated with length and diameter of the fiber, since these parameters influence their deposition, clearance, and biopersistence (durability) in the respiratory tract. It has been hypothesized that the critical lengths of fibers may differ for each certain disease, i.e., 2 μm for asbestosis, 5 μm for mesothelioma, and 10 μm for respiratory tract cancer.

After a contaminant has entered the airways, it must first be deposited in a significant dose onto susceptible tissue to produce an effect. Figure 3 illustrates the five mechanisms by which particle deposition can occur: impaction, sedimentation, Brownian diffusion, interception, and electrostatic precipitation. Electrostatic attraction of particles to the walls of respiratory airways is a minor mechanism and is not important for the inhalation deposition of most environmental contaminants. Interception is important for fiber deposition.

Larger particles are removed from the inhaled air by the mechanism of impaction at the various bifurcations. Impaction onto an airway surface occurs when a particle's momentum prevents it from changing course in an area where there is a rapid change in the direction of the airflow. It is the main mechanism for the deposition of particles having an aerodynamic size of $\geq 2.0$ μm. The largest particles are deposited in the anterior regions of the nose and are then most effectively cleared by mechanical means, including sneezing and nose-blowing. If the deposited substances are highly soluble, they can be absorbed readily into the blood and then transported to extrapulmonary sites. The less soluble substances are cleared from the upper airways by an effective mucociliary system, which can transport the deposited chemical from the airways to the throat, where it can be swallowed.

The probability of impaction increases with increased air velocity, rate of breathing, particle size, and density. Sedimentation is deposition due to gravity and is important for particles with an aerodynamic size of $\geq 0.5$ μm in medium and small airways where the air velocity is relatively low. In this case, the particles will fall

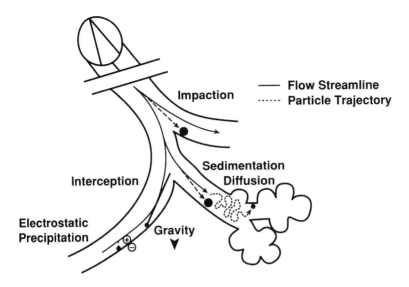

**Figure 3** Mechanisms for particle deposition in the respiratory tract.

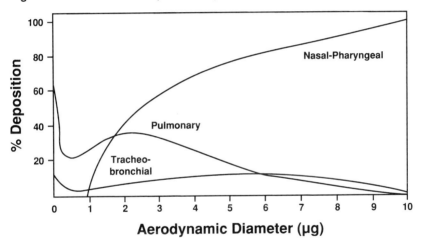

**Figure 4** Predicted regional deposition of inhaled aerosols as a function of particle size. (From Phalen, R. F., *Inhalation Studies: Foundation and Techniques*, CRC Press, Boca Raton, FL, 1984. With permission.)

out of the air stream at a constant rate when the gravitational forces on the airborne particle are balanced by the total force due to air buoyancy and air resistance. The smallest particles ($\leq 0.5$ $\mu$m) may be deposited onto the airway walls due to their bombardment by surrounding air molecules. Brownian diffusion is a major mechanism for deposition in airways where the airflow is low, such as are found in the alveoli and bronchiole. Figure 4 depicts the regional deposition of inhaled aerosols as a function of aerodynamic diameter.

## DOSIMETRY FOR GASES

While aerodynamic properties provide a commonality for examining particulate deposition, gaseous absorption is more complex and is more specific to the physico-chemical properties of the individual gas as well as to the morphology of the respiratory tract, the route of breathing, and the depth and rate of breathing. By considering all the relevant physical, chemical, and biological factors, conceptual models of transport and absorption have been developed to allow reasonable estimates of site-specific deposition for a few gases.

The major physicochemical properties influencing gaseous absorption are solubility and reactivity. The more soluble a gas, the more proximal the deposition. For example, the soluble formaldehyde deposits primarily in the nasopharyngeal region; the relatively insoluble ozone deposits throughout the respiratory tract, but the greatest tissue dose is primarily predicted to occur at the junctions of the TB and pulmonary regions. Reactivity with the liquid layers lining the respiratory tract (mucus or surfactant, depending upon the region) will influence the mass of the parent compound and, if applicable, the reaction products that may reach the tissue.

Gases may also be categorized as being stable, reactive, or metabolizable. An inhalation toxicologist should determine the stability of the gas upon inhalation. This may be determined or calculated by examining the spontaneous reactions likely to occur in a physical system.[17] If a gas is stable, like xenon, only physical or reversible effects, such as asphyxiation or anesthesia would be expected to occur. If it is not stable, it must be further categorized as to whether it is reactive or metabolizable. Reactive gases, such as formaldehyde and ruthenium tetroxide, are quickly converted to nonvolatile products in mucus and will have effects that are largely confined to the nasal tissues. The toxic action of moderately reactive gases, such as $O_3$ and chlorine, may be largely confined to the respiratory tract. Slowly reactive gases, such as many epoxides, tend to travel beyond the respiratory tract to become systemic toxicants, but due to their reactivity, may produce a concentration gradient between the respiratory tract and distal anatomic tissues.

Gases that have slow uncatalyzed reaction rates will either be excreted unchanged or will be metabolized. If metabolized, the metabolites are the potential toxicants, and it is the dose of these metabolites rather than of the parent compound that is the toxicologically relevant measurement. The toxicity of these metabolites also depends upon their reactive kinetics. Such metabolites may become a substrate for further metabolism. For the inhalation toxicologist, it is important to have a complete toxicokinetic profile of the fate of the inhaled toxicant up to the point at which the metabolites are either eliminated, fall into the category of stable compounds such as $CO_2$, or are metabolized to products that are incorporated into biomolecules such as acetic acid.[17,18]

## DOSIMETRY AND EXTRAPOLATION

When evaluating health effects due to toxicant exposure, there is often a lack of sufficient information on which to base quantitative assessments. However, the nature of information needed to respond to governmental regulations on worker and product safety, environmental pollution, etc. has placed an increased emphasis on quantifying risks from exposure to or use of various chemicals. As discussed in the previous section, health effects data are mostly available from animal toxicology studies, with more limited data being available from human clinical and epidemiological studies. Quantitative extrapolation modeling provides an integrated approach for combining animal and human data for making an overall assessment of toxicity. This process takes advantage of the strengths of both data bases while diminishing limitations. To accomplish quantitative extrapolations of any given health effect, dosimetry and species sensitivity information are needed. Species sensitivity relates to possible variations in response of different species receiving equivalent doses of a toxicant at a target site. Thus, even when the dosimetry of a compound is known, the biological response to an airborne toxicant at a given dose may be different in various species. Integrating

Table 3 **Guidelines for Extrapolation of Animal Data to Humans**

1. Use the species whose biological and physiological handling of the material most closely resembles humans.
2. Use several animal species.
3. Use several exposure levels and a variety of exposure patterns.
4. Whenever possible, try to determine the dose delivered to the target organ.

dosimetry and species sensitivity data is critical for being able to quantitatively extrapolate effective concentrations of a chemical in comparisons between animals and man. When judging the reasonableness of the extrapolation, the effect being extrapolated must be adequately understood, preferably at a mechanistic level.[17,18]

While the goal of defining human health risk is common to all quantitative extrapolation efforts, the nature of the process has additional advantages. In addition to characterizing risk based on currently available data, extrapolation modelling involves conceptual processes that allow one to identify research gaps, thereby facilitating the design of future experiments, which in turn will broaden the data base available for subsequent risk assessment. When integrating data on airborne toxicity and exposure, a number of extrapolation issues can arise, such as high to low exposure, intermittent to continuous exposure, acute to chronic effects, single compounds to mixtures, and route to route. Within these broad extrapolation issues, there can be additional complexities that are unique to whether a carcinogenic, behavioral, teratogenic, or cardiopulmonary endpoint is being extrapolated.

Further complications occur when one must extrapolate inhalation data from an animal species to man. Suppose, for example, that acute to chronic effects data need to be extrapolated. This requires developing experimental correlations between the known acute effects in animals and man from chemical exposure and also examining the relationship between acute and chronic effects seen in animals. Better understanding these relationships allows one to make more informed judgments when extrapolating from chronic effects in animals to possible chronic effects in man or when evaluating potential relationships between acute and chronic effects in man. Table 3 summarizes some scientific guidelines that animal toxicologists should consider when designing their research investigations so that the extrapolation of their inhalation data to humans can be scientifically sound and defensible.

# RESPIRATORY RESPONSES TO INHALED CHEMICALS

## IRRITATION RESPONSES

Irritation is one of the first functional responses to an acute exposure to many airborne chemicals. Effects such as a burning sensation in the eyes and nose, tearing, headaches, cough, and difficulty in breathing due to airway constriction are the early effects from numerous direct-acting chemicals. Examples of various airborne irritants and disease states that may be induced or exacerbated by such exposure are summarized in Table 4. It should be noted that, at present, there are no direct links between irritation and the causation of these diseases, especially since many of these diseases are chronic. But while irritation alone may not cause these debilitating diseases, their interaction with other chemicals in the air may be significant. Such responses have been successfully used as the basis for setting many occupational safety values (Threshold Limit Values). Those chemicals eliciting upper respiratory tract irritation are usually highly water soluble and include such chemicals as sulfur dioxide, ammonia, acrolein, chromium, photochemical smog, and formaldehyde.

Table 4 **Irritants and Disease States Which May Be Induced by or Exacerbated by Exposure to Irritants**

| Respiratory irritants | Health effects |
|---|---|
| • Formaldehyde | • Acute asphyxiation |
| • Dimethylamine | • Chronic bronchitis |
| • Methyl bromide | • Emphysema |
| • Chlorine | • Pulmonary fibrosis |
| • Cigarette smoke | • Pneumoconioses |
| • Nitrogen dioxide ($NO_2$) | • Cancer |
| • Bromotrichloromethane | • Allergic sensitization |
| • Ozone ($O_3$) | |
| • Nitrosamines | |
| • Oxygen | |
| • Phosgene | |
| • Methyl isocyanate | |
| • Ammonia | |
| • Halothane | |

Some individuals, such as asthmatics and the young, may be especially sensitive to such exposures and could be expected to experience adverse effects at much lower concentrations than the rest of the population. Gases and particles (e.g., ozone, nitrogen dioxide, chlorine, isocyanates, cadmium, beryllium, mercury) that are less water soluble tend to penetrate to the deeper regions of the lung, causing dyspnea, breathlessness, and chest tightness.

## INFLAMMATORY RESPONSES

An inflammatory response is characterized by changes in the permeability of the alveolar-capillary barrier, which in turn leads to an effusion of serous fluid (edema) and blood cells from the capillary side into the alveolar spaces. When this occurs, bacterial infection often compounds the damage. Together with this edema, alveolar macrophages and neutrophils infiltrate the lung. This is a normal response and may be the lung's first response against the insult. However, these defense cells may also be causally related to certain chronic lung diseases. As defense cells, they are capable of producing a number of potent agents, such as cytokines, chemotactic factors, prostaglandins, lysosomal enzymes, and active oxygen species. If the inhaled substance causes lysis of these defense cells, the highly active cellular products would be released into the lung, where they could act directly on the pulmonary tissue. For example, macrophages have been shown to release proteolytic enzymes that can degrade intercellular components of lung connective tissue and also interact with certain constituents of serum, such as complement. These agents may, alone or in combination, cause functional impairment of epithelial cells, mesothelial cells, and fibroblasts, resulting in disease.

A relatively common investigative method is to lavage the lungs of animals and humans to obtain samples to analyze as a sensitive biomarker of pulmonary injury. The analysis of this broncho-alveolar lavage fluid is an effective means to detect inflammatory responses in the lung because cell counts and cell distributions can be determined, along with measurements of protein and bioactive mediators. Table 5 summarizes some of these studies where lavage fluid has been used to measure the lung's response to various types of pulmonary irritants.

Table 5  **Pulmonary Toxicity Detected by Analysis of Bronchoalveolar Lavage Fluid**

| Chemical | Changes in Lavage Fluid[a] |
|---|---|
| Pneumotoxic agents | |
| Paraquat | PMN, AM, fibronectin |
| Bleomycin | PMN, AM, lymphocytes, albumin |
| Oxidant gases | |
| Nitrogen dioxide | Protein, PMN, phospholipid, LDH, enzymes |
| Ozone | Protein, albumin, LDH, PMN |
| Increased $O_2$ | Protein, hydroxyproline, PMN, albumin, phospholipids |
| Metallic compounds | |
| BeO | Protein, phospholipids, AM, lymphocytes, enzymes |
| $V_2O_5$ | PMN |
| $NiSO_4$, $NiCl_2$ | PMN, AM, protein, LDH |
| $CdCl_2$ | PMN, protein, LDH |
| Insoluble dusts and fibers | |
| Asbestos | PMN, AM, protein, LDH, interleukin-1, enzymes |
| Silica | PMN, AM, enzymes |
| Fly Ash | PMN, LDH |
| $Fe_2O_3$ | PMN, LDH, enzymes |
| $Al_2O_3$ | PMN, LDH, enzymes |

[a](AM) alveolar macrophage; (PMN) polymorphonuclear leukocytes; (LDH) lactate dehydrogenase.
Adapted from *Toxicology of the Lung,* Target Organ Toxicology Series, Gardner, D. E., Crapo, J. D., and Massaro, E. J., Eds., Raven Press, New York, 1988. With permission.

## ASPHYXIATION

Exposure to certain substances may also deprive the body of oxygen. Simple asphyxiants such as hydrogen, helium, methane, and nitrogen are physiologically inert gases that, in sufficient quantities, preclude the adequate supply of oxygen to the body. Chemical asphyxiants such as carbon monoxide, cyanide, hydrogen sulfide, and nitrites can render the body incapable of utilizing the oxygen available and are toxic at concentrations significantly below the levels needed for damage from simple asphyxiants. The effects of carbon monoxide inhalation have been documented extensively in animal models and humans and provide a good example of such an asphyxiant. Carbon monoxide is an odorless, colorless, tasteless gas that is principally encountered as a waste product of incomplete combustion of carbonaceous materials. Its toxic effects are related to the chemical interaction between carbon monoxide and hemoglobin, which forms carboxyhemoglobin (COHb) The binding affinity between carbon monoxide and hemoglobin is more than 200 times that of oxygen. Thus, a small amount of inhaled carbon monoxide can effectively block a large proportion of the hemoglobin from transporting oxygen. Carbon monoxide is not a cumulative poison, since COHb can be fully dissociated, and once the exposure ceases, the carbon monoxide is eliminated via the exhaled air. Many individuals who are occupationally exposed to carbon monoxide, such as traffic police and garage workers, may suffer from repeated low levels of exposure to carbon monoxide. Patients with angina pectoris, when exposed to low levels of carbon monoxide, will exhibit reduced

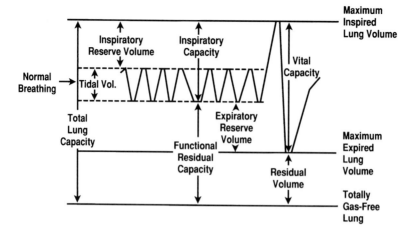

**Figure 5** Measurements of the volume of air moved during normal or forced inspiration or expiration.

time to angina attacks. At high levels, death may occur even in normal individuals. Cyanide, a chemical asphyxiant, which is rapidly absorbed through the respiratory tract, acts differently. It does not interfere with the transfer of oxygen to the tissue, but instead it prevents the tissue from using oxygen by inhibiting the tissue enzyme cytochrome oxidase that is needed for energy production.

## ACUTE STRUCTURAL CHANGES

An early cellular response to an acute assault is the damage to various epithelial cells of the respiratory tract. Epithelial cells can respond to toxicants differently based on their structure, function, and location in the airway. In general, the cells that are the most sensitive to direct-acting chemicals are Type 1 cells and ciliated epithelial cells; secretory bronchiolar cells are of intermediate sensitivity, and Type 2 alveolar cells are the most resistant. The impact of such damage depends upon the cell type affected and the degree of damage. For example, oxidants can cause a loss of ciliated epithelial cells, which are replaced by Clara cells, and Type 1 cells are replaced with Type 2 cells. The function of these lost cells is impacted, as modified by the function of the replacement cells. In the example above, sufficient damage would be expected to reduce tracheobronchial clearance (through the loss of ciliated cells) and also alter diffusing capacity of the pulmonary region through the replacement with thicker cells.

Epithelial cells also produce many different types of inflammatory mediators that can alter the permeability of the alveolar capillary barrier, act as chemotactic factors, and disrupt normal regulatory mechanism of smooth muscle contraction.

## PULMONARY FUNCTION CHANGES

An advanced assessment of the health effects of airborne contaminants on the respiratory system generally includes measurements of lung mechanics. Pulmonary function tests provide quantitative measurements useful in determining the nature and extent of dysfunction. A large number of tests have been developed to measure subtle changes in breathing patterns, lung mechanical properties, lung volumes, flow rates, gas exchange efficiencies, and airway resistances. Figure 5 shows a diagram of the various physiological subdivisions into which the air volume of the lung may be divided. Each of these measurements is defined in the Definitions section below.

Pulmonary function data have been used successfully (1) as a sensitive indicator of toxic response, (2) to diagnose and characterize the pathogenesis of obstructive and restrictive lung disease, and (3) to provide information useful in extrapolating to humans the effects seen in animals. When functional changes are identified, they may often provide an indicator of either a transient or permanent alteration in lung structure. Unfortunately, such functional tests can only indicate that a lesion exists when the lesion is large enough to impair normal function. However, this has inherent value in assessing the degree of adversity of an effect, since sensitive measurements of small cellular changes are often difficult to interpret in terms of public health impact. Although the exact techniques used in humans and animals may differ, the results to date suggest that similar functional changes reported in humans and animals would indicate similar structural changes. Examples of the pulmonary function measurements that are useful in evaluating effects of inhaled substances are given in Table 6. Exposure to air pollutants, such as sulfur dioxide and sulfuric acid, can produce increases in airway resistance, which is indicative of airway constriction. Ozone exposure produces decrements in exhaled volumes and flowrates and increases in airway reactivity. Most of these pulmonary functional changes tend to increase with increased concentrations and duration of exposure. Various aldehydes, such as formaldehyde and acrolein, can produce pulmonary changes including increases in pulmonary flow resistance, increases in tidal volume, and decreases in respiratory frequency. Effects on lung functions have also been associated with exposure to complex mixtures, like auto and diesel exhaust. Those with preexisting respiratory diseases (e.g., asthma, chronic bronchitis, and emphysema) are often more susceptible and sensitive to low exposure levels to many of these chemicals. There are a number of excellent review articles available that describe in more detail the methodology, physiological principles, and interpretation of pulmonary function testing in human and animals.[3-5,8,10]

## PULMONARY HOST DEFENSE SYSTEM EFFECTS

As discussed above, the respiratory system is highly dependent on maintaining a responsive defense system that protects the lung and the body from the constant assault of viable and nonviable airborne substances. Both human and animal studies have clearly demonstrated that when the efficiency of these pulmonary defenses is comprised, the potential risk of infectious pulmonary disease is significantly increased.[7,11] Various host defense parameters are used to examine problems associated with lung disease, including the rate of mucociliary clearance,[4] alveolar macrophages activity,[12] increased susceptibility to disease,[11] and impaired immune system function.[3,13]

The effectiveness of mucociliary clearance can be determined by measuring the rate of transport, the frequency of ciliary beating, and the structure and function of mucus-secreting cells. Any adverse effect causing a dysfunction of this system would be likely to increase the retention time of inhaled substances deposited in the respiratory system, thereby increasing the risk of disease. Various pollutants, such as ozone, nitrogen dioxide, nickel, cadmium, sulfuric acid, and complex mixtures such as tobacco smoke, have been shown to impair the body's ability to remove these contaminants through the depression of the mucociliary escalator. Effects noted were alterations in type and rate of mucus secretion, damage to ciliated epithelia, reduction in ciliary beating activity, and slowing of the clearance of tracer particles.

The functioning of the resident population of alveolar macrophages is crucial in defense against inhaled environmental pollutants. These cells are responsible for a variety of important activities, including the removal of inhaled nonviable airborne particles by phagocytosis, maintenance of pulmonary sterility by their bactericidal activities, and interaction with lymphoid cells for immunological protection. To adequately function, these cells must maintain mobility, a high degree of phagocytic activity, integrated membrane structure, and a well-developed and functioning

Table 6  **Techniques Available for Assessing Changes in Pulmonary Function**

| Parameter | Test Available | Measurements |
|---|---|---|
| Ventilatory exchange | Respiration rate | Frequency of breathing |
| | Minute ventilation | Total inspired/expired volume per minute |
| | Tidal volume | Volume of breathing |
| Static lung volumes | Vital capacity | ⎫ |
| | Total lung capancity | Elasticity of the lung and thorax |
| | Residual volume | |
| | Functional residual capacity | |
| | Expiratory reserve volume | Expiratory force |
| | Inspiratory capacity | Inspiratory force |
| Respiratory mechanics | Air flow resistance | Flow resistance of airways |
| | Total lung flow resistance | |
| | Static lung compliance | Stiffness of the lung |
| | Dynamic lung compliance | |
| | Lung and thoracic cage flow resistance | Changes in total respiratory system |
| | Flow-volume inspiratory | |
| | Maximum inspiratory flow | |
| | Volume curves | Overall mechanical function of lungs and thoracic wall |
| | Flow-volume maximum expiratory flow volume | |
| Distribution of ventilation | Closing volume | Closure of dependent airways |
| | Single and multiple breath Nitrogen washout | Distribution of ventilation |
| Diffusion | Blood-gas electrodes and tracer gases | Evaluates alveolar membrane thickness |
| Blood gases | Measurements of $O_2$, $CO_2$, $CO$, and pH | Total alveolar ventilation |
| Pulmonary circulation | Edema. Use of radioisotopes, wet/dry lung weights, gas transfer | Fluid accumulation in the lung |
| | Cardiovascular pressures | Hyper- and hypotension in vascular system |
| | Cardiovascular volumes, flow resistance, and distribution perfusion | Cardiovascular performance |

enzyme system. Exposure to trace metals, such as cadmium, nickel, and manganese, to complex aerosols (e.g., coal fly ash, diesel exhaust), and to numerous gases, such as ozone and nitrogen dioxide, can cause significant alterations in many of these macrophage functions. As would be expected, these pollutants also result in an increase in the host's risk of pulmonary infectious disease and perhaps to other chronic pulmonary diseases. Both human and animal studies have demonstrated that when the efficiency of these pulmonary defenses is compromised (e.g., reduced phagocytosis, bactericidal activity, or changes in numbers) inhaled viable infectious microorganisms can be expected to multiply and infect the lungs, causing serious respiratory disease.[11]

A number of effects on the host's immune system have been identified following exposure to foreign substances. Immunotoxicity is defined as an adverse effect in the structure and/or function of the normal immune system. The pulmonary cells having an immune function are heterogeneous, and external stimuli can modify their behavior both qualitatively and quantitatively. The lung's immune system can be viewed as having three distinct compartments, each containing immunocompetent cells (i.e., lymphocytes and macrophages). Morphologically, these three compartments include cells associated with the (1) bronchoalveolar airspace, (2) submucosal or secretory antibody system that lies beneath the tissue of the tracheobronchial tree, and (3) complex network of lymphatic ducts and lymph nodes associated with the tracheobronchial tree. A number of different kinds of responses have been associated with inhaled chemicals, including their ability to (1) reduce the normal immune response (immunosuppression), resulting in an increase incidence of infection or tumors, (2) overactivate or exaggerate the immune system, causing hypersensitivity reactions, or (3) cause an autoimmune reaction, which is the failure of the body to distinguish between "self" and "nonself." Sometimes chemicals cause more than just one effect on the immune system, depending on the concentration and duration of exposure. The specific type of response elicited depends on the specific chemical, as well as on the concentration and duration of exposure. Inhaled substances have been shown to cause or exacerbate various immune mediated-disorders, including some types of asthma, hypersensitivity pneumonitis, allergic rhinitis, bronchopulmonary aspergillosis, asbestosis, silicosis, coal workers' pneumoconiosis, bagassosis, and possibly byssinosis. See Table 7 for a summary of these types of effects.

## CHRONIC DISEASES

Identifying the pollutant(s) causing chronic pulmonary disease can be a problem because of the difficulty of performing chronic studies and the long latency periods sometimes involved. In many cases, the disease symptoms may not become evident until after 20 or more years of exposure. Chronic pulmonary diseases are often classified as restrictive lung disease, chronic obstructive lung disease, and cancer.

Both restrictive and obstructive lung diseases cause serious impairment of the flow of gases into the gas exchange regions of the lung. Analysis of lung volumes and the comparison to normal values can be useful to differentiate between obstructive and restrictive pulmonary disease. Chronic obstructive pulmonary disease (COPD) refers to a disease of uncertain etiology characterized by persistent slowing of airflow during forced expiration. Individuals with COPD have difficulty performing more than light to moderate exercise and frequently have associated cardiovascular disease and chronic cough. COPD may be of two major types—emphysema and chronic bronchitis. In both cases, there is small airway obstruction due airway restriction, increased mucus thickness, and increased airway smooth muscle tone, all leading to gas trapping, which makes expiration difficult.

Restrictive lung disease is characterized by a decrease in lung volumes, such as vital and total lung capacities. Such restriction decreases the lung's ability to expand, making inflation of the lungs more difficult. Restrictive defects may occur when the elastic properties of the lung are decreased and the lung becomes stiff, as in fibrotic disease, such as silicosis, pneumonia, and asbestosis.

Table 7   **Examples of Immunomodulation by Various Inhaled Chemicals**

| Classification | Symptoms | Chemical Agents |
|---|---|---|
| Immediate (type I) hypersensitivity | Bronchial asthma, asthmatic bronchitis, urticara, rhinitis, atopy | Beryllium, chloramine, ethylenediamine, ethylene oxide formaldehyde, isocyanates, platinum, nickel |
| Cytolytic (type II) hypersensitivity | Chemically induced hemolytic anemia, bone marrow depression, thrombocytopenia | Trimellitic anhydride, mercury |
| Arthus-immune complex (type III) hypersensitivity | Hypersensitivity pneumonitis, rheumatoid disease, sarcoidosis, vasculitis | Trimellitic anhydride, mercury |
| Cell-mediated (type IV) hypersensitivity | Contact dermatitis, sarcoidosis, anergy, delayed hypersensitivity | Beryllium, chromium, isocyanates, mercury, phthalic anhydride, trimellitic anhydride |
| Immunosuppression | Altered immune responses and host resistance following inhalation exposure | Asbestos, silica, metals, toluene, oxidant gases, tobacco smoke, benzene |
| Irritancy or nonimmunological | Pseudoallergic symptoms of bronchial asthma and asthmatic bronchitis | Formaldehyde, isocyanates, ethylenediamine |

## Chronic Obstructive Pulmonary Disease

The problem of chronic bronchitis is of great interest in industrial medicine. It is characterized by persistently increased secretion of mucus from bronchial glands and goblet cells lining the conducting airways and can be recognized by the presence of a chronic cough and recurrent expectoration. The symptoms of bronchitis are commonly associated with long-term cigarette smoking, dusty environments, such as in coal mines and grain elevators, and the exposure to ambient air heavily polluted with sulfur oxides and combustion products.

Emphysema is an anatomical alteration of the lung characterized by abnormal, uneven, permanent enlargement of the airspaces distal to the terminal bronchiole, accompanied by the destruction of the alveolar walls. Emphysema has been associated with exposure to coal dusts, cigarette smoke, osmium tetroxide, cadmium oxide, and some atmospheric pollutants. These substances cause cell injury and subsequent inflammatory responses in the lungs. During this process, proteases, lysosomal enzymes, and oxidants are released during phagocytosis, cell injury, and cell death. To maintain structural integrity, the lung has developed an effective antiprotease system. The main antiprotease in the lung is alpha-l-antiprotease, which is derived from serum. To maintain the structural integrity of the lung, a careful balance between lung proteases and antiproteases is critical. When this balance is upset, these reactive substances can begin to degrade pulmonary elastin and collagen, which is a key event in the destruction of the alveoli. This destruction of the lung connective tissue results in a subsequent loss in total lung surface area, reducing the lung's ability for gas exchange. Such structural changes are associated with various pulmonary functional abnormalities related to loss in lung elasticity and to decreases in normal diffusion capacity, lung compliance, and forced expiratory volume.

## Chronic Restrictive Lung Disease

In restrictive lung disease, inspiration is limited by reduced compliance of the lung or chest wall or a weakening of the inspiratory muscles. Both the forced expiratory volume (FEV) and the vital capacity (VC) are reduced, but characteristically the FEV/VC% is normal or increased. This is different than in obstructive lung disease where the FEV is reduced much more than the VC, giving a low FEV/VC%. Pulmonary fibrosis, a restrictive lung disease, appears to be increasing in the U.S. Toxic interaction of some chemical agents with the lung can result directly or indirectly in cell injury or death. This may be accompanied by an inflammatory response that may extend the damage to adjacent normal cells and connective tissue. Healing after cell injury or necrosis is achieved by fibrosis. Fibrogenesis is the process by which collagen is laid down as a structure or repair protein. Fibrosis may be nodular in character, as with silicosis, or diffuse, as in asbestosis. This reaction usually occurs at the expense of normal tissue function. During this repair, macrophages attach to the injured site, producing fibronectin and other growth factors, which attract fibroblasts and stimulate their proliferation. In a normal healing situation, such as in cases of a surgical wound, the healing by a fibrous scar is completed within a short time, and no further tissue disruption occurs. However, with continuous exposure to certain dusts (e.g., silica, asbestos, aluminum, coal dust, and kaolin), this fibrogenic reaction in the lungs becomes chronic and self-perpetuating. After several years, such fibrosis can become massive, causing lung stiffening, and can seriously impair breathing, even causing death.

Silicosis is a proliferative fibronodular lung disease occurring after sufficient high exposure to crystalline silica for 5 or more years. The silicotic nodule in the lung is a positive identification of silicosis. These nodules are made up of bundles of collagen fibers surrounded by macrophages, lymphocytes, and fibroblasts. It is believed that the lesion involves macrophages. When the number of particles phagocytosed becomes excessive, macrophages die and are lysed in large numbers. Certain mediators (fibronectin, interleukin-1, growth factors) within the macrophages are then released from the lysed macrophages, promoting fibroblast proliferation. The initial symptom of silicosis is shortness of breath accompanied by a fever caused by excessive cellular destruction. Silicosis can be further complicated by tuberculosis and nocardiosis (fungal disease).

Asbestosis is induced by high industrial exposure to asbestos dust, which was present before effective dust control standards were enforced. This disease appears to be declining, probably related to more effective control measures. Asbestosis is characterized by progressive lung stiffening, calcification, impaired gas exchange, and disability. Studies have shown that all urban dwellers retain mineral fibers in their lungs, but there is no evidence that these individuals are at greater health risk. Fibrosis from asbestos is most common in the lower lobes of the lungs. Studies have shown that lowering the total number of fibers in the air and decreasing the exposure frequency will significantly decrease the incidence of fibrosis.

Besides crystalline free silica, asbestos, and coal dust, other materials can incite pulmonary tissue reactions, sometimes with serious clinical effects. Substances that can cause pulmonary reactions of varying severity include graphite, fuller's earth, kaolin, iron, mixed dusts, and tungsten. Not all dusts cause disabling pulmonary disease; some dusts, when inhaled even in high concentrations, are retained in the lungs but cause little or no tissue reactions.

Granulomas are focal lesions that form as a result of an inflammatory reaction, which can be caused by biological, chemical, or physical agents. Chronic lung granulomatous disease may develop without a symptomatic acute phase and appears years after the initial exposure. One form is caused by exposure to beryllium fumes or dust.

**40**

Chronic beryllium disease is a progressive, debilitating granulomatous disorder that is primarily limited to the pulmonary interstitium. Exposure to beryllium fumes may also cause acute pulmonary edema and pneumonia. Berylliosis is believed to involve a beryllium-specific immune response. The most common symptom of the disease is dyspnea, which may be followed by weight loss, chest pain, cough, and fatigue.

## Cancer

Chemical carcinogenesis is often characterized by a long latency period, which in humans may be as long as 30 years. Because of this, there are multiple opportunities for individuals to be exposed to a wide range of other carcinogenic substances, making it most difficult, if not impossible, to identify the causative agent.

In spite of the dramatic decline in mortality and incidence of certain types of cancers (Hodgkin's disease, cervical and stomach cancer), lung cancer, which was rare in 1900, is now the current leading cause of cancer-related deaths and has been increasing at about 15% per year since 1973. While the mortality rates of lung cancer have leveled off for men, the incidence in women continues to climb. Most lung cancers are attributed to cigarette smoke. However, at the present time, there is considerable controversy about whether there is an increased risk of lung cancer directly related to living in an urban environment. But, occupational exposure to many industrial chemicals, such as formaldehyde, arsenic, benzene, and asbestos, are clearly associated with the development of cancer. For several human carcinogens (*bis* (chloro-methyl) ether, mustard gas, vinyl chloride), evidence for carcinogenicity was first found in experimental animals, but in general, the majority of carcinogens have been identified through epidemiological studies. However, such an association does not firmly establish that all chemicals that cause cancer in animals will also cause cancer in humans. Of all the agents for which "sufficient evidence" of carcinogenicity in animals have been shown, less than 15% of these chemicals have been clearly demonstrated to be definitely carcinogenic in humans. During the cancer process, there are many host factors that are involved that have the potential to either enhance or retard this process.

Many scientists are questioning the methods presently being used in assessing the cancer risk of various chemicals. Typically, animals are exposed to the maximum tolerable doses (MTD) of the test substances to see if such a dose elicits cancer. The results are then extrapolated to effects of minuscule doses in humans. Exposure to such huge doses may (1) overwhelm and inhibit the animal's natural detoxification mechanisms, (2) alter the normal metabolism of the test chemical, or (3) stimulate cell turnover rates, which may well be the driving force resulting in an incidence of mutation and cancer due to exposure to substances such as chloroform and saccharin. In such cases, the effects produced at the MTD may not predict the effects that would be expected in a target population exposed to lower doses, as found in the environment. In fact, more that half of the chemicals that were judged to be carcinogenic in the National Toxicology Program would not have been so classified if the results obtained with the MTD were excluded. An example is that exposure to 15 ppm formaldehyde causes nasal lesions that include cytolethality, squamous metaplasia, nasal epithelial ulcerations, regenerative cell proliferation, and cancer. But, an exposure to 0.7 ppm produces none of these lesions, probably because of natural defense mechanisms, such as mucociliary clearance and regional metabolism, which can protect the cells from formaldehyde accumulation.

Researchers have identified a wide variety of vapors, dusts, and aerosols that are classified as carcinogens. Table 8 provides examples of a number of airborne chemicals that are associated with increases in the incidence of tumors in humans and animals. A detailed description of the scientific bases of such classification for these

Table 8  **EPA Weight-of-Evidence Classification for Airborne Substances Associated with Increased Tumor Incidence in Laboratory Animals and Man**

| Organic Chemicals | |
| --- | --- |
| **Chemical** | **Group**[a] |
| Bis(chloromethyl)ether | A |
| Acrolein | C |
| Acrylonitrile | B1 |
| Formaldehyde | B1 |
| Benzene | A |
| Vinyl chloride | A |
| Inorganic Chemicals | |
| **Chemical** | **Group**[a] |
| Arsenic | A |
| Antimony | B2 |
| Cadmium chloride | B1 |
| Nickel compounds | B1 |
| Asbestos fiber | A |

[a]As more data become available, these classifications are reevaluated.

*Notes:* (A) is carcinogenic to humans (sufficient evidence from epidemiological studies to establish a causal association between exposure to the substance and cancer). (B) is probably carcinogenic to humans. This is divided into B1 (limited evidence of carcinogenicity to humans from epidemiological studies) and B2 (inadequate human data, but sufficient animal study evidence). (C) is possibly carcinogenic to humans (limited evidence of carcinogenicity in animals in the absence of human data).

carcinogenic chemicals is available.[14,15] Carcinogens can be divided conveniently into three general chemical classes: organic chemicals, inorganic chemicals, and radioactive material. In addition, certain complex mixtures, such as diesel engine exhaust, coke oven emissions, cigarette smoke, and coal tar aerosols, have also been shown to contain carcinogens. Evidence of carcinogenicity can be based on (1) an increase in the incidence of a specific tumor type, (2) the development of tumors earlier than seen in controls, (3) the presence of types of tumors normally not seen in the control group, and (4) an increase in multiplicity of tumors.

Various types of lung cancers have been noted in humans. All known human pulmonary carcinogens are taken into the body by the inhalation route; however, inhalation exposure to carcinogens can also result in neoplasms elsewhere in the body. For example, workers in the rubber industry, exposed to vinyl chloride by inhalation, may experience a significant increase in the incidence of liver cancer. Some inhaled chemicals are direct-acting pulmonary carcinogens, while other chemicals, such as benzo(a)pyrene and acrylonitrile, require metabolism in the respiratory tract to become carcinogenic. The three most common pulmonary cancers are those arising in the region of the tracheobronchial epithelium (bronchogenic carcinomas), alveolar epithelial carcinomas, and malignant mesotheliomas. Mesothelioma is normally a rare pleural or peritoneal tumor but is relatively common in individuals exposed to asbestos. The data would indicate that most cases of mesothelioma are related to crocidolite exposure. Both asbestosis and respiratory tract cancer are most likely to be caused by mineral fibers retained in the lungs that are not adequately cleared by alveolar macrophages or dissolved, i.e., those with diameters >0.15 μm. In contrast, mesothelioma is associated with asbestos fibers that are sufficiently fine to migrate from the pulmonary deposition site to the pleura and peritoneum (i.e., diameters <0.1 μm).

To prove that a suspected chemical actually causes cancer in humans is very difficult. The induction of cancer in humans and in animals by inhaled chemicals is extremely complex and is subject to and controlled by many factors.[14,16,17] Although test animals may respond differently than humans when exposed to a carcinogen, experience has shown that almost all human carcinogens also cause cancer in test animals; the one exception may be arsenic. This is understandable, since cancer starts in an individual cell, and all mammalian cells are comparable in structure and function. The basic premise of carcinogenesis research is that a substance that affects the animal cells in such a way as to cause cancer may affect human cells in the same way. Our understanding of the effect of human exposure to chemicals is incomplete. Some carcinogens have always been present in the environment, and their concentrations vary depending on geographical location and other environmental conditions. Each year, human activity may add more and more carcinogens to the environment. Logically, this increase in amount and number of carcinogens in the air, water, food, and soil is likely to lead to more human cancers.

## DEFINITIONS

**Absorption**—penetration of a substance into the bulk of a solid or liquid (cf. adsorption).

**Additivity**—the effects of a mixture are equal to the sum of the effects of the individual chemicals.

**Aerodynamic diameter**—the diameter of a unit density sphere having the same settling speed (under gravity) as the particle in question of whatever shape and density.

**Aerosol**—a suspension of liquid or solid particles in a gas.

**Air spaces**—all alveolar ducts, alveolar sacs, and alveoli. To be contrasted with airways.

**Airway conductance ($G_{aw}$)**—gas flow rate in the airway per unit of pressure difference between the airway opening and the alveoli, reciprocal of airway resistance. $G_{aw} = (1/R_{aw})$.

**Airway resistance ($R_{aw}$)**—the resistance to airflow afforded by the airways between the airway opening at the mouth and the alveoli.

**Airways**—all passageways of the respiratory tract from mouth or nares down to and including respiratory bronchioles. To be contrasted with air spaces.

**Allergen**—a material that, as a result of coming into contact with appropriate tissues of an animal body, induces a state of allergy or hypersensitivity, generally associated with idiosyncratic hypersensitivities.

**Alveolar macrophage (AM)**—a large, mononuclear, phagocytic cell found on the alveolar surface, responsible for particle clearance from the deep lung and for viral and bacterial killing.

**Alveolar septum**—a thin tissue partition between two adjacent pulmonary alveoli, consisting of a close-meshed capillary network and interstitium covered on both surfaces by alveolar epithelial cells.

**Bronchoconstriction**—reduction in the caliber (diameter) of the bronchi.

**Bronchoconstrictor**—an agent that causes a reduction in the caliber (diameter) of airways.

**Bronchodilator**—an agent that causes an increase in the caliber (diameter) of airways.

**Bronchus**—one of the subdivisions of the trachea serving to convey air to and from the lungs. The trachea divides into right and left main bronchi, which in turn form lobar, segmental, and subsegmental bronchi.

**Cilia**—motile, often hair-like extensions of a cell surface.

**Clara cell**—a nonciliated cell in the epithelium of the respiratory tract.

**Collagen**—the major protein of the white fibers of connective tissue, cartilage, and bone.

**Deposition**—the depositing of inhaled pollutants within the respiratory tract, which depends on breathing patterns, airway geometry, and the physical and chemical properties of the inhaled pollutants.

**Diffusing capacity of the lung ($D_L$, $D_LO_2$, $D_LCO$)**—amount of gas ($O_2$, CO, $CO_2$), commonly expressed as milliliters of gas (STPD), diffusing between alveolar gas and pulmonary capillary blood per torr mean gas pressure difference per minute, such as ml $O_2$/(minn-torr).

**Emphysema**—a condition of the lung characterized by abnormal, permanent enlargement of air spaces distal to the terminal bronchiole, accompanied by the destruction of their walls, and without obvious fibrosis.

**Endothelium**—a layer of flat cells, especially those lining blood and lymphatic vessels.

**Enzyme**—any of numerous proteins produced by living cells that catalyze biological reactions.

**Epithelium**—a primary animal tissue, distinguished by closely packed cells with little intercellular substance; covers free surfaces and lines body cavities and ducts, such as in the respiratory tract.

**Fibrosis**—the formation of fibrous tissue, usually as a repair or reactive process and not as a normal constituent of an organ or tissue; a restrictive lung disease.

**Fine particles**—airborne particles smaller than 2 to 3 μm in aerodynamic diameter.

**Forced expiratory volume (FEV)**—denotes the volume of gas that is exhaled in a given time interval from the beginning of the execution of a forced vital capacity.

Conventionally, the times used are 0.5, 0.75, or 1 s, symbolized $FEV_{0.5}$, $FEV_{0.75}$, $FEV_{1.0}$. These values are often expressed as a percent of the forced vital capacity; for example $(FEV_{1.0}/FVC) \times 100$.

**Gas exchange**—movement of oxygen from the alveoli into the pulmonary capillary blood as carbon dioxide enters the alveoli from the blood. In broader terms, the exchange of gases between alveoli and lung capillaries.

**Histamine**—a depressor amine derived from the amino acid histidine and found in all body tissues, with the highest concentration in the lungs, a constrictor of bronchial smooth muscle, and a vasodilator that causes a drop in blood pressure.

**Inspiratory reserve volume (IRV)**—the maximal volume of air inhaled from the normal end-tidal inspiratory level.

**Irritant potency**—the relative strength of an agent that produces irritation.

**$LC_{50}$**—concentration of a substance lethal to 50% of tested species.

**Leukocyte**—any of the white blood cells.

**Lung volume ($V_L$)**—actual volume of the lung, including the volume of the conducting airways.

**Mist**—suspension of liquid droplets formed by condensation of vapor or atomization; the droplet diameters exceed 10 μm.

**Morbidity**—The quantity or state of being diseased; also used in reference to the ratio of the number of sick individuals to the total population of a community (i.e., morbidity rate).

**Morphology**—structure and form of an organism at any stage of its life history.

**Morphometry**—the quantitative measurement of structure (morphology).

**Mortality rate**—for a given period of time, the ratio of the number of deaths occurring per 1000 population.

**Mucociliary action**—ciliary action of the mucous membranes lining respiratory tract airways that aids in removing particles from the lungs.

**Mucociliary transport**—the process by which mucus is transported, by ciliary action, from the lungs.

**Mucus**—the clear, viscid secretion of mucous membranes, consisting of mucin, epithelial cells, leukocytes, and various inorganic salts suspended in water.

**Nasopharyngeal**—relating to the nose or the nasal cavity and the pharynx (throat).

**Oronasal breathing**—breathing through the nose and mouth simultaneously; typical human breathing pattern at moderate to high levels of exercise vs. normally predominant nasal breathing while at rest.

**Oxidant**—a chemical compound that has the ability to remove, accept, or share electrons from another chemical species, thereby oxidizing it.

**Particle**—any object, solid or liquid, having definite physical boundaries in all directions; includes, for example, fine solid particles such as dust, smoke, fumes, or smog; found in the air or in emissions.

**Pathogen**—any virus, bacterium, or other microorganism capable of causing disease.

**Phagocytosis**—a mechanism by which alveolar macrophages and polymorphonuclear leukocytes engulf particles; one of several lung defense mechanisms by which foreign agents (biological and nonbiological) are removed from the respiratory tract.

**Pharyngeal regions**—the chamber at the oral end of the vertebrate alimentary canal, leading to the esophagus.

**Polymorphonuclear leukocytes (PMN)**—a type of leukocyte that phagocytizes particles; represents a secondary, nonspecific cellular defense mechanism.

**Potentiation**—the action of increasing one chemical by the action of another.

**Pulmonary edema**—an accumulation of excessive amounts of fluid in the lung extravascular tissue and air spaces.

**Pulmonary measurements**—measurements of the volume of air moved during a normal or forced inspiration or expiration.
Lung volume measurements = tidal volume, inspiratory reserve volume, expiratory reserve volume, residual volume (four basic independent volumes).
Capacities = combinations of basic volumes.
Total lung capacity (TLC) = tidal volume + inspiratory reserve volume + expiratory reserve volume + residual volume; the volume of gas in the lungs at the time of maximal inspiration or the sum of all volume compartments.
Vital capacity (VC) = tidal volume + inspiratory reserve volume + expiratory reserve volume; the greatest volume of gas that can be expelled by voluntary effort after maximal inspiration.
Functional residual capacity (FRC) = residual volume + expiratory reserve volume; the volume of gas remaining in the lungs at the resting, end-tidal expiratory position.
Inspiratory capacity (IC) = tidal volume + inspiratory reserve volume.
Inspiratory vital capacity (IVC) =the maximal volume that can be inspired from the resting end-expiratory position.
Expiratory reserve volume (ERV) = the maximal volume that can be exhaled from the resting end-tidal expiratory position.
Residual volume (RV) = that volume of air remaining in the lungs after maximal exhalation.
Residual volume (RV) = that volume of air remaining in the lungs after maximal exhalation.
Tidal volume = that volume of air inhaled or exhaled with each breath during quiet breathing.
Minute ventilation (MV) = the volume of gas exchanged per minute at rest or during any stated activity; the tidal volume times the number of respirations per minute.

**Respiratory cycle**—constituted by the inspiration followed by the expiration of a given volume of gas, called tidal volume. The duration of the respiratory cycle is the respiratory or ventilatory period, whose reciprocal is the ventilatory frequency.

**Respiratory frequency ($f_R$)**—the number of breathing cycles per unit of time, synonymous with breathing frequency ($f_B$).

**Synergism**—the joint action of agents so that their combined effect is greater than the sum of their individual effects.

**Systemic**—pertaining to or affecting the body as a whole; in air pollution toxicology, systemic generally refers to the body other than the respiratory tract.

**Thoracic**—of or pertaining to the chest.

**Trachea**—commonly known as the windpipe; a cartilaginous air tube extending from the larynx (voice box) into the thorax (chest), where it divides into left and right branches.

**Tracheobronchial region**—the area encompassed by the trachea to the gas exchange region of the lung; the conducting airways.

**Type 1 cells**—thin, alveolar surface, epithelial cells across which gas exchange occurs.

**Type 2 cells**—thicker, alveolar surface, epithelial cells that produce surfactant and serve as progenitor cells for Type 1 cell replacement.

**Ventilation**—physiological process by which gas is exchanged between the outside air and the lung.

## REFERENCES

1. National Research Council, Human Exposure Assessment for Airborne Pollutants: Advances and Opportunities, National Academy Press, Washington, D.C., 1991.
2. **Bates, D. V., Dungworth, D. L., Lee, P. N., McClellan, R. O., and Roe, F. J. C.,** *Assessment of Inhalation Hazards,* Springer-Verlag, New York, 1989.
3. *Toxicology of the Lung,* Gardner, D. E., Crapo, J. D., and Massaro, E. J., Eds., Target Organ Toxicology Series, Raven Press, New York, 1988.
4. *Concepts in Inhalation Toxicology,* McClellan, R. O. and Henderson, R. F., Eds., Hemisphere Publishing Co., New York, 1989.
5. *The Lung: Scientific Foundations,* Crystal, R. G., West, J. B., Barnes, P. J., Cherniack, N. S., and Weibel, E. R., Eds., Raven Press, New York, 1991.
6. **Gross, E. A. and Morgan, K. T.,** Morphometry of nasal passages and larynx, in *Comparative Biology of the Normal Lung,* Vol. 1, Plopper, C. G. and Pinkerton, K. E., Eds., CRC Press, Boca Raton, FL, 1992, Chap. 1.
7. *The Lung in the Transition Between Health and Disease,* Macklem, P. T. and Permutt, S., Eds., Marcel Dekker, New York, 1979.
8. **Phalen, R. F.,** *Inhalation Studies: Foundation and Techniques,* CRC Press, Boca Raton, FL, 1984.
9. Committee on the Institutional Means of Assessment of Risks to Public Health, National Research Council, Risk Assessment in the Federal Government: Managing the Process, National Academy Press, Washington, D.C., 1983.

10. **Frank, R., O'Neil, J. J., Utell, M. J., et al.,** *Inhalation Toxicology of Air Pollution: Clinical Research Considerations,* ASTMA, Philadelphia, 1985.
11. **Gardner, D. E.,** The use of experimental infections to monitor improvements in pulmonary defense, *J. Appl. Toxicol,* 6, 385, 1988.
12. **Gardner, D. E.,** Alterations in alveolar macrophage function by environmental chemicals, *Environ. Health Perspect.,* 55, 343, 1984.
13. **Gardner, D. E. and Graham, J. A.,** Attend immune functions and host resistance following exposure to common airborne pollutants, *Immunotoxicology,* Berlin, A., Dean, J., Draper, M. H., Smith, E. M. B., and Spreafico, F. Eds., Martinus-Nijhoff Publishers, Hingham, MA, 1987, 234.
14. *Air Toxic and Risk Assessment,* Calabrese, E. J. and Kenyon, E. M., Eds., Lewis Publishers, Boca Raton, FL, 1991.
15. Chemical carcinogens: a review of the science and its associated principles, *Fed. Reg.,* 50, 10372, 1986.
16. **Vainio, H., Hemminki, K., and Wilbourn, J.,** Data on the carcinogenicity of chemicals in the IRAC Monograph Programme, *Carcinogenesis,* 6, 1653, 1985.
17. **Furst, A.,** Yes, but is it a human carcinogen? *J. Am. Coll. Toxicol.,* 9, 1, 1990.
18. **Dahl, A. R.,** Dose concepts for inhaled vapors and gases, *Toxicol. Applied Pharmacol.,* 103, 185, 1990.
19. *Extrapolation Modeling of Inhaled Toxicants,* Miller, F. J. and Menzel, D. B., Eds., Hemisphere, New York, 1993.

# Chapter 3

# Carcinogenesis and Lung Cancer

*Steven H. Krasnow, M.D.*

## CONTENTS

## INTRODUCTION

Carcinogenesis is the process by which malignant cells arise from normal ones. While the notion that environmental agents could induce the formation of malignant tumors dates to the eighteenth century,[1] most of our current understanding of carcinogenesis has occurred in the last decade. This chapter addresses the principles of carcinogenesis, particularly as they apply to lung cancer. Our current concepts of carcinogenesis were developed largely from studies of chemical carcinogens, which play a major role in lung cancer development. However other carcinogens, including ionizing radiation, asbestos, and viruses, play important roles in human carcinogenesis. Recent developments in our understanding of oncogenes and their activation will also be discussed, as well as several host factors that influence susceptibility to cancer. Finally, it is important to understand some of the controversies surrounding the evaluation of carcinogens and their potential application to clinical decision-making.

Table 1  **Clinical Features of Lung Cancer Subtypes**

| | Nonsmall Cell | | | |
| | Adenocarcinoma | Large Cell | Squamous | Small Cell |
|---|---|---|---|---|
| % Cases | 35 | 15 | 30 | 20 |
| Location in lung | Peripheral | Peripheral | Central | Central |
| % In nonsmokers | 30 | <5 | <5 | <5 |
| % Operable | 50 | 50 | 50 | <10 |
| Response to chemotherapy or radiotherapy | Poor | Poor | Poor | Good |

## THE BRONCHIAL MUCOSA

Lung cancers usually arise from epithelial cells in the mucosa of the pulmonary airways. At least nine different cell types have been identified in the mucosa and six in submucosal glands.[2] Their relative numbers vary considerably from proximal to distal areas in the airways, and the ratios are also altered in various disease states.

The larger airways consist of pseudostratified columnar epithelium; all cells are in contact with the basement membrane; however, all are not in contact with the lumen. The latter include the basal cells, which actively divide and whose daughter cells differentiate into the other cell types. Another important cell not in contact with the lumen is the pulmonary neuroendocrine cell, which is believed to be of neural crest origin and exhibits amine precursor uptake and decarboxylation (APUD) properties. Ciliated cells reaching the lumen constitute the most common cells of the bronchial epithelium. The distal airways consist of a single layer of low cuboidal cells. Most are ciliated. Of the nonciliated cells, the most common is the Clara cell, a secretory cell of uncertain function. Goblet cells, which secrete mucus, are scattered throughout both large and small airways.

It is clear that the bronchial mucosa is a highly complex organ, with absorptive, secretory, and endocrine functions.[2] The cilia formed by the luminal cells act to clear particulate matter from the lung. The goblet cells bathe the luminal surface in protective mucus. These functions, as well as proliferation and differentiation of the basal cells, are under neurohumoral control mechanisms that may be disrupted by various diseases and toxins.

## LUNG CANCER

Lung cancer is the major cause of cancer death in both men and women (recently exceeding breast cancer in the latter) in the U.S.[3] At least 90% of the approximately 150,000 new cases annually are due to avoidable environmental exposures, particularly tobacco smoke.[4] Because the major cause of lung cancer is known, this malignancy provides a model for the study of carcinogenesis and of the other factors involved in cancer causation. It is now apparent that many host factors play a role in determining an individual's risk of developing lung cancer; these will be discussed below.

There are four major histologic types of bronchogenic carcinoma: epidermoid (or squamous), adenocarcinoma, large cell anaplastic, and small cell (Table 1). The latter, comprising about 20% of lung cancers, is believed to arise from pulmonary neuroendocrine cells. They are characterized by rapid growth, early metastasis, and moderate sensitivity to both radiotherapy and chemotherapy.

The other three major histologic subtypes, comprising about 80% of lung cancers, are similar in their growth kinetics and poor response to radiotherapy and/or chemotherapy. Thus, they are referred to collectively as ''non-small cell carcinoma.''

Epidermoid cancer is believed to arise from ciliated epithelial cells in the bronchial mucosa; because these cells are most numerous in the larger airways, squamous cell cancers tend to arise more centrally in the lung. They tend to cause extensive local tissue invasion; about 50% never metastasize beyond the thorax. Adenocarcinomas are believed to arise either from bronchiolar nonciliated cells, called Clara cells, or from alveolar lining cells. Thus, they tend to arise peripherally in the lung. They are characterized by a higher frequency of distant metastatic spread than with squamous cell tumors. Large cell anaplastic tumors are undifferentiated non-small cell tumors that cannot be readily classified as squamous or adenocarcinoma. They usually behave clinically like adenocarcinomas. However, the frequent presence of tumors with mixed histologic features, including small cell and non-small cell elements, suggests a probable common stem cell lineage for all the cell types.[5]

Tobacco use increases the risk for all the common histological subtypes. However, most of the approximately 10% of lung cancers not attributable to tobacco use are adenocarcinomas. For unknown reasons, adenocarcinomas also occur more frequently in women, while squamous cell carcinomas are more common in men.[6] With the rising incidence of lung cancer in women, adenocarcinoma is overtaking squamous cell carcinoma as the most common histologic subtype.[7]

Whether specific carcinogenic stimuli determine the histologic subtype that arises is unknown. Lung cancers induced in rodents by inhalation or intratracheal instillation of tobacco extracts or specific known carcinogens usually result in squamous cell or adenocarcinomas.[8] Nitrosamine-induced lung cancers are associated with hyperplasia of pulmonary neuroendocrine cells but few small cell carcinomas. Recently, Schuller and co-workers[9] found that lung cancers induced in hamsters with injected diethynitrosamine in combination with chronic hyperoxia (70% oxygen) resulted in a high frequency of neuroendocrine tumors. While the basis for this phenomenon was speculative, an early increase in bombesin and calcitonin levels in serum and hyperplastic pulmonary nodules prior to detection of malignancy suggests a role for these hormones in the development of these tumors.[10]

Factors regulating the growth of lung cancer cells have been extensively explored *in vitro*. Small cell lung cancer cells were difficult to grow in culture until it was recognized that they required defined media containing various growth factors, such as transferrin, hydrocortisone, and insulin,[11] and that the culture media in which these cells grew ("conditioned media") stimulated the growth of these cells.

Subsequently, it was shown that small cell carcinoma cell lines secreted a wide variety of hormones and growth factors. One of the most extensively studied factors, secreted by most small cell carcinoma cells in culture, is gastrin-releasing peptide (GRP), from which the neuropeptide bombesin is derived. This peptide stimulates the secretion of other hormones from a variety of cells. It is found in normal human lung, and increases in immunoreactive GRP occur in smokers compared with nonsmokers, a factor with possible potential in predicting lung cancer risk.[12] The receptor for GRP is a glycoprotein that binds a G protein. Many small cell lung cancer cells both secrete bombesin and express receptors for it, suggesting the possibility of autocrine growth stimulation as a mechanism of malignant transformation. Supporting this concept is the finding that antibodies to bombesin inhibit replication of small cell carcinoma cells in culture.[13]

Epidermal growth factor (EGF) has complex effects on the growth and differentiation of normal cells; it binds to a transmembrane receptor with tyrosine kinase activity. Epidermal growth factor may also serve as an autocrine growth factor in lung cancer. A marked increase in EGF receptors was first noted in squamous carcinoma cells,[14] and later, both nonsmall cell and small cell cancers were shown both to express EGF receptors and to secrete EGF.[15]

Expression of opioid receptors on both non-small cell and small cell lung cancer lines has been described.[16] Growth inhibition by opioids in these lines suggests a possible negative autocrine loop.

As will be discussed below, carcinogenesis often results in alterations of many genes encoding growth factors and/or their receptors. This illustrates the importance of these factors both in normal growth and differentiation as well as in malignant transformation.

## THE CANCER PHENOTYPE

Malignant cells share many of their morphologic features, metabolic functions, and growth characteristics with normal cells. It is this similarity that has made systemic therapy of most malignancies so difficult, since effective treatment necessarily would selectively destroy tumor cells and spare normal ones. Observed phenotypic alterations in malignant cells are frequently qualitative or quantitative distortions of the normal phenotype. However, several features allow us to distinguish neoplasms from normal tissues and malignant neoplasms from benign ones.

### Unregulated Cell Growth

Normal tissues are highly organized; cell production equals cell loss, except during growth and development, or during periods of tissue repair following injury. In tissue culture, normal cells exhibit contact inhibition, with growth slowing down as cells reach confluency.

In the bronchial mucosa, the basal cells comprise the generative compartment, providing daughter cells that differentiate into the other bronchial mucosal cell types.[2] Chronic injury to the bronchial mucosa, such as results from regular tobacco smoking, causes squamous metaplasia, wherein the number of ciliated and goblet cells are reduced and a stratified squamous epithelium prevails. The deeper layers contain dividing cells that result in more differentiated squamous cells toward the surface; most superficial layers are terminally differentiated and cannot divide, eventually sloughing off. This epithelial arrangement of multiple layers with terminally differentiated cells at the surface presumably minimizes injury to the dividing cells beneath. Nevertheless, despite an increased rate of cell turnover, a constant state of equilibrium is maintained, with cell production equal to cell loss.

In contrast, cell production in malignancy exceeds cell loss. *In vitro,* malignant cells continue to divide, piling up on one another. In malignant tumors *in vivo,* including lung cancers, failure of contact inhibition and other regulatory signals result in distortion of tissue architecture; accumulations of cells outstrip their own nutrient (blood) supply, often resulting in hypoxia and necrosis.

### Failure of Differentiation

While mitoses in normal tissues usually lead to a more differentiated phenotype of at least some of the daughter cells with a corresponding loss of mitotic capability, mitoses in malignant cells produce more cells of the same phenotype that retain their mitotic function, further promoting the imbalance of cell production and cell loss. Thus, an important feature of malignancy is the failure of cells to respond to normal signals that induce terminal differentiation.

Control of differentiation is complex. Studies of cultured human bronchial epithelial cells revealed that serum inhibited the growth of these cells and induced a squamous-appearing morphology.[17] The principle serum factor responsible was found to be transforming growth factor (TGF) beta-1,[18] one of a large group of peptides that act through membrane-bound receptors. The transforming growth factors are implicated in the regulation of growth and differentiation in many tissues.

In cultured normal human bronchial epithelial cells, terminal squamous differentiation is also induced by tumor promotors such as 12-O-tetradecanoyl-13-phorbolacetate (TPA), tobacco smoke condensate, and aldehydes.[19] Reduced responsiveness to these agents appears to be a necessary (but insufficient) aspect of malignant transformation.

Much attention has recently been directed to retinoids, which include vitamin A (retinol). These compounds induce terminal differentiation in epithelial tissues[20] and have been shown to protect against malignancies caused by environmental carcinogens.[21–23] Furthermore, vitamin A deprivation renders animals more susceptible to carcinogenesis with tobacco extracts.[24] In epidemiological studies, serum retinol levels have been inversely correlated with overall cancer risk[25] and with squamous cell lung cancer risk.[26] Recent chemoprevention trials have demonstrated that cis-retinoic acid can prevent head and neck cancers in high risk individuals.[27] Studies of dysplastic and malignant cultured oral cavity epithelial cell lines have shown a loss or dysregulation of retinoid-binding receptor activity in malignancy.[28]

### Immortalization

Normal cell lines have a limited life span; human embryonic cells in culture senesce after about 50 doublings; cells from an adult stop dividing after 30 to 40 divisions.[29] On the other hand, malignant transformation results in cells that can divide indefinitely. Many malignant cell lines have been maintained in culture for decades. Malignant transformation of hamster embryo cells in culture induced by various carcinogens suggests that immortalization appears early in transformation.[30] Immortalization appears to be necessary but not sufficient for malignant transformation; many immortalized cell lines exist that lack the other features of malignant cells and are nontumorigenic *in vivo*. Both failure of differentiation and failure of normal cell senescence in cancer imply that regulatory factors triggered by specific stimuli in normal cells are lost during malignant transformation.

### Invasion and Metastasis

The ability of tumor cells to invade or infiltrate normal tissues requires that the malignant cells produce proteolytic enzymes that disrupt normal tissue matrices[31] and angiogenesis factors to establish their own blood supply.[32,33] Tissue invasion and metastasis are the factors that distinguish most malignant neoplasms from benign ones. Benign tumors exhibit unregulated growth and may distort normal tissue architecture, without invading adjacent tissues.

## PRINCIPLES OF CARCINOGENESIS

The following principles relate the basic concepts underlying our current understanding of carcinogenesis. Some of these concepts are oversimplified but apply to the majority of experimental systems studied.

### CANCER IS CAUSED BY GENETIC ALTERATIONS

The genotypic (and most phenotypic) characteristics of a cell are passed on to daughter cells through mitosis, during which DNA is replicated. Each daughter cell usually resembles and behaves like cells in the line from which the parent cell arose because of heritable traits encoded in the cells, genomes. The altered characteristics of malignant cells are also passed on to their daughter cells during mitosis. Thus, cancer arises when a genomic change occurs in normal cellular DNA to allow the emergence of the cancer phenotype. This concept is referred to as the ''somatic mutation theory'' of cancer and is the most fundamental concept of carcinogenesis.

Table 2 **Stages in Carcinogenesis**

|  | **Initiation** | **Promotion** | **Conversion** |
|---|---|---|---|
| Implication | Initial step in carcinogenesis | Clonal expansion of initiated cells | Emergence of transformed phenotype |
| DNA damage | Yes | No | Yes |
| Reversibility | No | Yes | No |
| Latency | Short | Long | Short |
| Threshold? | No | Yes | No |

Studies cited throughout this chapter will support a link between carcinogenesis and DNA damage and/or failure to repair DNA damage. The inheritability of some cancers and the clonal nature of cancer (see below) further support this concept.

Our genome has developed through evolution over millions of years. Random mutations are ''experiments'' in evolution; most are detrimental and result in cell death or a growth disadvantage leading to eventual elimination. For mutations to result in cells with viable progeny, they must eventually confer a growth advantage over normal cells. As we shall see below, a single mutation is insufficient to cause malignant transformation; there is evidence that some steps leading to malignant transformation may result in a temporary growth disadvantage.[34]

## THE MULTISTAGE NATURE OF CARCINOGENESIS

Both laboratory models of carcinogenesis and epidemiologic data suggest that several sequential events must precede malignant transformation (Table 2). These include inheritable genetic alterations as well as epigenetic changes that ensure the survival and propogation of genetically altered cells.

It is now well recognized that the ''single hit'' concept, whereby a single genetic mutation results in malignancy, is untenable. Based on studies of mutation rates in the hgprt gene, it has been estimated that there are $1.4 \times 10^{-10}$ mutations/base pair/cell generation.[35] If this is representative of the rate in other loci, $2.8 \times 10^{15}$ mutated cells would arise in a lifetime. Despite the presence of DNA repair mechanisms and the fact that most mutations do not result in viable progeny, cancer would likely be much more frequent if a single mutation were sufficient to induce it.

However, the most cogent argument in favor of a multistage model of carcinogenesis in humans is that the incidence of most cancers rises exponentially late in life. If single random mutations resulted in malignant transformation, then cancer rates should be constant independent of age. A hypothesis requiring at least two genetic changes has been borne out thus far in all well-studied animal models and in some pediatric malignancies in humans. Furthermore, the presence of several stable genetic changes in most human cancers suggest that at least three to seven changes may be required in most malignancies, including lung cancer.[36]

Most currently accepted models of carcinogenesis include stages referred to as ''initiation,'' ''promotion,'' and ''conversion.''[37] These models derive from studies of the application of chemical carcinogens to animals. The most common models involve tumor induction in the skin of rodents. In these studies, the skin of the back is shaved, and prospective carcinogens are painted on the exposed skin. Tumors are histologically evaluated and quantitated over time.

A classic set of experiments employs a known carcinogen, such as benzo(a)pyrene (BP) or dimethylbenzanthracine (DMBA), painted on the skin of a mouse. A single high-dose exposure or multiple low-dose exposures result in formation of benign papillomas, some of which progress to malignant tumors. A single low-dose exposure will not result in tumor formation; however, subsequent to painting the same skin repeatedly with the phorbol ester 12–0-tetradecanoylphorbol-13-acetate (TPA; a derivative of croton oil), benign papillomas develop, some of which progress to frank malignancies. If TPA applications cease at the time of papilloma formation, some regress, but others persist. If the skin is treated repeatedly with TPA alone or prior to application of BP or DMBA, no papillomas or malignancies develop; thus, TPA is not itself a carcinogen (and, as noted above, may induce terminal differentiation in normal cells).

The changes following application of subcarcinogenic doses of BP or DMBA in these models are referred to as "initiation." Tumor formation results from application of TPA applied as long as 1 year after an initiator in this model. Thus, initiation is essentially irreversible. Furthermore, there is no threshold dose of initiator; the risk of initiation is proportional to the dose. Most initiators are mutagenic in bacterial assays (see below). Thus, initiation appears to result in permanent changes in the genome, which represent the first steps in malignant transformation.

The papilloma formation caused by TPA is referred to as "promotion." Most promotors are not mutagenic (and thus, not genotoxic). This explains why promotors must be applied after an initiator to exert its transforming effects and why its effects can be reversible. Promotors are, therefore, agents that facilitate clonal expansion of initiated cells.

The malignant changes that occur in some the papillomas after initiation and promotion are referred to as "conversion." Conversion is not reversible, implying that further genomic changes have occurred at this stage. Thus, carcinogenesis begins with the development of heritable genetic changes (initiation); the altered cells undergo clonal expansion (promotion), terminating in further genetic changes resulting in the malignant phenotype (conversion). Harris[37] also refers to a fourth phase called "progression" in which the features of invasion and metastasis become clinically evident. Progression is associated with its own distinct cellular characteristics.

It is now recognized that the promoting action of TPA is mediated by protein kinase C (PKC), a calcium-dependent protein kinase that serves as a transducer of signals from a variety of hormones and growth factors.[38] Activation of PKC results in phosphorylation of proteins involved in the regulation of cell growth and differentiation, such as epidermal growth factor receptor (EGFR). Other pathways leading to PKC stimulation (via formation of diacylglycerol and increases in intracellular calcium ions) are stimulated by certain oncogenes (see below); PKC, in turn, may stimulate the expression of other oncogenes.[39]

Other promotors act through diverse mechanisms, not all of which are fully understood. Potent promoting compounds, such as okadaic acid, do not activate PKC.[40] Many (but not all) promotors, including TPA, cause an inflammatory reaction in skin that results in a proliferative response that may itself accelerate neoplastic transformation in initiated cells (see below). The promoting action of asbestos in tobacco abusers is probably mediated through oxygen radical formation (see below). Hormones also constitute an important class of promotors that act through tissue-specific receptors to modulate cellular functions.[41]

The above idealized model of multistage carcinogenesis is surely oversimplified. Large doses or repeated applications of many initiators are sufficient to cause malignancy. Such compounds exhibit both initiating and promoting properties and are called "complete" carcinogens. Many nongenotoxic carcinogens act through diverse mechanisms and have not been well studied. Furthermore, relatively few studies have confirmed the multistage concept of carcinogenesis in tissues other than skin. The ability of certain retroviruses to induce cell transformation upon incorporation into a cell's genome challenges the multistage concept. However, it has been argued that the infection induces numerous genomic changes and does not constitute a "single hit" event (see below). Despite these disparities, the multistage concept of carcinogenesis is strongly supported by epidemiologic data in humans and constitutes a cornerstone of our understanding of carcinogenesis.

## CLONALITY OF CANCER

A clone is a population of cells arising from a single cell through successive mitoses. Most cancers are believed to represent clonal expansion of a single genetically altered cell. We have already alluded to the hypothesis that the development of the cancer phenotype requires multiple genetic changes, each of which occurs in low frequency. Therefore, the probability that these changes would occur simultaneously in numerous adjacent cells of the same tissue is extremely small, even when that tissue is exposed to carcinogenic stimuli.

The clonality of cancer has been evaluated in several ways, including expression of glucose 6-phosphate dehydrogenase (G6PD) isoenzymes in heterozygotes, cytogenetic analysis, and viral integration analysis. Expression of immunoglobulin light chains has also been useful in hematologic malignancies. Caution must be used in interpreting results of such analyses. Genetic instability in a clonal tumor may give rise to subclones with genomic differences; these may be mistakenly attributed to a polyclonal origin of the tumor. Alternatively, a clone with a selective growth advantage may arise from a polyclonal tumor and become predominant, leading to a misimpression that the tumor arose from one clone.

In females heterozygous for G6PD deficiency (an X-linked trait) or expressing polymorphism at the G6PD locus, tissues exhibit mosaicism for G6PD expression, with individual cells containing only normal or variant (or deficient) G6PD, due to inactivation of one X chromosome in each cell. Both benign and malignant tumors in such individuals have been observed to express only one or the other trait, indicating that they arose from one cell and that normal cells are not "recruited" by growing tumors.[42]

More recently, cytogenetics (including newer molecular genetic methods) has proven to be a more widely applicable approach to the evaluation of the clonality of tumors. The first cytogenetic abnormality found almost uniformly in a given tumor type was the Philadelphia chromosome in chronic myelogenous leukemia.[43] This was subsequently shown to be caused by a translocation t(9;22)(9[34];9[11]) resulting in hybridization of the abl oncogene on chromosome 9 with the bcr gene on chromosome 22.[44] Nonrandom chromosome abnormalities have now been described in a wide variety of hematologic and solid tumor malignancies.[45] More sophisticated banding techniques have revealed nonrandom chromosome defects in most tumors.

In non-small cell lung cancer patients, trisomy 7 or other abnormalities of chromosome 7 were described in eight of ten tumors.[46] Loss of alleles in 5q21 contiguous with the MCC gene, a defect associated with familial polyposis, has recently been found in many lung and (sporadic) colorectal carcinomas.[47] Frequent deletions from several chromosomes and double minute chromosomes have also been described.[48]

In small cell lung cancer, a deletion in the short arm of chromosome 3 has been described consistently.[49] The 3p21 locus is most consistently deleted, and a putative recessive gene at that locus has been postulated.[50] Miura et al.[51] have also reported frequent 5q deletions in small cell lung cancer.

## ONCOGENES AND SUPPRESSOR GENES

Many genes encode for proteins that are involved in cell growth and differentiation. It is reasonable to suggest that carcinogenesis may involve alterations in the function or control of such genes. In the past two decades, many examples of this phenomenon have been documented. Such altered genes that contribute to the cancer phenotype are called oncogenes. Their normal counterparts are usually essential for normal cell growth and differentiation and are referred to as protooncogenes.

## ONCOGENIC RETROVIRUSES

It has long been recognized that some viruses are tumorigenic when introduced into animals. The first of these to be studied was the Rous Avian Sarcoma virus.[52] Infection results in production of fibrosarcomas in chickens and transformation of chicken fibroblasts *in vitro*. The virus was subsequently found to carry a gene, called v-src, that was responsible for this oncogenic property. Later it was recognized that v-src exhibited striking homology to a normal cellular gene.[53]

The Avian Sarcoma virus is an example of a "retrovirus," so named because it encodes for the enzyme reverse transcriptase, allowing it to create a DNA copy using viral RNA as a template. The DNA transcripts become incorporated into the host genome, which produces more viral RNA as the cell replicates. Aside from this "pol" gene, the virus contains two other essential genes, the "gag" gene and the "env" gene, encoding core and envelope proteins, respectively. RNA sequences called "long terminal repeats" (LTRs) flank these three essential genes and provide regulatory signals. The fact that mutant Avian Sarcoma viruses lacking the v-src gene are fully competent to infect fibroblasts and to replicate but lack tumorigenicity suggests that the v-src may be acquired from c-src, perhaps incidentally, during infection. The viral gene responsible for malignant transformation is referred to as an "oncogene" and its normal cellular counterpart is referred to as a "protooncogene."

In the case of most other oncogenic retroviruses, the oncogene replaces part of the essential viral genome (pol, gag, or env genes), rendering such viruses incapable of replication without the aid of "helper" viruses. This lends further support to the concept that viral oncogenes are acquired from cellular protooncogenes. Because the Rous Avian Sarcoma virus retains all the genes essential for replication, it provides an unusually easy model to work with, and src is one of the most extensively studied of oncogenes.

## CELLULAR ONCOGENES

Over 40 cellular oncogenes have been discovered, most of whose products have been identified.[37,54] These findings have contributed greatly to our understanding both of the processes governing normal cell growth as well as those of carcinogenesis. Interestingly, almost all the oncogenes identified fall into one of four categories based on the function and location within the cell of their gene products. This suggests that relatively few types of processes are involved in cell growth and differentiation. While many more oncogenes are likely to be identified, it is unlikely that many more categories of oncogenes will be found. The following are categories of oncogenes.

### Protein Kinases

These are membrane-bound proteins constituting the largest class of oncogenes thus identified. They include src, abl, erb-B, v-kit (tyrosine kinases), raf, and mos (serine kinases). These proteins are similar to several growth factor receptors, which may explain their role in mitogenesis. For example, erb-B1 is homologous with the receptor for epidermal growth factor.[55] The fms oncogene exhibits homology with the monocyte colony-stimulating factor receptor (CSF-1).[56] The v-kit oncogene codes for a portion of platelet-derived growth factor (PDGF).[57] The oncogene c-raf is a cytoplasmic serine kinase that exhibits similarities to PKC; the v-raf product lacks an amino terminus negative regulatory domain and appears to be consitutively activated.[58]

## GTP Binding Proteins

These are the ras genes, a family of oncogenes whose products are 21 kDa (p21) proteins that bind guanosine nucleotides.[59] Such proteins, known as G proteins, bind to membrane receptors and act as signal transducers. When stimulated, the G proteins exchange guanosine diphosphate (GDP) for guanosine triphosphate (GTP). This results in activation of phospholipase C, which generates "second messengers" (diacylglycerol and inositol triphosphate); these, in turn, activate PKC (see above). The signal is normally abrogated by hydrolysis of the GTP moiety to GDP. Many ras oncogenes exhibit decreased hydrolytic activity, maintaining the complex in a prolonged state of stimulation to mitogenesis. Three separate genes encoding similar p21 products have been identified: H-ras, K-ras and N-ras. Oncogene activation almost always results from point mutations at amino acid positions 12, 13, or 61.

## Growth Factors

This class is characterized by v-sis, a retroviral oncogene found in the simian sarcoma virus (SSV), which causes fibrosarcomas and gliomas in monkeys. The oncogene encodes part of the PDGF B chain (the 5' end is supplied by the env gene of the virus). PDGF B acts as an autocrine growth factor, stimulating its own receptors, which are tyrosine kinases. These, in turn, stimulate signal transducing substrates that activate the nuclear protooncogenes c-fos and c-myc. The biology of this oncogene has recently been reviewed.[60]

## Nuclear Regulatory Proteins

These include c-myc, c-myb, c-fos, and jun oncogenes. The activities of these genes vary during the cell cycle, being most active prior to or during mitosis. The protooncogene products are involved in the regulation of gene transcription and probably bind directly to nucleic acids. The genes and their products are usually identical to those of the normal protooncogene; however, they tend to be overexpressed, either by gene amplification or by translocation of the gene, resulting in dysregulation.

The myc gene family has been most extensively studied; at least three genes have been well characterized in human tumors (c-myc, N-myc, and L-myc), and others have been identified.[61] c-myc is located on chromosome 8 and is often involved in the breakpoint region of the 8:14 translocation in Burkitt's lymphoma. All myc genes are highly homologous in their second and third exons but show considerable variation in their first (noncoding) exon.

The products of the jun and fos genes form heterodimers that constitute the AP-1 transcription factor.[62] AP-1 binds specific nucleotide sequences and activates the transcription of several proteins involved in growth and differentiation. The heterodimer forms an alpha helix held together by the hydrophobic interaction of regularly spaced leucine molecules along each protein strand—the so-called "leucine zipper"—and this configuration is essential for dimerization.[63]

## SUPPRESSOR GENES

While growth deregulation may result from activation of a protooncogene to an oncogene, it may also result from loss of genes that normally exert a negative effect on cell growth. Initially considered to be recessive, it quickly became apparent that mutated suppressor genes may act in a dominant fashion, often in concert with other oncogenes (see below).

Evidence for suppressor genes first derived from observations on retinoblastoma, a rare pediatric malignancy that may occur in familial and sporadic forms. In the former instance, tumors occur in infancy and usually affect both eyes; in the latter instance, they occur later and usually affect only one eye. These features led to the theory that individuals afflicted with the familial form were born heterozygous for a tumor suppressor gene and that loss of the normal allele resulted in malignancy.[64] In the sporadic form, the individual was born with a normal complement of these genes, and both had to be subsequently lost before malignancy occurred. The finding that retinoblastoma was frequently associated with deletions at the 13q14 locus led to the discovery of the "Rb" gene, whose product has since been identified as a 110kd nuclear phosphoprotein expressed in the retina and other tissues.[65] Deletion or mutation at the Rb locus has subsequently been shown to be associated with several other malignancies, including sarcomas, melanoma, lung and bladder cancers.[66] Reintroduction of Rb gene into malignant cells lacking it using a retroviral vector has resulted in reversion of many phenotypic features of malignancy.[67]

It has been postulated that the Rb gene product suppresses the exit of cells from the G1 phase of the cell cycle.[68] In normal cells, phosphorylation of the Rb gene product temporarily inactivates it, allowing progression of mitosis. Several DNA oncogene products inactivate the Rb gene product by binding to it, deregulating cell growth.[69,70]

At the present time, the p53 gene (chromosome 17p13) is the most intensively studied suppressor gene. It was originally discovered in SV40 transformed cells, in which its product complexed to the SV40 large T antigen.[71] Its half-life in transformed cells was much longer than in normal cells;[72] thus, total p53 levels were much higher in transformed cells. It was originally considered an oncogene because transfection of p53 from transformed cells into normal cells resulted in immortalization. However, it was later found that all the p53 proteins found in transformed cells were mutants[73] and that the wild type p53 gene product acted as a downregulator of transcription, inhibiting cell replication.[74] Introduction of wild type p53 gene into transformed cells by transfection may cause reversion to normal phenotype.[75,76] The precise mode of action of p53 is unknown, but a DNA binding domain recognizing a specific nucleotide sequence has been identified.[77]

Loss or mutations at the p53 locus have been associated with malignancies, including those of lung, breast, ovary, brain, and gastrointestinal tract. A variety of mutations have been described, but they tend to occur in one of a few regions of the protein; codons 175, 248, and 273 are most frequently involved. The mutation sites and specific base substitutions involved tend to be specific for certain malignancies.[77]

A rare familial cancer syndrome called Li Fraumeni syndrome, in which numerous malignancies, especially sarcomas, occur early in life, has been found to be associated with a germ line inheritance of a mutant p53.[78] Affected individuals may develop multiple primary cancers and may be particularly susceptible to posttherapeutic malignancies induced by radiotherapy and/or chemotherapy (see below).

Identification of tumor suppressor genes is more difficult than finding oncogenes because, in the case of the latter, one needs to identify an abnormal protein, but in the former, one is usually looking for a gene product that is not present or not functional. One clue to the presence of a tumor suppressor gene in a chromosome is loss of heterozygosity (LOH). If one allele of a tumor suppressor gene is inherited as a germ line mutation, the second may be lost by deletion. In these cases, the cell may be hemizygous for the chromosome bearing the gene, the chromosome may be lost and reduplication of the chromosome with the mutant allele may occur, or mitotic recombination may result in duplication of the mutant allele. Alternatively, a second mutation may occur in the normal allele. Studies in retinoblastoma cells revealed that all these events occur.[79] In the first three instances, heterozygosity in the region of the gene of interest (or in the entire chromosone), detectable by analyses of restriction fragment length polymorphisms (RFLPs), is lost. In the case of p53 gene, LOH in the region of chromosome 17p13 has frequently been detected. While heterozygosities do not establish the presence of tumor suppressor genes, they frequently provide a clue to their locations.

Several other suppressor genes have been identified and others have been predicted, including the DCC gene (DCC = deleted in colon cancer) on chromosome 18q, which has been associated with the late stages of colon carcinogenesis.[80] Wilms tumor is also associated with a putative tumor suppressor gene on chromosome 11p13.[81] Recently, LOH on chromosome 8p was described in 40% of non-small cell lung cancer surgical specimens and in other tumors.[82]

## ONCOGENE ACTIVATION

Oncogenes may be activated by several different mechanisms. If a retrovirus is incorporated into the cellular genome near a protooncogene, conversion to an oncogene may result from displacement of the gene away from its normal regulatory signals in the genome, from upregulating signals provided by retroviral LTRs, or from disruption of the normal structure of the protooncogene or incorporation of the protooncogene into the viral genome.[83] Mutations in protooncogenes due to chemical carcinogens[84] or radiation[85] may similarly uncouple a protooncogene from its normal regulatory signals.

As a general rule, protooncogenes whose protein products act in the cytoplasm become activated by changes in structure.[86] The best studied examples are the ras genes, where point mutations occurring almost exclusively in codons 12, 13, or 61 result in oncogene activation. Even low levels of the resulting mutant protein can transform cells. For src, abl, and neu genes, overexpression of the normal protein product has little or no ability to transform cells, whereas mutant proteins have powerful transforming ability.

In contrast, protooncogenes, whose protein products act in the nucleus, are usually activated by deregulation.[86] Amplification of the gene products for myc genes may result from chromosomal translocation, separating the gene from its normal regulatory signals or from nearby insertion of a retrovirus. The p53 gene product, a protein with a short half-life, may become activated by binding to the SV40 large T antigen, which stabilizes it.

## DNA VIRAL ONCOGENES

Only a small percentage of RNA retroviruses are oncogenic. In contrast, most categories of DNA viruses include oncogenic strains.[87] Unlike oncogenes associated with retroviruses, those associated with DNA viruses do not appear to be homologous with recognized cellular protooncogenes. However, their action is mediated by interaction with cellular oncogene products. For example, several DNA viral oncogene products, including the adenovirus E1A protein and the polyoma virus SV40 large T antigen, contain homologous sequences that bind to the products of tumor suppressor genes.[87] The resulting complex disrupts the downregulatory function of these gene products. Some human papillomaviruses, implicated in several human malignancies, encode for a protein that binds the p53 gene product and promotes its degradation.[88]

## COOPERATION AMONG ONCOGENES

It has been stressed above that carcinogenesis is a multistage process, and no single genetic alteration is likely to result in the malignant phenotype. This concept is supported by findings suggesting that oncogenic cell transformation requires cooperation

Table 3  **Oncogene Products in Lung Cancer**

| Class | Small Cell | Non-Small Cell |
|---|---|---|
| Protein kinases | fms, raf, kit | fur, fes |
| GTP binding proteins | — | ras |
| Growth factors | — | — |
| Nuclear proteins | myc, myb | jun, fos |
| Tumor suppressor genes | Rb, p53 | p53 |

among different classes of oncogenes. With certain exceptions, oncogenes whose products act at the plasma membrane or cytoplasm (e.g., the ras family, src and erbB oncogenes) induce changes in cell shape, anchorage-dependent growth, and requirements for growth factors but do not immortalize cells. In contrast, oncogenes whose products are located in the nucleus (e.g., the myc family, fos and jun oncogenes) immortalize cells but do not induce the other phenotypic changes.[86,89] Since both processes must occur in carcinogenesis, cooperation between two or more oncogenes must occur in most circumstances.

As an example of the above, Pfeifer et al.[90] found that simultaneous introduction of c-raf and c-myc oncogenes into human bronchial cell line BEAS-2B resulted in malignant transformation with tumors in nude mice resembling large cell carcinomas expressing neuron-specific enolase. Transfection within either oncogene alone did not result in tumorigenic transformation.

Cooperation between dominant oncogenes and mutant suppressor genes has also been well documented. While a mutant p53 was able to immortalize murine cells, malignant transformation required the presence of ras gene expression as well.[91]

## ONCOGENES IN LUNG CANCER (Table 3)
Activated oncogenes play prominent roles in lung carcinogenesis, and several recent comprehensive reviews exist.[54,92,93] Only a brief overview will be presented here.

### Small Cell Lung Cancer
The myc oncogene (located on chromosome 8) has been most extensively studied in small cell lung cancer. This family of oncogenes encodes for nuclear proteins involved in transcription regulation. Deregulation results from gene amplification and from upregulation of promotor genes. c-myc was the first oncogene identified in small cell lung cancer.[94] It was found mainly in morphologically variant cell lines with aggressive behavior and rarely in cell lines with classic morphology. Its amplification is a negative prognostic factor.[95] N-myc (on chromosome 2), an oncogene first identified in neuroblastoma cells, has been identified in about 20% of small cell cancer cell lines studied,[96] both in classic and variant forms. L-myc (on chromosome 1) has been found in about 10% of small cell tumors studied, both in classic and variant histologies.[97]

Almost all examined small cell lung cancer lines overexpress c-myb, another oncogene (chromosome 6) involved in nuclear transcription regulation.[98,99] This oncogene has not been found in non-small cell cancers.

The fms (chromosome 5) has also been found in many variant small cell lung cancer lines.[98] The normal gene product encodes for a tyrosine kinase that is a monocyte colony-stimulating factor. However, the c-fms gene product in small cell lung cancer is not homologous at the 5′ terminus and might exert a different function.

Most small cell lung cancers also exhibit activation of c-raf and LOH at the c-raf oncogene locus on 3p25.[100] This is somewhat distal to the 3p21–3 deletion site classically associated with small cell lung cancer.[49]

The oncogene c-kit encodes for a tyrosine kinase, which serves as the receptor for stem cell factor (SCF). This growth factor is produced by marrow stromal cells and stimulates the growth and differentiation of early marrow progenitors. It has been proposed as a therapeutic agent to enhance marrow recovery after intensive chemotherapy. However, some solid tumors, including small cell lung cancers, also produce SCF as well as c-kit, suggesting that SCF may be an autocrine growth factor for small cell lung cancer.[101,102]

Rb gene deletions and/or mutations have been detected in most small cell lung cancers.[103-105] Inactivation of p53 gene through various types of mutation have been described.[106] As with Rb gene product, malignant transformation is associated with prolonged expression, which has been demonstrated in one study by increased immunohistochemical staining of the gene product.[107]

### Non-Small Cell Lung Cancer

The most extensively studied oncogenes in non-small lung cancer are in the ras family. K-ras activation has been identified in 15 to 50% of lung adenocarcinomas[108,109] and is associated with a poor prognosis, particularly when the mutation occurs at codon 12.[110] ras mutations are rarely found in other non-small cell lung cancers, nor in small cell carcinomas. A similar spectrum of K-ras activation has been found in dogs dying of spontaneous lung cancers.[111]

The erbB-1 oncogene, which codes for EGFR, is frequently active in squamous cell lung cancers.[112] The erbB-2 (neu) oncogene, which exhibits homology to EGFR but is distinct from it, is frequently overexpressed in breast adenocarcinoma and constitutes a negative prognostic factor. It has also been found to be expressed in 28% of adenocarcinomas of the lung but rarely in squamous cell lung carcinomas.[113]

The c-fur oncogene, which encodes for a membrane-bound protein, was found to be overexpressed in most non-small cell lung cancer surgical samples studied, but not in small cell carcinomas.[114] It is located near the locus for the fes protooncogene, which encodes for a membrane-bound tyrosine kinase. Oncogenic activation of fes has also been found in many non-small cell cancers, especially adenocarcinomas.[115] Coactivation of these protooncogenes may occur in some lung cancers. Cooperation between fes and H-ras genes have also been suggested by their frequent coactivation in adenocarcinomas.[115]

A wide variety of p53 gene deletions and mutations were described in 74% of non-small cell lung cancer cell lines of all histologies.[116] In other studies, about half of fresh surgical specimens of human non-small cell lung cancers of various histologies bore p53 mutations.[117-119] These also included a variety of mutations and deletions. Accumulation of p53 protein in surgical specimens of non-small cell lung cancers detectable by a monoclonal antibody linked immunoperoxidase assay was associated with a poor prognosis.[120]

## DNA DAMAGE AND REPAIR MECHANISMS

Carcinogenesis requires genomic changes, which may arise from damage to DNA. Various mechanisms of DNA damage have been described that are positively correlated with malignant transformation. Adducts are covalent complexes between DNA bases and carcinogens or their metabolites. DNA strand breakage and base pair mismatches are caused by free radical damage. Viruses and activated oncogenes (see above) may result in major chromosomal alterations and/or rearrangements of genetic material within chromosomes.

The frequency of cancer and/or fatal mutations would be unacceptably high if the cell had no mechanism by which to defend itself from these insults or to repair damaged DNA. Several mechanisms have evolved to prevent DNA damage; in addition, there are enzymes to repair specific DNA abnormalities as they occur.

## DNA DAMAGE

### DNA Adducts

DNA adducts are the products of covalent bonding between DNA bases and activated carcinogens or their metabolites. Not all DNA adducts predispose to carcinogenesis. Those that are likely to do so render the involved codon more susceptible to inheritable replication errors. The adduct itself cannot induce a mutation because it is not an inheritable DNA alteration. It is an error in the opposite strand that must occur during replication. Since DNA repair mechanisms exist to excise adducts or mismatched base pairs, the rate at which these repair processes occur and the rate of DNA replication affect mutation incidences resulting from adducts.

One of the most important classes of compounds forming DNA adducts is the nitrosamines. These are formed endogenously from the nitrosation of heterocyclic amines. Such compounds exist in many cooked food products.[121] Tobacco smoke also contains "tobacco-specific nitrosamines," derived from nitrosation of nicotine and other alkaloids in tobacco.[122]

One of the most clinically important types of carcinogen-DNA adducts is the $O^6$-alkylguanines, which result from exposure to nitrosamines such as 4-(methylnitrosamino)-1-(3-pyridyl)-1-butanone (NNK) and other alkylating agents.[123] While the most common adducts formed by nitrosamines are $N^7$-alkylguanines, these adducts and others formed by nitrosamines correlate poorly with mutagenicity. In contrast, the mutagenic $O^6$-alkylguanines probably confer conformational changes in DNA that disrupt base pairing with cytosine. Their concentrations in target rat bronchial mucosal cells is linearly correlated with tumor induction.[124]

Alkylguanine moieties adjacent to a 3′ purine seem to form more readily and are repaired more slowly than those adjacent to a pyrimidine.[125] This suggests that certain nucleotide sequences are more sensitive to mutagenesis following adduct formation. A relevant example is methylnitroso-urea (MNU)-induced activation of H-ras oncogenes (see above); these almost always occur at the second guanine of codon 12 (GGC).[126]

Several cytotoxic drugs used in cancer chemotherapy, have been shown to be carcinogenic (see below) and form alkylguanine adducts in DNA, including mechlorethamine and cyclophosphamide.[127] Cis-diamminodichloroplatinum(II) (cisplatinum) forms intrastrand adducts with adjacent guanine moieties, particularly at the N-7 positions.[128] Many other classes of chemotherapy agents form adducts with DNA, but the significance of these adducts in the drugs' mechanisms of action and/or in carcinogenesis are less certain.

Another important group of adducts is formed by the activated byproducts of polycyclic aromatic hydrocarbons (PAHs), such as BP. An important BP intermediate, benzo(a)pyrene-7,8-diol-9,10-epoxide (BPDE), forms DNA adducts that are frequently detected in tissues of workers exposed to PAHs, such as coke oven workers and roofers.[129]

Many adducts, including the ones discussed above, often result in instability of N-glycosidic bonds linking bases to the deoxyribose sugar moiety. The ensuing loss of the bases from the DNA molecule creates apurinic or apyrimidinic (AP) sites that can result in copying errors during transcription.[130] The mutagenic potential of such AP sites has been demonstrated in bacterial cells and, in a few studies, in mammalian cells; in one such study, the H-ras protooncogene was shown to be activated by depurination.[131] The significance of mutagenesis caused by AP sites may be evidenced by the presence of a diverse class of enzymes to repair these defects (see below).

Some adducts associated with increased risk of carcinogenesis form in the absence of specific chemical carcinogens, such as 8-hydroxyguanine. This adduct appears to be formed from compounds that result in free radical formation because it is not associated with carcinogenic analogues that do not induce free radical formation.[132]

### Assessment of DNA Adducts

The dosimetry of DNA adduct formation is complex. Clearly, there is a poor correlation between the concentration of carcinogen in the environment or duration of carcinogen exposure and risk of carcinogenesis. This is due to the multiple other factors affecting rates of carcinogenesis, including enzymatic activation of carcinogens, DNA repair rates, and DNA replication rates. The rates of each of these processes may vary widely among both tissues and individuals and are governed, in turn, by both genetic and epigenetic influences (examples of the latter include enzyme induction and toxicity resulting in increased cell turnover). The "molecular dose" of a carcinogen,[133] the concentration at the level of DNA, may be quite different from that to which the whole organism is exposed and is probably better correlated with cancer risk. Therefore, assessment of DNA adduct formation is important for identifying the sites of action of particular carcinogens and for evaluating the risks of particular carcinogens to individual subjects.

The molecular dose of carcinogens detected as DNA adducts has been assessed by several types of methods. Recent technologic advances have allowed accurate measurement of specific DNA adducts of extremely low incidence.[134] As examples, an ultrasensitive enzyme radioimmunoassay (USERIA) can detect as few as one PAH-DNA adduct in $10^7$ nucleotides, and high pressure liquid chromatography-linked synchronous fluorescence spectrophotometry (HPLC-SFS) can specifically detect as few as one BPDE-DNA adduct in $5 \times 10^6$ nucleotides.[134,135] $^{32}$P-postlabeling, combined with thin layer chromatography and autoradiography of DNA hydrolysates, has been used to screen for several adducts in various tissues.[136] Examination of exfoliated bladder cells by this method may have practical application in monitoring individuals at risk for bladder cancer.[137]

Hemoglobin adducts have been used to assess long-term exposure to carcinogens. Activated carcinogens form adducts with proteins, as well as with nucleic acids. Furthermore, while nucleic acid adducts are repaired, protein adducts are not. The stability of such protein adducts may, therefore, provide an indirect measurement of long-term carcinogen exposure, which may be relevant to carcinogenesis risk. As an example, 4-aminobiphenyl (4ABP), a potent bladder carcinogen found in tobacco smoke, is metabolized to an electrophilic intermediate that forms DNA adducts. Hemoglobin adducts of 4ABP have been measured in smokers and, despite great interindividual variations, changes in levels correlated with smoking cessation.[138] Metabolites of other tobacco carcinogens, such as ethylene oxide, form adducts with hemoglobin whose concentrations correlate well with smoking history.[139] Hemoglobin is a convenient protein to study because it has a long half-life and is readily available in gram quantities without invasive methods.

While carcinogen testing cannot be done in humans, studies performed in *in vitro* explants of human tissues or cultured human cells[140] confirm that DNA adduct formation is qualitatively similar to that seen *in vivo* and that cancer risk is quantitatively correlated with the adduct formation. This confirms the relevance of these models for studying carcinogen exposure risks in humans.

Table 4  **Biologically Important Oxygen Active Species**

| Free Radicals | Structure | Comment |
|---|---|---|
| Superoxide | $O_2\cdot^-$ | Most common free radical in biological systems |
| Hydroxyl | $OH\cdot$ | May induce lipid peroxidation |
| Perhydroxyl | $HO_2\cdot^-$ | Exists in equilibrium with superoxide |
| Peroxyl | $ROO\cdot^-$ | Organic group replaces hydrogen in perhydroxyl radical; long half-life |
| **Other Species** | | |
| Hydrogen Peroxide | $H_2O_2$ | Converted to hydroxyl radicals, especially in presence of superoxide |
| Delta Singlet Oxygen | $O_2$ | Long half-life |

## Reactive Oxygen Species

Oxygen is ubiquitous on Earth and constitutes 21% of the atmosphere we breathe. Oxygen is the major electron acceptor in aerobic respiration; thus, it is essential to the life of most eukaryotes. However, our dependence on oxygen is precarious; its major metabolic byproducts are chemically reactive species that are highly toxic to most macromolecules. Evidence of this is found in the numerous mechanisms that have evolved to protect us from oxidative damage, including enzymes (e.g., superoxide dismutase, peroxidase, and catalase) and antioxidants (e.g., vitamins C and E, glutathione). The forces of oxidative stress and antioxidant protective mechanisms are normally in equilibrium, but the former can overwhelm the latter under appropriate circumstances.

Molecular oxygen in the atmosphere is relatively inert because the two electrons in the outermost orbitals are paired. Oxygen species with unpaired electrons in the outer orbitals (''free radicals'') and other reactive forms (Table 4) are unstable, and their chemistry has been reviewed.[141,142] These include superoxide anion ($O_2\cdot^-$), the form most commonly produced by oxidative stress in biological systems, and hydroxyl radical ($OH\cdot$). In addition, reactive oxygen species that are not free radicals result from oxidative stress, including hydrogen peroxide ($H_2O_2$) and singlet oxygen. Superoxide anion exists in equilibrium with a protonated form, the perhydroxyl radical ($HO_2\cdot^-$), which is in low concentrations at neutral pH. When the hydrogen atom is replaced by an organic group, peroxyl radicals ($ROO\cdot^-$) are formed. They commonly result from lipid peroxidation.

Oxygen-reactive species in cells may arise endogenously from mitochondrial or microsomal electron transport reactions, from reactions catalyzed by oxidant enzymes (e.g., oxidases, cyclooxygenase, and lipoxygenase), from products of phagocytic cells such as neutrophils and monocytes, and from nonenzymatic oxidation of substances such as iron and epinephrine. Exogenous sources of such compounds derive from ionizing radiation, products of tobacco smoke, diesel fuel exhaust, and other air pollutants, and from numerous oxidant toxins (e.g., paraquat) and other compounds.

The location of nucleic acids within the nucleus of the cell, a site relatively remote from most sources of oxygen-reactive species, may not be coincidental. Similarly, the central location of biologically important metal ions within large molecules (e.g., iron within globin and ferritin) may reduce their access to oxygen reactive species. Most superoxide anions and hydrogen peroxide are generated extracellularly or in the cytoplasm. Hydroxyl ions must be formed close to DNA to damage it because of their short half-life. While superoxide anions and hydrogen peroxide are not highly reactive, the latter is converted into hydroxyl radical, a highly reactive species, in the presence of metal ions; this reaction is accelerated by the presence of the super-oxide anion. In addition, hydroxyl radicals can be formed by the reaction of ionizing radiation on water. Hydroxyl ions formed extracellularly may cause lipid peroxidation; these species are relatively stable and may migrate to the nucleus. Marnett has summarized data suggesting a role for peroxyl free radicals in malignant transformation.[143]

Exposure of bacteria, viral plasmids, or mammalian cells to activated neutrophils or macrophages or to hydrogen peroxide may result in DNA damage, including single strand breaks, base substitutions, and base deletions. At least three adducts are associated with oxidative DNA damage: 5-hydroxymethyluracil, thymine glycol, and 8-hydroxyguanine (discussed above). Activated neutrophils and macrophages from normal individuals but not from those with chronic granulomatous disease may induce mutations and chromosomal abnormalities.[144] Hydrogen peroxide can induce malignant transformation in mouse fibroblasts[145] and induces squamous metaplasia in mammalian bronchial epithelium.[146]

The role of reactive oxygen species in carcinogenesis is not limited to DNA damage. Oxidants also induce growth and alter gene expression through some of the same signalling mechanisms used by growth factors and phorbol esters, including modulation of PKC activity[147] and activation of c-fos and c-myc oncogenes.[148,149]

The discussion above suggests that oxygen-reactive species may form a common final pathway in lung carcinogenesis from many diverse sources. Tobacco smoke is rich not only in oxidizing chemicals, but also in transitional metal ions and aldehydes which may induce lipid peroxidation, and radioactive isotopes such as polonium-210. Polyaromatic hydrocarbons form free radicals, and their reactivity is correlated with carcinogenic potential. Moreover, tobacco smoke, as well as other lung carcinogens such as asbestos, induce inflammation of the bronchial mucosa with accumulation of activated neutrophils and macrophages. Finally, whereas ionizing radiation may directly damage DNA, most of its energy is absorbed by water, with the consequent production of hydroxyl ions and other unstable oxygen species. Thus, oxygen intermediates may mediate lung carcinogenesis from apparently diverse insults.

## Altered Cytosine Methylation

A small percentage of cytosine residues of DNA in vertebrate cells undergoes post-transcriptional methylation at position 5.[150] The extent of methylation is species-specific and tissue-specific and is under genetic control. In most systems studied, DNA of actively expressed genes is less methylated than the corresponding DNA in tissues not expressing that gene. Many viruses, including some oncogenic viruses, exhibit virtually no DNA methylation. Several mammalian tumors exhibit less DNA methylation than corresponding host tissue. These observations suggest that DNA methylation may serve to downregulate DNA transcription. Furthermore, methylation of specific sequences might commit cells to differentiation.

It has also been suggested that malignant transformation may be triggered by changes in DNA methylation.[151] If methylated residues are excised in the process of repairing damaged DNA in a protooncogene, activation might result. Methylcytosine residues have been noted to exhibit increased susceptibility to spontaneous mutations through deamination or demethylation reactions.[152] Carcinogen-induced hypomethylation of cultured human bronchial cells has also been demonstrated.[153]

## DNA REPAIR AND PROTECTION

The histologic features of the bronchial mucosa serve to minimize exposure to DNA-damaging agents. The basal cells, which are the progenitors of the lining epithelium, are located in the deepest layers of the mucosa, farthest from contact with exogenous carcinogens. The mucosal surface is lined with ciliated cells that trap and expel foreign airborne substances from the lung. The mucous lining of the mucosa also protects it from vapors and particles containing carcinogens (e.g., tobacco smoke). Finally, the alveolar macrophages engulf foreign substances and aid in their clearance from the lung, while providing immune surveillance against viruses and, possibly, malignantly transformed cells. Defects in mucociliary clearance, both hereditary (e.g., ''immotile cilia syndromes'' such as Kartagener's disease) and acquired (e.g., from chronic exposure to tobacco smoke) may affect an individual's susceptibility to lung carcinogenesis.

Several enzymes recognize abnormal DNA base pairing or posttranslational DNA alterations, such as adducts, and function to repair or excise them.[130] The enzymes probably recognize distortions in the conformation of the nucleic acid helix rather than specific structural abnormalities. Studies in Chinese hamster ovary cells suggest that DNA repair capacities are conserved by preferential activity in expressed genes.[154]

An important enzyme involved in repair of DNA due to alkylating agents is $O^6$-methylguanine-DNA methyltransferase ($O^6$-MT), an enzyme that transfers the alkyl group to one of its own cysteine moieties, inactivating that enzyme molecule in the process. While this enzyme is inducible in bacteria, it appears not to be inducible in vertebrates. Therefore, this repair mechanism might be overwhelmed in the face of large or chronic exposure to carcinogens such as nitrosamines. Large differences in $O^6$-MT activity among tissues and among individuals have been noted.[155] Whether these differences are genetic or epigenetic is not clear.

Glycosylases are a group of at least six enzymes that recognize specific adducts and remove the affected base by cleaving the N-glycosylic bond with deoxyribose, creating an AP site.[130] The AP site is subsequently repaired by removal of the unbound deoxyribose moiety by an endonuclease, followed by reconstitution with DNA polymerase, using the opposite strand as a template. A ligase is required to incorporate the reconstituted strand into the DNA. Larger lesions, such as pyrimidine dimers formed by ultraviolet light damage, can be repaired by excision of up to 30 bases by an endonuclease, followed by repair with polymerase and ligase. Some large lesions are ignored during DNA replication and are repaired later; in some cases, the abnormal region is not replicated, leaving a gap that is repaired after transcription.

Protection of DNA and other macromolecules from oxidative damage is afforded by several enzymes, antioxidants such as vitamins C and E, which scavenge for free radicals, and the reducing substrate glutathione. The latter, a small molecule with a sulfhydryl side group, is ubiquitous in mammalian tissues. The reduced sulfhydryl group becomes oxidized by reactive oxygen species. It can be regenerated by acquiring a hydrogen atom from NADPH, which is, in turn, generated from the hexose monophosphate shunt. A deficiency of glucose 6-phosphate dehydrogenase (G6PD), the initial enzyme in this pathway, results in hemolytic anemia under conditions of oxidative stress, because depletion of reduced glutathione in erythrocytes results in oxidative damage to the red cell membrane. The principle enzymes involved in protection from oxidative damage are superoxide dismutase, which catalyzes the formation of superoxide to hydrogen peroxide, catalase, and peroxidase, which convert hydrogen peroxide to water, the latter using glutathione as a substrate.

Epidemiologic studies examining a link between anticarcinogenesis and dietary intake of antioxidants such as vitamins C and E are controversial and frequently

68

Table 5  **Hereditary Defects in DNA Repair**

| Disease | Clinical Features | Defect |
|---|---|---|
| Xeroderma pigmentosum | Skin malignancies (multiple) | Failure to excise pyrimidine dimers |
| Ataxia telangiectasia | Cerebellar ataxia, oculocutaneous telangiectasia, immunodeficiency | Abnormality at chromosome 11q22–23 |
| Bloom syndrome | Short stature, photosensitivity, leukemias | DNA ligase deficiency |
| Fanconi's anemia | Renal and osseous defects, skin hyperpigmentation, aplastic anemia, leukemia | Sensitivity to DNA crosslinking agents |

flawed by the need to obtain retrospective dietary histories. In one large, carefully conducted prospective study,[156] the risk of lung cancer in nonsmokers was inversely related to the intake of these micronutrients and also of margarine and fruits. The contributions of other unmeasured dietary factors could not be ruled out.

The importance of DNA repair mechanisms may be illustrated by autosomal recessive disorders characterized both by defects in the ability to repair or defend against DNA damage and by susceptibility to various malignancies.[37,157] These include xeroderma pigmentosum (XP), ataxia telangiectasia (AT), Bloom syndrome (BS), and Fanconi's anemia (FA) (Table 5).

Patients with XP are highly sensitive to DNA damage from ultraviolet light and develop high rates of skin malignancies, including melanoma. Ultraviolet light induces pyrimidine dimerization in DNA that is normally excised and repaired. In XP, an early step in the excision process does not occur.[158]

In AT, patients exhibit cerebellar ataxia, oculocutaneous telangiectasias, immunocompromise with frequent pulmonary infections, and endocrine abnormalities. Homozygotes are highly sensitive to ionizing radiation and exhibit high rates of lymphoma and other malignancies.[159] The biochemical defect in AT is unknown, but a chromosomal defect at 11q22–23 has been suggested by genetic linkage analysis.[160]

Bloom syndrome is characterized by short stature, photosensitivity, and a high rate of leukemia. Cells from BS patients exhibit a high rate of sister chromatid exchange[161], and a DNA ligase defect has been detected.[162,163]

Fanconi's anemia is characterized by renal and osseous defects, skin hyperpigmentation, aplastic anemia, and a high rate of leukemia. Frequent chromosome breakages are observed. Patients are highly sensitive to DNA crosslinking agents.[164] The defect has not been identified.

The above disorders illustrate extremes in interindividual variation in cancer susceptibility, probably due to alterations in DNA repair capabilities. More subtle variations have been detected by quantitatively assaying bleomycin-induced chromosome breakage in cultured lymphocytes from patients with and without cancer exposed to tobacco and/or ethanol.[165] The subject of interindividual variations in cancer susceptibility is discussed in a separate section below.

## ENVIRONMENTAL LUNG CARCINOGENS

While there is overwhelming evidence that lung cancer is caused by exogenous carcinogens, principally from tobacco smoke, the body is also constantly exposed to endogenous carcinogens. Nitrosamines are formed *in vivo* from many food products.[166] Oxygen reactive species are constantly generated by neutrophils and macrophages; their production may be greatly increased during inflammatory processes.[167] Hormones act as cocarcinogens in several common tumors, probably by increasing cell turnover and, thus, the risk of mutation.[168] Estrogens clearly influence the development and growth of breast and endometrial cancers; likewise, the growth of prostate cancer is strongly influenced by androgens. While the role of endogenous carcinogens in human cancer formation has not been established, *in vitro* studies and the presence of DNA repair mechanisms described above suggest that they must be added to any superimposed environmental challenge.

## CHEMICAL CARCINOGENS

### Tobacco

An association between the rising popularity of cigarette smoking and increasing lung cancer incidence was first noted in the 1930s in the U.S. By the 1950s, case control studies[169–171] had established a close relationship. In 1976, an important study by Doll and Peto[172] in British physicians confirmed this relationship and also demonstrated that smoking cessation was associated, after 3 years, with a fall in lung cancer incidence to nearly baseline levels within 15 years. Such data, along with animal and *in vitro* studies, have unequivocally demonstrated that tobacco causes lung cancer. Tobacco abuse accounts for about 95% of the lung cancer incidence and about 30% of all cancer deaths in the U.S.[173,174]

Epidemiological studies demonstrate that changes in lung cancer incidence lag about 20 years behind changes in tobacco use.[3] This interval includes the period of carcinogenesis and the duration of subclinical cancer growth. Cell kinetic studies suggest that about 30 doublings (representing a mass 1cm in diameter and containing $10^9$ cells) occur before most cancers can be clinically detected. Estimates of tumor doubling time in patients with measurable lung cancers have yielded mean values of 87 d for squamous cell tumors, 134 d for adenocarcinomas.[175] These figures suggest that the first malignant lung cancer cells arise about 7 to 11 years before they can be detected. (The actual period is probably shorter because growth is more rapid in subclinical cancers.) The greater part of the 20-year period, therefore, must represent the period of carcinogenesis.

In the U.S., the lung cancer incidence has leveled off in men, reflecting a decline in smoking that began in the 1960s, coinciding with increasing awareness of tobacco health hazards. Unfortunately, lung cancer incidence is rising in women, a trend reflecting their increased smoking beginning in the 1960s, coincident with the rise of gender activism. Lung cancer is now the leading cause of cancer death in women, surpassing breast cancer.[3]

Tobacco smoke and its condensates act as complete carcinogens in the two-stage mouse skin carcinogenesis model.[37] However, there are literally thousands of chemicals in cigarette smoke. The vapor phase of cigarette smoke, defined as the fraction that largely passes through a standard fiberglass filter, contains mainly initiating compounds, such as oxygen active species and aldehydes resulting from combustion. These include acrolein, 1,3-butadiene, and benzene. The substances largely trapped by fiberglass filters, referred to as cigarette smoke condensate (CSC), contain a wide variety of both initiating and promoting compounds. These include polycyclic aromatic hydrocarbons (PAHs), such as benz(a)anthracene and benzo(a)pyrene, tobacco-specific nitrosamines, such as 4-(methylnitrosamino)-1-(3-pyridyl)-1-butanone (NNK), heavy metals, such as nickel, arsenic, and lead, and the alpha-emitting radioisotope, polonium-210. Nicotine, the major addictive substance in tobacco, can be nitrosated to several N-nitrosamines[176] and may act itself as a cofactor in dimethylbenz(a)anthracene (DMBA)-induced cancers in the hamster cheek pouch.[177]

### Environmental Tobacco Smoke

As noted above, about 10% of lung cancer occurs in nonsmokers. It is reasonable to hypothesize that at least some of those cases arise from "passive smoking," the inhalation of others' tobacco smoke. Sidestream smoke (that arising from the lit end of a cigarette) contains qualitatively similar carcinogens to mainstream smoke (that inhaled and exhaled by the smoker). Many carcinogens in sidestream smoke are present in higher quantity than in mainstream smoke because they are not removed by contact with the smoker's mucosal lining and by cigarette filters, which are present on most cigarettes smoked in the U.S. Because there is a dose–response relationship between carcinogen exposure and cancer incidence in smokers, and because there is no minimum threshold of exposure for cancer incidence, it is plausible that some lung cancers may be caused by sidestream smoke.

Objective evidence that sidestream smoke results in significant exposure of nonsmokers derives from direct measurements of nicotine levels in tissues of exposed nonsmokers. For example, elevated nicotine levels have been detected in cervical lavages of nonsmokers exposed to sidestream smoke[178] and increased urinary cotinine (a metabolite of nicotine) has been measured in nonsmoking spouses of smokers.[179] In another study, nonsmokers exposed to cigarette smoke under experimental conditions for 3 h exhibited elevated numbers of sensitized peripheral blood neutrophils that produced increased amounts of oxygen-active species.[180] It was hypothesized that oxidant-mediated tissue damage might correlate with an increased lung cancer risk in nonsmokers.

Several studies have addressed the role of "passive" smoking in the incidence of lung cancer in nonsmokers, and the major studies were recently reviewed.[181] In general, they demonstrate a trend toward increased lung cancer risk from environmental tobacco exposure; however, all contain methodological flaws and have produced conflicting results. One of the most recent and best designed studies[182] revealed that only exposure to >25 smoker-years during childhood and adolescence increased the risk of lung cancer in nonsmokers and that this accounted for about 17% of lung cancer incidence in nonsmokers. Exposure to a spouse's smoking in the home did not increase the risk of lung cancer. The study was subsequently criticized for failing to control for several potential variables[183-185] and on statistical grounds.[186]

Despite the methodological difficulties in unequivocally demonstrating an increased lung cancer risk in passive smokers, the suggestive data, combined with a burgeoning literature on other adverse health effects of environmental smoking, have led to extensive legislative measures to minimize the exposure of nonsmokers to tobacco smoke. At the federal level, these have included restricting tobacco advertising, mandating warning labels on tobacco products, restricting smoking on domestic airline flights, and restricting smoking in many federal buildings and agencies,[187] including the National Institutes of Health and the Department of Veterans Affairs. Most states now have laws restricting smoking in public places and/or in the worksite. Local initiatives have also succeeded in limiting exposure to tobacco smoke.[188] The tobacco industry has also experienced increasing numbers of lawsuits from individuals resulting from the health consequences of smoking[189] although, to date, most of these have been unsuccessful.

### Other Air Pollutants and Industrial Exposures

In contrast to the carcinogenic effects of tobacco products, which are relatively easily demonstrated and constitute a major topic of this volume, the contribution of air pollution to cancer incidence, particularly of the lung, is more difficult to assess. It has been well demonstrated that lung cancer incidence is higher in urban than in rural areas,[190] even when adjusted for tobacco abuse. Furthermore, many substances, proven to be lung carcinogens either in animals or from occupational exposures,[191] have been detected in polluted air. Nevertheless, it is risky to attribute excess lung cancer mortality in urban areas to air pollution because other lifestyle variables, e,g., occupation, passive smoking, or other undefined factors, may play a role.

Another problem in assessing the carcinogenic risk of air pollution is the considerable geographic and temporal variation in specific pollutants present. This is exemplified by the polycyclic aromatic hydrocarbons (PAHs), a well-studied class of air pollutants, which includes benzo(a)pyrene. The major sources of these substances in air are coal burning and coke production. Use of coal as a major source of fuel in the U.S. has declined since the introduction of oil and gas in the 1940s. This factor must be taken into consideration in epidemiologic studies seeking a link between lung cancer and air pollutants. It has also been pointed out[191] that the definitions of urban and rural areas may be blurred as industry expands out of cities into extensive suburban areas and fewer truly rural areas exist.

Automobile emissions constitute another major source of modern urban air pollution. Occupational exposures to various components of petroleum products were assessed by Siemiatycki et al.[192] They found modest positive correlations between exposure to mineral spirits, diesel fuel, and lubricating oils and lung cancer incidence (among other cancer sites). Several epidemiologic studies of lung cancer incidence from diesel exhaust exposure suggest a modest occupational risk but probably no significant risk to the general public.[193] Benzene, an organic solvent now a major component of unleaded gasoline, is a well-known leukemogenic carcinogen.[194] Chronic low-dose exposure in mice has also been demonstrated to result in a variety of solid neoplasms, including lung cancers.[195] It is probably too early to assess the hazards of benzene in unleaded gasoline on the general public.

A relationship between occupational chemical exposures and lung cancer is often easier to establish than between air pollution in general and lung cancer, because the populations of interest are better defined and the suspected carcinogens exist in relatively higher concentrations. As pointed out earlier in this chapter, the first association between chemicals and cancer risk were made from observations of occupational exposures. The clearest examples of occupational lung carcinogens are asbestos and radon, discussed elsewhere in this chapter.

Other agents associated with lung cancer include a variety of metal ions and/or their salts. Arsenic has been recognized as a lung carcinogen in the copper smelting industry[196,197] and in handlers of arsenic-containing pesticides.[198] The skin and hepatic carcinogenicity of arsenic-containing medicinal preparations such as Fowler's solution has also been established.[199] An excess of lung cancer deaths has been noted in workers in the chrome pigment industry;[200,201] however, complicated pharmacokinetics and different oxidation states of chromium have made recommendations about exposure limits particularly difficult.[202] Crude nickel ores (but not refined nickel compounds) have also been implicated in lung carcinogenesis.[203] Rat renal sarcomas induced by nickel subsulfide and iron are associated with K-ras oncogene activation characterized by GGT to GTT transversions in codon 12; this suggests that 8-hydroxyguanine adducts resulting from oxidative damage were responsible.[204] Both nickel and chromium compounds have also demonstrated pulmonary carcinogenicity in laboratory animals.[205] More tenuous associations with lung cancer incidence have been found for iron oxides[206] and for beryllium.[207]

*Bis*(chloromethyl)ether, a compound used in the manufacture of ion exchange resins, has been proven to cause lung cancers in exposed workers.[208] The carcinogenicity of PAHs in coal carbonization and coke workers was alluded to above. Lung cancers have also been linked to acrylonitrile[209] and to vinyl chloride,[210] a compound

Table 6    **Viruses Implicated in Human Cancers**

| Virus | Malignancy |
|---|---|
| **RNA viruses** | |
| HTLV-1 | Endemic T-cell leukemia |
| **DNA viruses** | |
| Human papillomavirus | Cervix, lung, esophagus |
| Epstein-Barr virus | Nasopharynx, Burkitt lymphoma |
| Hepatitis B virus | Hepatoma |

whose carcinogenicity was first noted by a rise in the incidence of usually rare hepatic angiosarcomas in exposed workers.

## Cancer Chemotherapy Drugs

Some cancer chemotherapy drugs form DNA adducts (see above) and are known carcinogens. Evaluation of the mutagenicity of various chemotherapy agents measured as the ability to induce sister chromatid exchanges in lymphocytes revealed strong and persistent mutagenicity only for alkylating agents.[211] These findings correlate with the relatively high incidence of malignancies seen in patients following treatment with alkylating agents.[212-215] Most of these posttherapeutic malignancies are leukemias. While many other classes of cytotoxic drugs are mutagenic and presumed to be carcinogenic, supporting clinical data are lacking.

A high incidence of second lung cancers has been seen in long-term survivors of small cell carcinoma of the lung treated with chemotherapy.[216-219] This is probably due to the same risk factors that gave rise to the small cell cancer rather than to the chemotherapy because other second malignancies seen in these patients are also tobacco-related. However, the increased incidence of leukemias in long-term survivors with small cell lung cancers[220-222] suggests that a contribution from chemotherapeutic agents and radiotherapy in producing second lung cancers cannot be ruled out.

Chemotherapy agents may also pose an occupational hazard. While no clear instance of carcinogenicity has been proven, an increase in chromosome abnormalities has been observed in nurses handling chemotherapeutic agents[223] and increased fetal loss has been reported in a similar population.[224]

## VIRUSES (Table 6)

The numerous oncogene-bearing retroviruses discussed above infect various animal models but not humans. Thus far, the only retrovirus implicated directly in human malignancy is HTLV-1 (HTLV = human tumor leukemia virus), which is associated with endemic T cell leukemia/lymphoma.[225] This clinically distinct malignancy, largely confined to the Caribbean basin, the southeastern U.S., and Japan, is characterized by a rapidly fatal leukemia, hypercalcemia, and bone metastases. The virus is closely related to HIV-1 (formerly HTLV-III), the causative agent of AIDS (acquired immunodeficiency syndrome). The HIV-1 virus is associated (but not necessarily causally) with Kaposi's sarcoma, non-Hodgkin lymphomas, and possibly, other malignancies.

DNA viruses associated with human malignancies include Epstein-Barr virus in nasopharyngeal carcinoma and Burkitt's lymphoma, hepatitis B virus in hepatoma, and human papillomaviruses in cervical and, perhaps, esophagus and bronchogenic carcinomas. Many more examples of DNA viral carcinogenesis in animal models exist.[226]

Viruses have been implicated both as initiators and promotors of carcinogenesis. The roles of retroviral and DNA viral oncogenes in the former process have been discussed above. Inflammation and increased cell turnover associated with chronic viral infection may increase the susceptibility of affected tissues to tumor initiators. One proposed example is the association of hepatoma with hepatitis B infection;[168] another example is accelerated chemical carcinogenesis in the chronically infected rat lung.[227] For most human viral-associated cancers, molecular genetic mechanisms have not been well elucidated.

## PHYSICAL AGENTS

### Asbestos

Asbestos is a commercial product widely used in construction for its tensile strength and resistance to corrosion. It is produced from a family of fibrous hydrated silicates whose average fiber length to width ratio is 3:1 but whose individual fiber dimensions vary widely.[228] Two classes of asbestos are recognized. The serpentines, represented by chrysotile, consist of sheets of pseudohexagonal arrays with fibrillar subunits of silicon oxide and magnesium. The amphiboles, represented by amosite, crocidolite, and tremoline, consist of straight, rodlike structures of silicon oxide linked by various cations. Asbestos is ubiquitous in the earth's crust, and mixed forms, along with other commercially useful minerals, may be found together.

Chrysotile is, by far, the most widely useful form in this country. Asbestos use increased dramatically during World War II, when it was widely used in the shipping industry. It has also been incorporated into cement pipes used as water and sewage conduits and is used in roofing, paneling, and tiling. Its thermal stability has made it useful in automotive brake linings. It is also found widely in plastics and textiles.

The health dangers of asbestos were first recognized in the 1960s.[228,229] This is attributable to the long latency period of its effects and to its greatly increased use in the 1940s. The major organ affected is the lung and pleural lining, because asbestos gains access to the body by inhalation of fibers released into the air. The typical lesion consists of diffuse pulmonary fibrosis, slowly progressive over many years. Fibrosis also occurs in the parietal pleura of the thorax and abdomen, resulting in calcified plaques.

Two tumors are clearly associated with asbestos exposure: mesothelioma and bronchogenic carcinoma. The former is an extremely rare malignancy, whose occurrence is considered pathognomonic of asbestos exposure. About 3% of workers chronically exposed to asbestos will develop mesothelioma. Its incidence is unaffected by concurrent tobacco use.

Bronchogenic carcinoma is far more common. Incidence varies greatly with the type of asbestos involved, crocodolite, for example, being more carcinogenic than chrysotile. It also varies with stage of production, with cancer incidences higher in those working with finished products than in miners of asbestos. Of greatest significance is the hyperadditive effect of tobacco use. In fact, the risk of bronchogenic carcinoma in nonsmoking asbestos workers is only slightly above that of nonsmokers not exposed to asbestos. In contrast, smoking asbestos workers incur a risk approaching 100 times that of nonsmoking, nonexposed individuals.[230]

A major factor in the carcinogenic potential of asbestos is fiber dimensions. Lung cancer is associated mainly with exposure to fibers >5µm in length. For mesotheliomas, a small fiber diameter is also critical, as only long, thin fibers are transported to the periphery of the lung. The long fiber lengths required for the adverse health effects of asbestos have led many investigators to propose that both the characteristic fibrotic reaction induced by asbestos and its carcinogenesis are mediated by activation of pulmonary macrophages.[231,232] Macrophages produce inflammatory mediators and growth factors, as well as active oxygen intermediates. While short asbestos fibers are phagocytosed, resulting in transient inflammation and few long-term effects, longer fibers cannot be completely phagocytosed and result in a persistent inflammatory reaction that is perpetuated over years by new macrophages as old ones die. Thus, inflammation and increased cell turnover in the presence of oxygen radicals may explain the mechanism of carcinogenesis. The greatly enhanced rate of lung carcinoma formation in smokers fits well with the concept that asbestos acts as a promoter in cells initiated by tobacco products. Browne[233] has extended this concept to mesothelioma formation. He notes that the rare cases of mesothelioma not associated with asbestos arise in the face of chronic inflammation (e.g., tuberculosis, emphysema), that long fiber length is also required for mesothelioma induction, and that the process appears to require a threshold level of exposure and a latent period of induction that is inversely proportional to exposure dose. These features, particularly the latter, are incompatible with initiation. Moreover, asbestos is one of the few well-documented carcinogens that is not mutagenic in short-term assays.

Nevertheless, other studies have demonstrated that asbestos can act as a complete carcinogen.[234] It induces tumors in animals by multiple routes of administration independent of age of first exposure (suggesting an effect on an early step in carcinogenesis). While it does not induce gene mutations, it does transform mesothelial cells and fibroblasts in culture and induces nonrandom chromosomal abnormalities in several models. Cytological studies suggest that phagocytosed fibers aggregate around the nucleus and may physically disrupt the mitotic process, resulting in chromosome loss and/or damage. Thus, asbestos carcinogenesis remains a complex and controversial issue.

## Radiation

### General Principles

Ionizing radiation and ultraviolet radiation fall in the spectrum of electromagnetic radiation, including visible and infrared light, radiofrequency, and electrical waves. All are part of our normal environment. Low levels of background radiation exist in the earth, largely in the form of radon and its decay products. Ultraviolet radiation is emitted by the sun. Artificial sources of radiation include harnessed nuclear energy or nuclear weapons, diagnostic and therapeutic radiation, and, possibly, the alpha particle-emitting $^{210}$Po in tobacco.

The energy of a photon is proportional to its frequency. Visible light, infrared light, and radio waves all have energies <2 eV; ultraviolet radiation exhibits energies between 2 to 10 eV, and ionizing radiation has energies >10 eV. The threshold for the ability of electrons in biological systems to absorb photon energy is about 10 eV; thus, higher energy photons eject electrons, resulting in ionization.

Ionizing radiation may arise from electromagnetic sources (gamma rays or X-rays) or from particulate sources (electrons, neutrons, alpha particles, and other heavy particles). What they have in common is their ability to release enough energy to break covalent bonds. However, different sources of ionizing radiation with equal energies do not exert the same effects in biological systems. This is due to differences in linear energy transfer (LET), the energy lost over the distance the particle or photon travels. A high LET source, such as an alpha particle, is densely ionizing but loses energy rapidly along its path and can travel only a short distance through tissues. On the other hand, gamma rays exhibit low LET, are sparsely ionizing, but exhibit great tissue penetration. Neutrons exhibit LETs between that of gamma rays and of alpha particles.

Doses of absorbed radiation used to be expressed in "rad" (1 rad = 100 erg/g). In the SI system, now widely accepted, the unit is the Gray (Gy), 1 Gy = 1 J/Kg. A centigray (cGy), 1/100 of a Gray, is equal to 1 rad. In order to assess the actual biological effect of a radiation dose, the absorbed dose (i.e., number of cGy) must be multiplied by a "quality factor" that takes LET into consideration. This resulting value estimates the "relative biological effectiveness" (RBE) and is expressed in units called "sieverts" in the SI system (the old unit was the rem; 100 rem = 1 sievert). For gamma radiation (or X-rays), the quality factor is unity; thus, the RBE can be expressed in cGy. Larger RBE values are obtained for high LET radiation.

Radiation carcinogenesis results from nucleic acid damage, causing gene mutations or deletions[235] and presumably resulting in the activation of oncogenes.[236] The dosimetry of radiation carcinogenesis in a particular tissue is complex, depending not only on the dose but on cell viability after radiation, the effectiveness of DNA repair mechanisms, and other initiating and promoting factors impinging on the target cells. The major target of low LET radiation is probably water, with resulting production of oxygen active species (see above). High LET particles may interact directly with nucleic acids.

Carcinogenicity is proportional to dose rate for low LET radiation.[237] Dose protraction (lower doses over longer periods of time) probably permits DNA repair mechanisms to operate. In contrast, high LET radiation carcinogenicity is enhanced by protracted, low doses, suggesting that DNA repair cannot take place due to the densely ionizing nature of each particle.

### Atom Bomb Experience

Much has been learned about carcinogenesis resulting from short, high-dose radiation exposure in humans by careful followup of the more than 120,000 survivors of the atomic bomb blasts in Hiroshima and Nagasaki at the conclusion of World War II. Survivors felt not to have been exposed to significant radiation doses serve as controls for those who were exposed. A high incidence of leukemias over the first 20 years in exposure victims has subsequently diminished.[237] This phenomenon probably reflects the relatively short latent period for leukemogenesis. Many other malignancies, including lung cancers, began to appear in excess numbers, but their onset coincided with the usual age of onset for these malignancies. In other words, the latency period was not shortened, but the incidence was increased. A large fraction of this population is still entering the ages of risk for solid tumors, so the ultimate risks and risk durations are not known.

Another more recent incident that will yield important data on short, high-dose radiation exposure is the Chernobyl disaster of 1986. There is no agreement on estimates of future excess cancers and cancer deaths at this time because the many possible variables determining dosimetry are unknown.[238]

### Radon-222

In recent years, it has become apparent that the greatest source of exposure to ionizing radiation for the general public is from radon-222 decay products. Radon-222 is a radioactive gas formed as a decay product of radium-226, which is ubiquitous in the Earth's surface. It was first recognized as a carcinogen when an excess of bone sarcomas and carcinomas of the paranasal sinuses were noted in workers painting radium dials on watches.[239] These workers licked their brushes to make a fine point, ingesting large quantities of radium. The paranasal sinus cancers were attributed to the generation of radon gas.

High concentrations of radon gas in poorly ventilated uranium mines have been causally linked to a high incidence of lung cancers in uranium miners.[240] Smoking greatly enhances the risk but does not account for it. Most studies indicate an increase in all lung cancer histological subtypes, with small cell carcinoma most greatly increased, but methodological problems make this latter point controversial.[241] A recent study of lung tumor tissue from miners exposed to high radon concentrations revealed a high frequency of p53 gene mutations.[242] Activation of several oncogenes by diverse mechanisms has been reported in animal models of radiation carcinogenesis.[236]

Assessing occupational exposure to radon is difficult. The effective concentration of radon gas in air is expressed as pCi/l (a pCi is 2.2 decays/min). However, exposure must also take into account that several decay products result from radon-222 and exist in mixtures of varying concentrations in different settings. This has led to the formulation of the working level (WL), the concentration of radon daughters per liter that yield $1.3 \times 10^5$ MeV of alpha energy. Exposure is expressed in working level months (WLM), the WL exposure in 170 h (the number of hours a miner works in an average month). There is no simple relationship between exposure in WLM and risk because the dosimetry is complex, being influenced by breathing patterns, flow rates, physiological factors such as mucociliary clearance, and smoking. Because accurate smoking histories are difficult to obtain retrospectively (in deceased individuals) and because smoking constitutes an even greater cause of lung cancer, this factor remains particularly problematic.

It is clear that smoking exerts a supraadditive effect on the incidence of radon-induced lung cancers. Epidemiologic studies suggest that occupational exposure to 4 WLM/year from ages 20 to 40 or a lifetime exposure of 1 WLM yearly will increase the risk of lung cancer by a factor of about 1.6 over rates in the general population. Most of this increased risk occurs in smokers.[243]

Nonoccupational exposure to high radon levels had occasionally been detected in areas where soil containing radioactive waste was used as foundation backfill for homes. However, over the last decade, it became apparent that some geographic areas (such as the Reading Prong in Pennsylvania) contained noncontaminated soil high in radon and that indoor concentrations could exceed those permissible in mines.[244] Moreover, there may be wide variations in radon levels from building to building within a given geographic area.

It is clear that ventilation plays a large role in determining radon levels, poorly ventilated areas containing much higher concentrations than well-ventilated areas. Ambient dust is another important factor, because radon daughters are adsorbed onto solid surfaces (a factor that may partially explain the synergism between tobacco smoking and radon in lung carcinogenesis). Because radon levels are generally much lower in indoor environments than in mines, and because of the many variables in assessing exposure noted above, risk estimates in the general population cannot be directly measured. The average nonoccupational lifetime exposure in the U.S. has been estimated as 10 to 20 WLM;[245] the increased lung cancer risk in miners begins to be detected at lifetime exposures of about 50 WLM. It is controversial whether extrapolation from occupational exposure studies to nonoccupational exposure settings is valid. Despite uncertainties in calculating actual increased risk from indoor radon exposure, common sense dictates that radon levels unusually higher than average are to be avoided, and recommendations have been made, based on extrapolation from the data in miners. The U.S. Environmental Protection Agency has published "A Citizen's Guide to Radon"[246] in which 4 pCi/l is cited as the maximum allowable average yearly concentration; the guide encouraged mass testing of homes for radon concentrations. This concentration has been extrapolated to predict a life-time excess lung cancer risk of .9 to 3.4% in different mathematical models.

Radon detection is now relatively inexpensive and widely available. The most commonly used devices are charcoal-containing canisters placed for a defined period of time in the lowest level of the house with doors and windows closed. Radon is adsorbed onto the charcoal, which is then eluted and measured in a laboratory. The effectiveness of methods to reduce radon accumulation in the home are controversial but center around covering dirt floors, sealing off pipes and cracks that expose the foundation soil, and most importantly, improving ventilation to outside air.[246]

### Polonium-210

Polonium-210, an alpha-emitting decay product of Pb-210, is a component of to-bacco. Its precise role in human lung carcinogenesis is unclear because its effects cannot be isolated from those of other carcinogenic components. Levels up to 1 pCi per cigarette have been detected, and peripheral lung tissue concentrations of up to .025 pCi/gm have been measured in smokers (compared to negligible concentrations in nonsmokers).[247] In the Syrian golden hamster, intratracheal instillation of large doses of Po-210 alone or in combination with benzo(a)pyrene exhibited additive tumorigenic effects.[248] The carcinogenicity of intratracheally instilled Po-210 is also enhanced by high oxygen concentrations,[249] which induce bronchial epithelium to proliferate.

### Medical Radiation Exposure

The most common manmade source of radiation exposure is from diagnostic and therapeutic applications. Millions of diagnostic X-ray and radionuclide studies are performed yearly in the U.S. In addition, radiotherapy may deliver high doses, usually to limited parts of the body. The potential for carcinogenesis varies widely among tissues and, as will be discussed below, probably among individuals.

In the past, high-dose exposures have resulted in measurable increases in cancer mortality. One example is the use of Thoratrast, an alpha-emitting radionuclide used in liver imaging, which caused an increased incidence of normally rare hepatobiliary carcinomas and other malignancies.[250] Another study demonstrated increased rates of breast carcinoma resulting from excessive use of chest fluoroscopy in women.[251] There is now an extensive literature on thyroid carcinogenesis resulting from expo-sure to low-dose therapeutic radiation in childhood for benign conditions.[252,253] An excess of cancer mortality in radiologists has also been reported prior to the devel-opment of appropriate shielding techniques and lower doses used in diagnostic studies.[254]

In many cases, cancers associated with therapeutic radiation increase with dose up to a point, then decrease at higher doses. This phenomenon is well demonstrated in thyroid carcinogenesis, where cumulative doses over 2000 cGy are not associated with cancer formation. The probable explanation is that the potentially transformed cells are destroyed at high doses.

One of the most widely publicized controversies about the potential carcinoge-nicity of radiation used in medical practice concerned diagnostic mammography for breast cancer. This concern originally arose at a time when the doses employed were far greater than those used currently.[255] Currently, doses of about 0.4 cGy per study are employed, and subsequent analysis has confirmed the benefits and very minimal risks associated with these doses in women over age 50, the age group usually screened.[256]

## PREDISPOSING HOST FACTORS

It has long been recognized that not all individuals with equal exposure to tobacco products are at equal risk of developing lung cancer. Both the observation of lifelong smokers who do not develop tobacco-related tumors and the observation of familial clustering of lung cancer cases suggest that inherited host factors may influence the rate of pulmonary carcinogenesis. In a recent study of families with lung cancer probands, evidence for Mendelian inheritance of a codominant autosomal gene predisposing to lung cancer at an early age was described.[257]

As discussed above, most carcinogens are really procarcinogens that must be activated to electrophilic forms. Likely metabolizing enzymes are the mixed function oxidases (cytochrome p450 enzymes). Variations in the expression of p450 enzymes can be identified by their ability to metabolize substrate pharmaceuticals. The antihypertensive drug debrisoquine is oxidized to 4-hydroxydebrisoquine by the cytochrome p450IID6. Extensive, poor, and intermediate hydroxylator phenotypes have been described; the ratios are consistent with autosomal recessive inheritance of the poor hydroxylator phenotype, which is also associated with marked sensitivity to substrate drugs.[258]

Recently, it has been reported that tobacco users who are extensive metabolizers of debrisoquine are also at greater risk of developing lung cancer compared with slow metabolizers.[259-261] The extensive metabolizer phenotype was three times more common in lung cancer patients than in healthy-smoker controls. Analogous findings have been described for susceptibility to bladder cancer[262] and to gastrointestinal malignancies.[263] The p450IID6 cytochrome is encoded for by the CYP2D6 gene and this gene has recently been cloned. It is possible that the debrisoquine oxidation phenotype is only a marker for susceptibility differences and not the responsible factor. Linkage of polymorphisms in the CYP2D6 gene must be correlated with differences in susceptibility to lung cancer to prove its direct involvement in lung cancer susceptibility. It should be noted that not all studies examining debrisoquine phenotype and lung cancer susceptibility have confirmed this relationship.[264]

In a similar line of analysis, Kawajiri et al.[265] have identified restriction fragment length polymorphisms (RFLPs) in the gene encoding the cytochrome p450IA1, consistent with two homozygous states and one heterozygous state at a single locus. These were associated with genetically determined differences in lung cancer susceptibility in smokers, the rarer homozygous state being found three times more frequently in smokers with lung cancer. Moreover, the distribution of risk associated with the three genotypes matched the previously described trimodal distribution of risk associated with aryl hydrocarbon hydroxylase (AHH) inducibility phenotypes.[266] Cytochrome p450AIA metabolizes hydrocarbons such as benzo(a)pyrene; the AHH inducibility phenotypes may be explained by this group of gene polymorphisms.

The cancer susceptibilities associated with the cytochrome p450 phenotypes described above are the first to be correlated with changes at the level of the gene. Since numerous enzymes are involved in carcinogen metabolism and DNA repair, it is likely that other genetic factors explaining variations in individual cancer susceptibility will be uncovered. In this regard, Seidegard et al.[267,268] have described an isozyme of glutathione transferase whose activity is decreased in lung cancer patients compared to smoking controls. The glutathione transferase activity appears to be strongly under genetic control but is also affected by extent of smoking history, suggesting an epigenetic influence. Rudiger et al.[269] have shown that fibroblasts cultured from lung cancer patients exhibit lower activity of $O^6$-methylguanine-DNA methyltransferase, and hence lower DNA repair rates, than those from normal controls or from patients with malignant melanoma.

# CARCINOGEN TESTING

As noted above, the first recognized carcinogens were identified by the high rates of cancer induced in occupational settings. These carcinogens were either especially potent or the occupational exposure to them was very great. In our industrialized society, we are all exposed to at least low levels of thousands of compounds in our air, water, and food. Many of these compounds may have low but significant carcinogenic potency below levels likely to be detected by epidemiologic studies. Furthermore, the effects of multiple compounds of low carcinogenic potency are difficult to assess. It is desirable to identify potential carcinogens in our environment so that we may restrict our exposure to them. However, no ideal method has been developed to accomplish this. Furthermore, there is little agreement on what levels of carcinogens are ''acceptable'' or clinically important.

Testing of potential carcinogens in humans is obviously untenable. Epidemiologic assessments of carcinogenesis in humans are invaluable, and many have been cited throughout this review. However, they also have serious limitations.[270] Prospective studies usually require large cohorts and years of follow-up. Retrospective case control studies must contend with many hidden variables and with possible imprecision in reported carcinogen exposure levels. Because of limitations in studying human carcinogenesis directly, most studies are done in animals or *in vitro* models. Extrapolation of data from these systems to humans is fraught with difficulties but has formed the basis for occupational regulatory guidelines.

Rodents are the commonly used models for *in vivo* carcinogen testing because they are relatively inexpensive and easy to work with and because many human carcinogens are also carcinogenic in rodents. While chronic administration of low doses of carcinogens simulating human exposure are often performed, these are obviously time-consuming. Therefore, more convenient ''short-term tests'' of carcinogenicity, that can be completed quickly, have been developed. Because carcinogenesis usually occurs over years of exposure to environmental or occupational carcinogens, endpoints other than tumor formation are often used. Carcinogen testing may be performed in both *in vitro* or *in vivo* models. The classic *in vitro* short-term test for carcinogenicity is the Ames mutagenicity assay,[271] which employs histidine-requiring mutants of *Salmonella typhimurium* that are sensitive to reversion in the presence of carcinogens. In the presence of suspected carcinogens, revertants are detected as colonies growing in histidine-deficient media. A critical component of the assay is the addition of rodent microsomal liver extract to provide cytochrome p450 enzymes, in view of the fact that most human carcinogens require metabolic activation (see above).

Early studies employing a narrow range of carcinogenic compounds revealed high specificity and sensitivity for the Ames assay. However, subsequent studies by the National Toxicology Program (NTP), employing a broader range of chemicals, demonstrated a sensitivity for proven carcinogens of only 48%.[272] Significantly, however, 89% of chemicals that were mutagenic in the Ames assay were demonstrated to be carcinogens. More recently, an NTP study of 301 chemicals classified according to presence or absence of electrophilic properties suggesting DNA toxicity (''structural alerts'') confirmed these features of the Ames assay for compounds with structural alerts; carcinogens without structural alerts (presumably nongenotoxic compounds) were not usually detected in the Ames assay.[273] However, it has been pointed out that most of the approximately 60 chemicals thus far identified as definite human carcinogens possess structures of genotoxic agents, validating the continued usefulness of these assays.[274]

Other *in vitro* short-term tests for carcinogenicity employing mammalian cells and liver extracts have been developed and have been reviewed.[275] These have as end-points either specific gene mutations or cytogenetic abnormalities (chromosome de-rangements or sister chromatid exchange). While these have varying specificity and sensitivity, none have greater accuracy than the Ames assay; furthermore, batteries of different *in vitro* assays have little greater predictability than a single assay.[272]

Short-term tests of carcinogenicity performed *in vivo* in rodents have become widely used to detect potential DNA-damaging compounds not detected in *in vitro* assays and studies in whole animals may provide more clinically relevant information about potential carcinogens in humans. Endpoints in these studies include assessment of chromosomal aberrations in bone marrow (selected because it is an easily acces-sible source of rapidly dividing cells), and assessment of unscheduled DNA synthesis (DNA turnover in normally nondividing cells by tritiated thymidine uptake) in the liver (a frequent site of carcinogenesis in rodents). Detection of DNA adducts in a variety of tissues, and induction of tumors in transgenic rodents bearing activated oncogenes[275] are also endpoints.

Different endpoints are usually needed for testing nongenotoxic carcinogens, such as tumor promoters, since mutational events are not induced by these compounds. A frequently used animal model for testing these agents is the SENCAR mouse,[276] a strain that is particularly sensitive to skin tumor-promoting activity of phorbol esters. Other assay endpoints may directly measure PKC activation.[38]

One of the major controversies in carcinogenicity testing in rodents, both in short-term tests and in long-term studies, is dosing. There is no general agreement on such issues as dose, route, and schedule of the compounds to be tested with respect to their relevance to human carcinogenesis. The NTP uses a wide variety of approaches, chosen selectively for each compound based on its chemistry, pharmacology, and known toxicology.[277] One of the most interesting controversies concerns the wide-spread practice of employing the ''maximum tolerated dose'' (MTD) as at least one of the dose levels. This is usually defined as the highest dose that does not cause lethality (other than from tumor induction). The problem is that toxic doses may cause tissue damage and inflammation, resulting in increased cell turnover, a factor that, in itself, may promote tumor formation and, thus, false positive results.

A major proponent of this criticism in recent years is Ames, the developer of the Ames mutagenesis assay, who has argued that most carcinogens are endogenously formed or are natural food components rather than synthetic, environmental agents.[278,279] He noted that about half of all compounds tested in rodents were mu-tagenic, whether or not they were carcinogens, and that, of the relatively small frac-tion that were natural substances, the same percentage proved mutagenic. He concluded that chronic testing at or near the MTD resulted in tissue damage leading to tumor promotion via increased cell turnover. He further noted that even potent proven carcinogens such as tobacco may exert much of their effects by the inflam-matory reaction they induce rather than through genotoxic mechanisms.

Ames' criticism challenges the concept of a linear dose relationship for carcinogen potency and, thus, the validity of extrapolating human risks of low-dose exposures from carcinogenicity of high doses in animals. He thus challenges the basis for reg-ulatory decisions about potential carcinogens in the environment and downplays the risks of trace contaminants of our food and water supply by synthetic carcinogens. As an example, he has argued that the recent banning of daminozide (Alar), a pes-ticide that had been widely used on commercially grown apple trees, may actually increase human exposure to fungal mutagens in apples that the pesticide sup-pressed.[280] His arguments have been supported by Cohen and Ellwein, who provide examples of both genotoxic and nongenotoxic agents that may exert their carcinogenic effects through increased cell turnover rather than mutational events.[281,282] They are also supported by the fact that other cancer risk factors, such as hormones (in breast, prostate, and endometrial cancers) and infectious agents (such as hepatitis B in hepatoma and human papillomavirus in cervical cancer) induce increased cell turnover.[168]

Some investigators have called for abandonment of the MTD in rodent carcinogenicity testing in favor of pharmacokinetic-based dosing schemes[283] or modifications to take cell proliferation into account in designing and interpreting these studies.[284] However, others have defended the practice, citing lack of direct proof that tumor induction results solely from excess cell proliferation in these assays, that the hypothesis ignores the multistage (and multifactorial) nature of carcinogenesis, and that many studies demonstrate a good correlation between carcinogenicity in humans and animals.[285,286] Hoel et al.[287] found that, among 99 carcinogenic and noncarcinogenic compounds tested at their MTDs in rodents, very few caused target organ toxicity leading to increased cell turnover.

Despite the difficulties in designing and interpreting the results of rodent carcinogenicity studies, they provide valuable information about human carcinogens and will continue to be used. It is clear, however, that they cannot be used to quantify the risk of a carcinogen in humans. First, as discussed above, the interindividual risk of carcinogenesis may vary by orders of magnitude. Second, humans are frequently exposed to mixtures of low doses of carcinogens, both chemical and physical; the risks of these are not usually subject to animal testing, although certain chemical mixtures are being tested by the NTP.[288] More precise methods for assessing individual human risks may lie in carcinogen adduct-DNA detection and other tests of possible DNA damage discussed earlier in this chapter.[37,270]

## SUMMARY AND CONCLUSIONS

Our current understanding is that carcinogenesis is a protracted, multistage process, evolving over many years in most cases, with each stage progressing a cell toward malignant transformation. Clinically detectable cancer probably constitutes only a small percentage of the natural history of this process. Sporn[289] has argued that the term "carcinogenesis" should replace the term "cancer" to emphasize this continuum. By analogy with syphilis, which refers to the protracted (and easily treated) latent phase of the disease as well as to the late (tertiary) phase, which is irreversible, he argues that directing treatment only to clinically apparent malignancy is analogous to treating syphilis only in its tertiary phase.

While the multistage nature of carcinogenesis has been recognized for decades, the virtual explosion of knowledge brought on by new technologies in the field of molecular genetics in the past decade may, indeed, make early cancer diagnosis and cancer prevention practical. While there may be hundreds of oncogenes yet to be identified, it appears that they will fall into only a few general categories, allowing our understanding of tumor cell growth and differentiation to be a realistic prospect. Moreover, it is clear that carcinogenesis results from various sources of insults to DNA (e.g., radiation, chemicals, asbestos) but that these are all mediated through mutations in these critical target genes.

The potential for advances in cancer diagnosis resulting from these findings is illustrated by some recent examples. The most widely used screening method for colon cancer relies on detection of occult blood in the stools; the value of screening by this approach in reducing colon cancer mortality is still controversial.[290,291] Recently, it was demonstrated that the polymerase chain reaction (PCR) could be used to detect mutated K-ras oncogenes in the stools of patients with subsequently proven colorectal cancers or large (premalignant) adenomas.[292] A similar strategy, using bronchoscopic lavage, has been proposed for lung cancer detection,[293] for which effective screening methods currently do not exist.

Understanding the factors responsible for cell growth and differentiation may also lead to strategies of cancer prevention. As cited above, retinoids have been shown to be useful in preventing oral cavity cancers. The hormone tamoxifen, which acts via hormone receptors in breast cancers, has recently been shown, in addition, to induce transforming growth factor-beta production by stromal cells in breast cancers;[294] this may explain tamoxifen's effectiveness in some hormone receptor negative tumors and may lead to further strategies to take advantage of this phenomenon. While classical epidemiology has been useful in defining cancer risks in populations, it is limited by the great interindividual variations in cancer risks among individuals. Recent technical advances in detecting DNA damage and cellular mutational events with great sensitivity, as discussed earlier in this chapter, have established the field of molecular epidemiology, which seeks to determine individual cancer risk. Together with classic epidemiology, this field may help to define the biologic dose of a carcinogen and allow risk assessments for individuals. This, in turn, may guide future studies in cancer prevention.

Finally, laboratory strategies in molecular genetic engineering may have clinical application in the treatment of cancer. The suppression of the malignant phenotype after introduction of wild type p53 gene or Rb gene has been cited above.[67,75,76] Viral vectors have also been used to transfect antisense RNA specific for oncogenes[295] and hormones regulating growth of lung cancer cells.[296] Monoclonal antibodies against portions of oncogene products have also transiently reversed the malignant phenotype.[297] It is only a matter of time before such laboratory approaches will be adapted for use in the clinical setting.

## NOTE ADDED DURING PROOF:

Two additional classes of oncogenes have been described and may have relevance in lung cancer. The cyclins are a group of genes involved in regulating passage through the cell cycle and have recently been reviewed.[298] One of these, PRAD1/cyclin D, has been implicated as a human protooncogene, whose activation may play a role in several malignancies including lung cancer. In addition, the bcl-2 gene, a protooncogene involved in protecting the cell from "programmed death" (apoptosis), is activated in the 14:18 translocation involving many B cell lymphomas. Recently, bcl-2 activation was reported as a favorable prognostic factor in some non-small cell lung cancers[299] and has also been found in small cell carcinoma cell lines.[300]

## REFERENCES

1. **Pott, P.,** Chirurgical Observations Relative to the Cataract, the Polypus of the Nose, the Cancer of the Scrotum, and Different Kinds of Ruptures and the Mortification of the Toes and Feet, Clark and Collins, London, 1775, reprinted in Shimkin, M. B., Some Classics in Experimental Oncology, NIH Publication #80-2150, Bethesda, MD, 1980.
2. **Becker, K. L. and Gazdar, A. F.,** *The Endocrine Lung in Health and Disease,* W. B. Saunders, Philadelphia, 1984, Chap. 3, 56–78.
3. **Boring, C. C., Squires, T. S., and Tong, T.** Cancer statistics, 1993, *CA Cancer J. Clinicians,* 43, 7, 1993.

4.  **Doll, R. and Peto, R.,** The causes of cancer: quantitative estimates of avoidable risks of cancer in the United States today, *J. Natl. Cancer Inst.,* 66, 1192, 1981.

5.  **Gazdar, A. F., Carney, D. N., Guccion, J. G., et al.,** Small cell carcinoma of the lung: cellular origin and relationship to other pulmonary tumors, in *Small Cell Lung Cancer,* Greco, F. A., Oldham, R. K., Punn, P. A., Jr., Eds., Grune and Stratton, 1981, Chap. 7, 152–3.

6.  **Devesa, S. S., Shaw, G. L., and Blot, W. J.,** Changing patterns of lung cancer incidence by histologic type, *Cancer Epidemiol. Biomarkers Prev.,* 1, 29, 1991.

7.  **Vincent, R. G., Pickren, J. W., Lane, W. W., et al.,** The changing histopathology of lung cancer: a review of 1682 cases, *Cancer,* 39, 1647, 1977.

8.  **Becker, K. L. and Gazdar, A. F.,** *The Endocrine Lung in Health and Disease,* W. B. Saunders, Philadelphia, 1984, 341.

9.  **Schuller, H. M., Becker, K. L., and Witschi, H. P.,** An animal model for neuroendocrine lung cancer, *Carcinogenesis,* 9, 293, 1988.

10. **Nylen, E. S., Becker, K. L., Joshi, P. A., et al.,** Pulmonary bombesin and calcitonin in hamsters during exposure to hyperoxia and diethylnitrosamine, *Am. J. Respir. Cell. Mol. Biol.,* 2, 25, 1990.

11. **Simms, E., Gazdar, A. F., Abrams, P. G., et al.,** Growth of human small cell (oat cell) carcinoma of the lung in serum-free growth factor-supplemented medium, *Cancer Res.,* 40, 4356, 1980.

12. **Aguayo, S. M., Kane, M. A., Kimg, T. E., et al.,** Increased levels of bombesin-like peptides in the lower respiratory tract of asymptomatic cigarette smokers, *J. Clin. Res.,* 84, 1105, 1989.

13. **Cuttita, F., Carney, D. N., Mulshine, J., et al.,** Bombesin-like peptides can function as autocrine growth factors in human small cell lung cancer, *Nature,* 316, 823, 1985.

14. **Hendler, F. J. and Ozanne, W.,** Human squamous cell lung cancers express increased epidermal growth factor receptors, *J. Clin. Invest.,* 74, 647, 1984.

15. **Haeder, M., Rotsch, M., Bepler, G., et al.,** Epidermoid growth factor receptor expression in human lung cancer cell lines, *Cancer Res.,* 48, 1132, 1988.

16. **Maneckjee, R. and Minna, D. J.,** Opioid and nicotine receptors affect growth regulation of human lung cancer cell lines, *Proc. Natl. Acad. Sci. U.S.A.,* 87, 3294, 1990.

17. **Lechner, J. F., McClendon, I. A., LaVeck, M. A., et al.,** Differential control by platelet factors of squamous differentiation in normal and malignant human bronchial epithelial cells, *Cancer Res.* 43, 5915, 1983.

18. **Masui, T., Wakefield, L. M., Lechner, J. F., et al.,** Type B transforming growth factor is the primary differentiation inducing serum factor for normal human bronchial epithelial cells, *Proc. Natl. Acad. Sci. U.S.A.,* 83, 2438, 1986.

19. **Pfeifer, M. A., Lechner, J. F., Masui, T., et al.,** Control of growth and squamous differentiation in normal human bronchial epithelial cells by chemical and biological modifiers and transferred genes, *Environ. Health Perspect.,* 80, 209, 1989.

20. **Sporn, M. B. and Roberts, A. B.,** Role of retinoids in differentiation and carcinogenesis, *Cancer Res.,* 43, 3034, 1983.

21. **Lasnitzki, I.,** The influence of A hyper-vitaminosis on the effect of 20-methylcholanthrene on mouse prostate glands grown *in vitro, Br. J. Cancer,* 9, 434, 1955.

22. **Sporn, M. B., Dunlop, N. M., Newton, D. L., et al.,** Prevention of chemical carcinogenesis by vitamin A and its synthetic analogs (retinoids), *Fed. Proc.* 35, 1332, 1976.

23. **McCormack, D. L., Burns, F. J., and Albert, R. E.,** Inhibition of benzo(a)pyrene-induced mammary carcinogenesis by retinyl acetate, *J. Natl. Cancer Inst.,* 66, 559, 1981.

24.  **Bhide, S. V., Ammigan, N., Nair, U. J., et al.,** Carcinogenicity studies of tobacco extract in vitamin A-deficient Sprague-Dawley rats, *Cancer Res.,* 51, 3018, 1991.

25.  **Willett, W. C., Polk, B. F., Underwood, B. A., et al.,** Relation of serum vitamins A and E and carotenoids to the risk of cancer, *N. Engl. J. Med.,* 310, 430, 1984.

26.  **Menkes, M. S., Comstock, G. W., Vuilleumier, J. P., et al.,** Serum beta-carotene, vitamins A and E, selenium, and the risk of lung cancer, *N. Engl. J. Med.,* 315, 1250, 1986.

27.  **Hong, W. K., Lippman, S. M., Itri, L., et al.,** Prevention of second primary tumors with isotretinoin in squamous-cell carcinoma of the head and neck, *N. Engl. J. Med.,* 323, 795, 1990.

28.  **Hu, L., Crowe, D. L., Rheinwald, J. G., et al.,** Abnormal expression of retinoic acid receptors and keratin 19 by human oral and epidermal squamous cell carcinoma cell lines, *Cancer Res.,* 51, 3972, 1991.

29.  **Hayflick, L.,** The limited in vitro lifetime of human diploid cell strains, *Exp. Cell Res.,* 37, 614, 1965.

30.  **Newbold, R. F., Overell, R. W., and Connell, J. R.,** Induction of immortality is an early event in malignant transformation of mammalian cells by carcinogens, *Nature,* 299, 633, 1982.

31.  **Jones, P. A., and DeClerk, Y. A.,** Extracellular matrix destruction by invasive tumor cells, *Cancer Metastasis Rev.,* 1, 289, 1982.

32.  **Folkman, J.,** Tumor angiogenesis: therapeutic implications, *N. Engl. J. Med.,* 285, 1182, 1971.

33.  **Folkman, J., Watson, K., Ingber, D., et al.,** Induction of angiogenesis during the transition from hyperplasia to neoplasia, *Nature,* 339, 58, 1989.

34.  **Temin, H. M.,** Evolution of cancer genes as a mutation-driven process, *Cancer Res.,* 48, 1697, 1988.

35.  **Loeb, L. A.,** Mutator phenotype may be required for multistage carcinogenesis, *Cancer Res.,* 51, 3075, 1991.

36.  **Armitage, P. and Doll, R.,** The age distribution of cancer and a multi-stage theory of carcinogenesis, *Br. J. Cancer,* 8, 1, 1954.

37.  **Willey, J. C., and Harris, C. C.,** Cellular and molecular biological aspects of human bronchogenic carcinogenesis, *CRC Crit. Rev. Oncol. Hematol.,* 10, 181, 1990.

38.  **Blumberg, P. M.,** Protein kinase C as the receptor for the phorbol ester tumor promoters: Sixth Rhoads Memorial Award Lecture, *Cancer Res.,* 48, 1, 1988.

39.  **Weinstein, I. B.,** Nonmutagenic mechanisms in carcinogenesis: role of protein kinase C in signal transduction and growth control, *Environ Health Perspect.,* 93, 175, 1991.

40.  **Ward, J. M. and Ito, N.,** Development of new medium-term bioassays for carcinogens, *Cancer Res.,* 48, 5051, 1988.

41.  **Henderson, B. E., Ross, R. K., Pike, M. C., et al.,** Endogenous hormones as a major factor in human cancer, *Cancer Res.,* 42, 3232, 1982.

42.  **Fialkow, P. J.,** Clonal origin of human tumors, *Biochim. Biophys. Acta,* 458, 283, 1976.

43.  **Nowell, P. C. and Hungerford, D. A.,** A minute chromosome in human granulocytic leukemia, *Science,* 132, 1497, 1960.

44.  **Rowley, J. D.,** Molecular cytogenetics: Rosetta Stone for understanding cancer — Twenty-Ninth GHA Clowes Memorial Award Lecture, *Cancer Res.,* 50, 3816, 1990.

45.  **Hart, J. S.,** Chromosome abnormalities in human neoplasia, *Cancer Treat. Rev.,* 10, 173, 1983.

46. **Lee, J. S., Pathak, S., Hopwood, V., et al.,** Involvement of chromosome 7 in primary lung tumor and nonmalignant normal lung tissue, *Cancer Res.,* 47, 6349, 1987.

47. **Ashton-Rickardt, P. G., et al.,** MCC, a candidate familial polyposis gene in 5q.21, shows frequent allele loss in colorectal and lung cancer, *Oncogene,* 6, 1881, 1991.

48. **Testa, J. R. and Siegfried, J. M.,** Chromosome abnormalities in human non-small cell lung cancer, *Cancer Res.,* Suppl. 52, 2702s, 1992.

49. **Whang-Peng, J., et al.,** A nonrandom chromosomal abnormality, del 3p(14–23), in human small cell lung cancer, *Genet. Cytogenet.,* 6, 119, 1982.

50. **Carritt, B., Kok, K., van den Berg, A., et al.,** A gene from human chromosome region 3p21 with reduced expression in small cell lung cancer, *Cancer Res.,* 52, 1536, 1992.

51. **Miura, I., Graziano, S. L., Cheng, J. Q., et al.,** Chromosome alterations in human small cell lung cancer: frequent involvement of 5q[1], *Cancer Res.,* 52, 1322, 1992.

52. **Rous, P.,** A sarcoma of the fowl transmissible by an agent separable from the tumor cells, *J. Exp. Med.,* 13, 397, 1911.

53. **Stehelin, D., Varmus, H. E., Bishop, J. M., et al.,** DNA related to the transforming gene(s) of Avian Sarcoma virus is present in normal avian DNA, *Nature,* 260, 170, 1976.

54. **Buchhagen, D. L.,** Molecular mechanisms in lung pathogenesis, *Biochim. Biophys. Acta,* 1072, 159, 1991.

55. **Downward, J., Yarden, Y., Mayes, E., et al.,** Close similarity of epidermal growth factor receptor and v-erb-B oncogene protein sequences, *Nature,* 307, 521, 1984.

56. **Sherr, C. J., Rettenmier, C. W., Sacca, R., et al.,** The c-fms protooncogene product is related to the receptor for the mononuclear phagocyte growth factor, CSF-1, *Cell,* 41, 665, 1985.

57. **Yarden, Y., Escobedo, J. A., Kuang, W. J., et al.,** Structure of the receptor for platelet-derived growth factor helps define a family of closely related growth factor receptors, *Nature,* 323, 226, 1986.

58. **Druker, B. J., Mamon, H. J., and Roberts, T. M.,** Oncogenes, growth factors and signal transduction, *N. Engl. J. Med.,* 321, 1383, 1989.

59. **Bos, J. L.,** The ras gene family and human carcinogenesis, *Mutat. Res.,* 195, 255, 1988.

60. **Westermark, B. and Heldin, C. H.,** Platelet-derived growth factor in autocrine function, *Cancer Res.,* 51, 5087, 1991.

61. **Ingvarsson, S., Asker, C., Axelson, H., et al.,** Structure and expression of B-myc, a new member of the myc family, *Mol. Cell. Biol.,* 8, 3168, 1988.

62. **Angel, P., Allegretto, E. A., Okino, S. T., et al.,** Oncogene jun encodes a sequence specific trans-activator similar to AP-1, *Nature,* 332, 166, 1988.

63. **Vogt, P. K.,** The jun oncogene: a transcription regulator, *Adv. Oncol.* 5, 5, 1989.

64. **Knudson, A. G.,** Mutation and cancer. Statistical studies of retinoblastoma, *Proc. Natl. Acad. Sci. U.S.A.,* 68, 820, 1971.

65. **Lee, W. H., Brookstein, R., Hong, F., et al.,** Human retinoblastoma susceptibility gene: cloning, identification and sequence, *Science,* 235, 1394, 1987.

66. **Marshall, C. J.,** Tumor suppressor genes, *Cell,* 64, 313, 1991.

67. **Huang, H. J. S., Yee, J. K., Shew, J. Y., et al.,** Suppression of the neoplastic phenotype by replacement of the Rb gene in human cancer cells, *Science,* 24, 1563, 1988.

68. **Laiho, M., DeCaprio, J. A., Ludlow, J. A., et al.,** Growth inhibition by TGF-beta linked to suppression of retinoblastoma protein phosphorylation, *Cell,* 62, 175, 1990.

69. **Whyte, P., Buchkovich, K. J., Horowitz, J. M., et al.,** Association between an oncogene and an antioncogene: the adenovirus E1A proteins bind to the retino-blastoma gene product, *Nature,* 334, 124, 1988.

70. **Dyson, N., Bernards, R., Friend, S. H., et al.,** Large T antigens and many polyomaviruses are able to form complexes with the retinoblastoma protein, *J. Virol.,* 64, 1353, 1990.

71. **Lane, D. P. and Crawford, L. V.,** T antigen is bound to a host protein in SV40-transformed cells, *Nature,* 278, 261, 1979.

72. **Oren, M., Maltzman, W., and Levine, A. J.,** Post-translational regulation of the 54K cellular tumor antigen in normal and transformed cells, *Mol. Cell. Biol.,* 1, 101, 1981.

73. **Hinds, P., Finlay, C., and Levine, A. J.,** Mutation is required to activate the p53 gene for cooperation with ras oncogene and transformation, *J. Virol.,* 63, 739, 1989.

74. **Finlay, C. A., Hinds, P. W., and Levine, A. J.,** The p53 protooncogene can act as a suppressor of transformation, *Cell,* 57, 1083, 1989.

75. **Baker, S. J., Markowitz, S., Fearon, E. R., et al.,** Suppression of human colo-rectal carcinoma cell growth by wild-type p53, *Science,* 249, 912, 1990.

76. **Chen, P. L., Chen, Y., Bookstein, R., et al.,** Genetic mechanisms of tumor sup-pression by the human p53 gene, *Science* 250, 1576, 1990.

77. **Levine, A. J.,** The role of p53 as a tumor suppressor in human cancers, *Adv. Oncol.,* 8, 3, 1992.

78. **Srivastava, S., Zou, Z., Pirollo, K., et al.,** Germ-line transmission of a mutated p53 gene in a cancer-prone family with Li-Fraumeni syndrome, *Nature,* 348, 747, 1990.

79. **Cavenee, W. K., Hansen, M. F., Nordenskjold, M., et al.,** Genetic origin of mutations predisposing to retinoblastoma, *Science,* 228, 501, 1985.

80. **Fearon, E. R., Cho, K. R., Nigro, J. M., et al.,** Identification of a chromosome 18q gene that is altered in colorectal cancers, *Science,* 247, 49, 1990.

81. **Call, K., Galser, T., Ito, C. Y., et al.,** Isolation and characterization of a zinc finger polypeptide gene at the human chromosome 11 Wilms tumor locus, *Cell,* 60, 509, 1990.

82. **Emi, M., Fujiwara, Y., Nakajima, T., et al.,** Frequent loss of heterozygosity for loci on chromosome 8p in hepatocellular carcinoma, colorectal cancer and lung cancer, *Cancer Res.,* 52, 5368, 1992.

83. **Gordon, H.,** Oncogenes, *Mayo Clin. Proc.,* 60, 697, 1985.

84. **Yuspa, S. H. and Poirer, M. C.,** Chemical carcinogenesis: from animal models to molecular models in one decade, *Adv. Cancer Res.,* 50, 25, 1988.

85. **Stegeimeier, B. L., Gillett, N. A., Rebar, A. H., et al.,** The molecular progression of plutonium-239-induced rat lung carcinogenesis: Ki-ras expression and activa-tion, *Mol. Carcinogenesis,* 4, 43, 1991.

86. **Weinberg, R. A.,** The action of oncogenes in the cytoplasm and nucleus, *Science,* 230, 770, 1985.

87. **Levine, A. J.,** Oncogenes of DNA tumor viruses, *Cancer Res.,* 48, 493, 1988.

88. **Scheffner, M., Werness, B. A., Huibregtse, J. M., et al.,** The E6 oncoprotein encoded by human papillomavirus types 16 and 18 promotes the degradation of p53, *Cell,* 63, 1129, 1990.

89. **Hunter, T.,** Cooperation between oncogenes, *Cell,* 64, 249, 1991.

90. **Pfeifer, A. M. A., Mark, G. E., Malan-Shibley, L., et al.,** Cooperation of c-raf and c-myc protooncogenes in the neoplastic transformation of SV40 T antigen immortalized human bronchial epithelial cells, *Proc. Natl. Acad. Sci. U.S.A.,* 86, 10075, 1989.

91. **Parada, L. F., Land, H., Weinberg, R. A., et al.,** Cooperation between gene encoding p53 tumor antigen and ras in cellular transformation, *Nature,* 312, 649, 1984.

92. **Harris, C. C., Reddel, R., Pfeifer, A., et al.,** Role of oncogenes and tumour suppressor genes in human lung carcinogenesis, in *Relevance to Human Cancer of N-Nitroso Compounds, Tobacco Smoke and Mycotoxins,* O'Neill, I. K., Chen, J., Bartsch, H., Eds., IARC, Lyon, 1991, 294.

93. **Iman, D. S. and Harris, C. C.,** Oncogenes and tumor suppressor genes in human lung carcinogenesis, *Crit. Rev. Oncogenesis,* 2, 161, 1991.

94. **Little, C. D., Nau, M. M., Carney, D. N., et al.,** Amplification and expression of the c-myc oncogene in human lung cancer cell lines, *Nature,* 306, 194, 1983.

95. **Johnson, B. E., Ihde, D. C., Makuch, R. W., et al.,** myc family oncogene amplification in tumor cell lines established from small cell lung cancer patients and its relationship to clinical status and course, *J. Clin. Invest.,* 79, 1629, 1987.

96. **Nau, M. M., Brooks, B. J., Carney, D. N., et al.,** Human small-cell lung cancers show amplification and expression of the N-myc gene, *Proc. Natl. Acad. Sci. U.S.A.,* 83, 1092, 1986.

97. **Nau, M. M., Brooks, B. J., Battey, J., et al.,** L-myc, a new myc-related gene amplified and expressed in human small cell lung cancer, *Nature,* 318, 69, 1985.

98. **Kiefer, P. E., Bepler, G., Kubasch, M., et al.,** Amplification and expression of protooncogenes in human small cell lung cancer cell lines, *Cancer Res.,* 47, 6236, 1987.

99. **Griffin, C. A. and Baylin, S. B.,** Expression of the c-myb oncogene in human small cell lung cancer, *Recent Results Cancer Res.,* 99, 237, 1985.

100. **Sithanandam, G., Dean, M., Brennscheidt, U., et al.,** Loss of heterozygosity at the c-raf locus in small cell lung carcinoma, *Oncogene,* 4, 451, 1989.

101. **Hibi, K., Takahashi, T., Sekido, Y., et al.,** Coexpression of the stem cell factor and the c-kit genes in small-cell lung cancer, *Oncogene,* 6, 2291, 1991.

102. **Turner, A. M., Zsebo, K. M., Martin, F., et al.,** Nonhematopoietic tumor cell lines express stem cell factor and display c-kit receptors, *Blood* 80, 374, 1992.

103. **Harbour, J. W., Lai, S. L., Whang-Peng, J., et al.,** Abnormalities of structure and expression of the retinoblastoma gene in SCLC, *Science,* 241, 353, 1988.

104. **Hensel, C. H., Hsieh, C. L., Gazdar, A. F., et al.,** Altered structure and expression of the human retinoblastoma susceptibility gene in small cell lung cancer, *Cancer Res.,* 50, 3067, 1990.

105. **Yokota, J., Akiyama, T., Fung, Y. K. T., et al.,** Altered expression of the retinoblastoma (RB) gene in small-cell carcinoma of the lung, *Oncogene,* 3, 471, 1988.

106. **Takahashi, T., Nau, M. M., Chiba, I., et al.,** p53: a frequent target for genetic abnormalities in lung cancer, *Science,* 246, 491, 1989.

107. **Iggo, R., Gatter, K., and Bartek, J.,** Increased expression of mutant forms of p53 oncogene in primary lung cancer, *Lancet,* 335, 675, 1990.

108. **Sugio, K., Ishida, T., and Yokoyama, H.,** ras gene mutations as a prognostic marker in adenocarcinoma of the human lung without lymph node metastases, *Cancer Res.,* 52, 2903, 1992.

109. **Rodenhuis, S., van de Wetering, M. L., Mooi, W. J., et al.,** Mutational activation of the K-ras oncogene: a possible pathogenetic factor in adenocarcinoma of the lung, *N. Engl. J. Med.,* 317, 929, 1987.

110. **Slebos, R. J. C., Kibbelaar, R. E., Dalesio, M. D. O., et al.,** K-ras oncogene activation as a prognostic marker in adenocarcinoma of the lung, *N. Engl. J. Med.,* 323, 561, 1990.

111. **Kraegel, S. A., Gumerlock, P. H., Dungworth, D. L., et al.,** K-ras activation in non-small cell lung cancer in the dog, *Cancer Res.,* 52, 4724, 1992.

112. **Cline, M. J. and Battifora, H.,** Abnormalities of protooncogenes in non-small cell lung cancer; correlation with tumor type and clinical characteristics, *Cancer,* 60, 2669, 1987.

113. **Tateishi, M., Ishida, T., Mitsudomi, T., et al.,** Prognostic value of c-erbB-2 protein expression in human lung adenocarcinoma and squamous cell carcinoma, *Eur. J. Cancer,* 27, 1372, 1991.

114. **Schalken, J. A., Roebroek, A. J. M., Oomen, P. P. C. A., et al.,** fur gene expression as a discriminating marker for small cell and nonsmall cell lung carcinomas, *J. Clin. Invest.,* 80, 1545, 1987.

115. **Nishio, H., Nakamura, A., Horai, T., et al.,** Clinical and histopathologic evaluation of the expression of Ha-ras and fes oncogene products in lung cancer, *Cancer,* 69, 1130, 1992.

116. **Mitsudomi, T., Steinberg, S. M., Nau, M. M., et al.,** p53 gene mutations in non-small-cell lung cancer cell lines and their correlation with the presence of ras mutations and clinical features, *Oncogene,* 6, 171, 1991.

117. **Suzuki, H., Takahashi, T., Kuroisha, T., et al.,** p53 mutations in nonsmall cell lung cancer in Japan: association between mutations and smoking, *Cancer Res.,* 52, 734, 1992.

118. **Chiba, I., Takahashi, T., Nau, M. M., et al.,** Mutations in the p53 gene are frequent in primary, resected non-small cell lung cancer, *Oncogene,* 5, 1603, 1990.

119. **Kishimoto, Y., Murakami, Y., Shiraishi, M., et al.,** Aberrations of the p53 tumor suppressor gene in human non-small cell carcinomas of the lung, *Cancer Res.,* 52, 4799, 1992.

120. **Quinlan, D. C., Davidson, A. G., Summers, C. L., et al.,** Accumulation of p53 protein correlates with a poor prognosis in human lung cancer, *Cancer Res.,* 52, 4828, 1992.

121. **Wakabayashi, K., Nagao, M., Esumi, H., et al.,** Food-derived mutagens and carcinogens, *Cancer Res.,* 52(suppl.), 2092, 1992.

122. **Hoffmann, D., Adams, J. D., Brunnemann, K. D., et al.,** Assessment of tobacco-specific N-nitrosamines in tobacco products, *Cancer Res.,* 39, 2505, 1979.

123. **Peterson, L. A. and Hecht, S. S.,** $O^6$-Methylguanine is a critical determinant of 4-(methylnitrosamino)-1-(3-pyridyl)-1-butanone tumorigenesis in A/J mouse lung, *Cancer Res.,* 51, 5557, 1991.

124. **Belinsky, S. A., Foley, J. F., White, C. M., et al.,** Dose–response relationship between $O^6$-methylguanine formation in Clara cells and induction of pulmonary neoplasia in the rat by 4-(methylnitrosamino)-1-(3-pyridyl)-1-butanone, *Cancer Res.,* 50, 3772, 1990.

125. **Dolan, M. E., Oplinger, M., and Pegg, A. E.,** Sequence specificity of guanine alkylation and repair, *Carcinogenesis,* 9, 2139, 1988.

126. **Zarbl, H., Sukumar, S., Arthur, A. V., et al.,** Direct mutagenesis of Ha-ras-1 oncogenes by N-nitroso-N-methylurea during initiation of mammary carcinogenesis in rats, *Nature,* 315, 382, 1985.

127. **Kubota, M.,** Generation of DNA damage by antineoplastic agents, *Anticancer Drugs,* 2, 531, 1991.

128. **Lippard, S. J.,** New chemistry of an old molecule: cis-[Pt(NH$_3$)$_2$Cl$_2$], *Science,* 218, 1075, 1982.

129. **Sawicki, E.,** Airborne carcinogens and allied compounds, *Arch. Environ. Health,* 14, 46, 1967.

130. **Loeb, L. A. and Preston, B. D.,** Mutagenesis by apurinic/apyrimidinic sites, *Annu. Rev. Genet.,* 20, 201, 1986.

131. **Vousden, K. H., Bos, J. L., Marshall, C. J., et al.,** Mutations activating human c-HA-ras-1 protooncogene induced by chemical carcinogens and depurination, *Proc. Natl. Acad. Sci. U.S.A.,* 83, 1222, 1986.

132. **Floyd, R. A.,** The role of 8-hydroxyguanine in carcinogenesis, *Carcinogenesis,* 11, 1447, 1990.

133. **Swenberg, J. A., et al.,** High- to low-dose extrapolation: critical determinants involved in the dose response of carcinogenic substances, *Environ. Health Perspect.,* 76, 57, 1987.

134. **Wilson, V. L., Weston, A., Manchester, D. K., et al.,** Alkyl and aryl carcinogen adducts detected in human peripheral lung, *Carcinogenesis,* 10, 2149, 1989.

135. **Wogan, G. N.,** Markers of exposure to carcinogens, *Environ. Health Perspect.,* 81, 9, 1989.

136. **Watson, W. P.,** Post-radiolabelling for detecting DNA damage, *Mutagenesis,* 2, 319, 1987.

137. **Talaska, G., Schamer, M., Skipper, P., et al.,** Detection of carcinogen-DNA adducts in exfoliated urothelial cells of cigarette smokers: association with smoking, hemoglobin adducts, and urinary mutagenicity, *Cancer Epidemiol Biomarkers Prev.,* 1, 61, 1991.

138. **Maclure, M., Bryant, M. S., Skipper, P. L., et al.,** Decline of the hemoglobin adduct of 4-aminobiphenyl during withdrawal from smoking, *Cancer Res.,* 50, 1810, 1990.

139. **Tornqvist, M., Osterman-Golkar, S., Kautiainen, A., et al.,** Tissue doses of ethylene oxide in cigarette smokers determined from adduct levels in hemoglobin, *Carcinogenesis,* 7, 1519, 1986.

140. **Harris, C. C.,** Human tissues and cells in carcinogenesis research, *Cancer Res.,* 47, 1, 1987.

141. **Southorn, P. A. and Powis, G.,** Free radicals in medicine. I. Chemical nature and biologic reactions, *Mayo Clin. Proc.,* 63, 381, 1988; II. Involvement in human disease, *Mayo Clin. Proc.,* 63, 390, 1988.

142. **Cross, C. E., Halliwell, B., Borish, E. T., et al.,** Oxygen radicals and human disease, *Ann. Int. Med.,* 107, 526, 1987.

143. **Marnett, L. J.,** Peroxyl free radicals: potential mediators of tumor initiation and promotion, *Carcinogenesis,* 8, 1365, 1987.

144. **Weitzman, S. A. and Stossel, T. P.,** Mutations caused by human phagocytes, *Science,* 212, 546, 1981.

145. **Nassi-Calo, L., Mello-Filho, A. C., Meneghini, R.,** o-Phenanthroline protects mammalian cells from hydrogen peroxide-induced gene mutation and morphologic transformation, *Carcinogenesis,* 10, 1055, 1989.

146. **Radosevich, C. A. and Weitzman, S. A.,** Hydrogen peroxide induces squamous metaplasia in hamster tracheal organ explant culture model, *Carcinogenesis,* 10, 1943, 1989.

147. **Larsson, R. and Cerutti, P.,** Translocation and enhancement of phosphotransferase activity of protein kinase C following exposure of mouse epidermal cells to oxidants, *Cancer Res.,* 49, 5627, 1989.

148. **Crawford, D., Zbinden, I., Amstad, P., et al.,** Oxidant stress induces the protooncogenes c-fos and c-myc in mouse epidermal cells, *Oncogene,* 3, 27, 1988.

149. **Amstad, P. A., Krupitza, G., and Cerutti, P. A.,** Mechanism of c-fos induction by active oxygen, *Cancer Res.,* 52, 3952, 1992.

150. **Ehrlich, M. and Wang, R. Y. H.,** 5-Methylcytosine in eukaryotic cells, *Science,* 212, 1350, 1981.

151. **Holliday, R.,** A new theory of carcinogenesis, *Br. J. Cancer,* 40, 513, 1979.

152. **Ehrlich, M., Zhang, X. Y., and Inamdar, N. M.,** Spontaneous deamination of cytosine and 5-methylcytosine residues in DNA and replacement of 5-methylcytosine residues with cytosine residues, *Mutat. Res.,* 238, 277, 1990.

153. **Wilson, V. L., Smith, R. A., Longoria, J., et al.,** Chemical carcinogen-induced decreases in genomic 5-methylcytidine content of normal human bronchial epithelial cells, *Proc. Natl. Acad. Sci. U.S.A.,* 84, 3298, 1987.

154. **Hanawalt, P. C.,** Preferential DNA repair in expressed genes, *Environ. Health Perspect.,* 76, 9, 1987.

155. **Myrnes, B., et al.,** Interindividual variation in the activity of $O^6$-methyl guanine-DNA methyltransferase and uracil-DNA glycosylase in human organs, *Carcinogenesis,* 4, 1565, 1983.

156. **Kneckt, P., Jarvinen, R., Seppanen, R., et al.,** Dietary antioxidants and the risk of lung cancer, *Am. J. Epidemiol.,* 134, 471, 1991.

157. **Smith, P. J.,** Carcinogenesis: molecular defences against carcinogens, *Br. Med. Bull.,* 47, 3, 1991.

158. **Cleaver, J. E.,** Defective repair replication of DNA in xeroderma pigmentosum, *Nature,* 218, 652, 1968.

159. **Morrell, D., Cromartie, E., and Swift, M.,** Mortality and cancer incidence in 263 patients with ataxia-telangiectasia, *J. Natl. Cancer Inst.,* 77, 89, 1986.

160. **Gatti, R. A., Berkel, I., Boder, E., et al.,** Localization of an ataxia-telangiectasia gene to chromosome 11q22–23, *Nature,* 336, 577, 1988.

161. **Kuhn, E. M. and Therman, E.,** Cytogenetics of Bloom's syndrome, *Cancer Genet. Cytogenet.,* 22, 1, 1986.

162. **Willis, A. E. and Lindahl, T.,** DNA ligase I deficiency in Bloom's syndrome, *Nature,* 325, 355, 1987.

163. **Chan, J. Y., Becker, F. F., German, J., et al.,** Altered DNA ligase I activity in Bloom's syndrome cells, *Nature,* 325, 357, 1987.

164. **Auerbach, A. D. and Wolman, S. R.,** Susceptibility of Fanconi's anaemia fibroblasts to chromosome damage by carcinogens, *Nature,* 261, 494, 1976.

165. **Hsu, T. C., Spitz, M. R., and Schantz, S. P.,** Mutagen sensitivity: a biologic marker of cancer susceptibility, *Cancer Epidemiol. Biomarkers Prev.,* 1, 83, 1991.

166. **Bartsch, H.,** N-nitroso compounds and human cancer: where do we stand?, in *Relevance to Human Cancer of N-Nitroso Compounds, Tobacco Smoke and Mycotoxins,* O'Neill, I. K., Chen, J., and Bartsch, H., Eds., IARC, Lyon, 1991.

167. **Weitzman, S. A. and Gordon, L. I.,** Inflammation and cancer: role of phagocyte-generated oxidants in carcinogenesis, *Blood,* 76, 655, 1990.

168. **Preston-Martin, S., Pike, M., Ross, R. K., et al.,** Increased cell division as a cause of human cancer, *Cancer Res.,* 50, 7415, 1990.

169. **Doll, R. and Hill, A. B.,** A study of the aetiology of carcinoma of the lung, *Br. Med. J.,* 2, 1271, 1952.

170. **Wynder, E. L. and Graham, E. A.,** Tobacco smoking as a possible etiologic factor in bronchogenic carcinoma: a study of six hundred and eighty-four proved cases, *JAMA,* 143, 329, 1950.

171. **Levin, M. L., Golstein, H., and Gerhardt, P. R.,** Cancer and tobacco smoking: a preliminary report, *JAMA,* 143, 336, 1950.

172. **Doll, R. and Peto, R.,** Mortality in relation to smoking: 20 years' observations on male british doctors, *Br. Med. J.,* 2, 1525, 1976.

173. **Pitot, H. C.,** Principles of cancer biology: chemical carcinogenesis, in *Cancer: Principles and Practice of Oncology,* 2nd ed., Devita, V. T., Jr., Hellman, S., and Rosenberg, S. A., Eds., Lippincott, Philadelphia, 1985, 79.

174. **Doll, R. and Peto, R.,** The causes of cancer: quantitative estimates of avoidable risks of cancer in the United States today, *J. Natl. Cancer Inst.,* 66, 1191, 1981.

175. **Shackney, S. E., McCormack, G. W., and Cucheral, G. J.,** Growth rate patterns of solid tumors and their relation to responsiveness to therapy, *Ann. Intern. Med.,* 89, 107, 1978.

176. **Brunnemann, K. D., Djordjevic, M. V., Feng, R., et al.,** Analysis and pyrolysis of some N-nitrosamino acids in tobacco and tobacco smoke, in *Relevance to Human Cancer of N-Nitroso Compounds, Tobacco Smoke and Mycotoxins,* O'Neill, I. K., Chen, J., and Bartsch, H., Eds., IARC, Lyon, 1991, 477.

177. **Chen, Y. P. and Squier, C. A.,** Effect of nicotine on 7, 12-dimethylbenz[a]-anthracene carcinogenesis in hamster cheek pouch, *J. Natl. Cancer Inst.,* 82, 861, 1990.

178. **Jones, C. J., Schiffman, M. H., Kurman, R., et al.,** Elevated nicotine levels in cervical lavages from passive smokers, *Am. J. Public Health,* 81, 378, 1991.

179. **Thompson, S. G., Store, R., Nanchahal, K., et al.,** Relation of urinary cotinine concentrations to cigarette smoking and to exposure to other people's smoke, *Thorax,* 45, 356, 1990.

180. **Anderson, R., Theron, A. J., Richards, G. A., et al.,** Passive smoking by humans sensitizes circulating neutrophils, *Am. Rev. Resp. Dis.,* 144, 570, 1991.

181. **Byrd, J. C.,** Environmental Tobacco Smoke, *Med. Clin. N. Am.* 76, 377, 1992.

182. **Janerich, D. T., Thompson, W. D., Varela, L. R., et al.,** Lung cancer and exposure to tobacco smoke in the household, *N. Engl. J. Med.,* 323, 632, 1990.

183. **Lee, P. N.,** *N. Engl. J. Med.,* 324, 412 (letter), 1991.

184. **van Poppei, G., Kok, F. J., Gorgels, W. J., et al.,** *N. Engl. J. Med.,* 324, 413 (letter), 1991.

185. **Sterling, T. D. and Weinkam, J. J.,** *N. Engl. J. Med.,* 324, 414 (letter), 1991.

186. **Borelli, T. J.,** *N. Engl. J. Med.,* 324, 414 (letter), 1991.

187. **Bierer, M. F. and Rigotti, N. A.,** Public policy for the control of tobacco-related disease, *Med. Clin. N. Am.,* 76, 515, 1992.

188. **Samuels, B. and Glantz, S. A.,** The politics of local tobacco control, *JAMA,* 266, 2110, 1991.

189. **Daynard, R. A.,** Tobacco liability litigation as a cancer control strategy, *J. Natl. Cancer Inst.,* 80, 9, 1988.

190. **Buell, P.,** Relative impact of smoking and air pollution on lung cancer, *Arch. Environ. Health,* 15, 291, 1967.

191. **Tomatis, L.,** Outdoor air pollution and lung cancer, *Ann. Oncol.,* 2, 265, 1991.

192. **Siemiatycki, J., Dewar, R., Nadon, L., et al.,** Associations between several sites of cancer and twelve petroleum-derived liquids, *Scand. J. Work Environ. Health,* 13, 493, 1987.

193. **Fraser, D.,** Lung cancer risk and diesel exhaust exposure, *Public Health Rev.,* 14, 139, 1986.

194. **Askoy, M., Erdem, S., and Dincol, G.,** Leukemia in shoe workers exposed chronically to benzene, *Blood,* 44, 837, 1974.

195. **Cronkite, E. P., Drew, R. T., Inoue, T., et al.,** Hematotoxicity and carcinogenicity of inhaled benzene, *Environ. Health Perspect.,* 82, 97, 1989.

196. **Lee, A. M. and Fraumeni, J. F., Dr.,** Arsenic and respiratory cancer in man: an occupational study, *J. Natl. Cancer Inst.,* 42, 1045, 1969.

197. **Pinto, S. S., Enterline, P. E., Henderson, V., et al.,** Mortality experience in relation to a measured arsenic trioxide exposure, *Environ. Health Perspect.,* 19, 127, 1977.

198. **Mabuchi, K., Lilianfeld, A. M., and Snell, L. M.,** Cancer and occupational exposure to arsenic: a study of pesticide workers, *Prev. Med.,* 9, 51, 1980.

199. **Pershagen, G.,** The carcinogenicity of arsenic, *Environ. Health Perspect.,* 40, 93, 1981.

200. **Langard, S. and Norseth, T.,** A cohort study of bronchial carcinomas in workers producing chromate pigments, *Br. J. Ind. Med.* 32, 62, 1975.

201. **Dalagar, N. A., Mason, T. J., Fraumeni, J. F., Jr., et al.,** Cancer mortality among workers exposed to zinc chromate paints, *JOM,* 22, 25, 1980.

202. **LaVelle, J. M.,** Mechanisms of toxicity/carcinogenicity and superfund decisions, *Environ. Health Perspect.,* 92, 127, 1991.

203. **Pederson, E., Hogetveit, A. C., and Anderson, A.,** Cancer of respiratory organs among workers at a nickel refinery in Norway, *Int. J. Cancer,* 12, 32, 1973.

204. **Higginbotham, K. G., Rice, J. M., Diwan, B. A., et al.,** GGT to GTT transversions in codon 12 of the K-ras oncogene in rat renal sarcomas induced with nickel subsulfide or nickel subsulfide/iron are consistent with oxidative damage to DNA, *Cancer Res.,* 52, 4747, 1992.

205. **Sunderman, F. W., Jr.,** Carcinogenic effects of metals, *Fed. Proc.,* 37, 40, 1978.

206. **Axelson, O. and Sjoberg, A.,** Cancer incidence and exposure to iron oxide dust, *JOM,* 21, 419, 1979.

207. **Wagoner, J. K., Infante, P. F., and Bayliss, D. L.,** An etiologic agent in the induction of lung cancer, nonneoplastic respiratory disease, and heart disease among industrially exposed workers, *Environ. Res.,* 21, 15, 1980.

208. **Pasternak, B. S., Shore, R. E., and Albert, R. E.,** Occupational exposure to chloromethyl ethers, *JOM,* 19, 741, 1977.

209. **O'Berg, M. T.,** Epidemiologic study of workers exposed to acrylonitrile, *JOM,* 22, 245, 1980.

210. **Buffler, P. A., Wood, S., Eifler, C., et al.,** Mortality experience of workers in a vinyl chloride monomer production plant, *JOM,* 21, 195, 1979.

211. **Abe, T., Tsuda, S., Maekawa, T., et al.,** Sister chromatid exchanges induced by cancer chemotherapeutic agents in vitro and in vivo: consideration of the hazard of drugs as possible mutagens and carcinogens causing second malignancies, *Cancer Treat. Rep.,* 69, 505, 1985.

212. **Kaldor, J. M., Day, N. E., Pettersson, F., et al.,** Leukemia following chemotherapy for ovarian cancer, *N. Engl. J. Med.,* 322, 1, 1990.

213. **Kaldor, J. M., Day, N. E., and Clarke, E. A.,** Leukemia following Hodgkin's disease, *N. Engl. J. Med.,* 322, 7, 1990.

214. **Witherspoon, R. P., Fisher, L. D., Schoch, G., et al.,** Secondary cancers after bone marrow transplantation for leukemia or aplastic anemia, *N. Engl. J. Med.,* 321, 784, 1989.

215. **Baker, G. L., Kahl, L. E., Zee, B. C., et al.,** Malignancy following treatment of rheumatoid arthritis with cyclophosphamide, *Am. J. Med.,* 83, 1, 1987.

216. **Craig, J., Powell, B., Muss, H. B., et al.,** Second primary bronchogenic carcinomas after small cell carcinoma, *Am. J. Med.,* 76, 1013, 1984.

217. **Johnson, B. E., Ihde, D. C., Matthews, M. J., et al.,** Non-small-cell lung cancer: major cause of late morbidity in patients with small cell lung cancer, *Am. J. Med.,* 80, 1103, 1986.

218. **Heyne, K. H., Lippman, S. M., Lee, J. J., et al.,** The incidence of second primary tumors in long-term survivors of small-cell lung cancer, *J. Clin. Oncol.,* 10, 1519, 1992.

219. **Sagman, U., Lishner, M., Maki, E., et al.,** Second primary malignancies following diagnosis of small-cell lung cancer, *J. Clin. Oncol.,* 10, 1525, 1992.

220. **Chak, L. Y., Sikic, B. I., Tucker, M. A., et al.,** Increased incidence of acute nonlymphocytic leukemia following therapy in patients with small cell carcinoma of the lung, *J. Clin. Oncol.,* 2, 385, 1984.

221. **Pederson-Bjergaard, J., Osterlind, K., Hansen, M., et al.,** Acute nonlymphocytic leukemia, preleukemia and solid tumors following intensive chemotherapy of small cell carcinoma of the lung, *Blood,* 66, 1393, 1985.

222. **Johnson, D. H., Porter, L. L., List, A. F., et al.,** Acute nonlymphocytic leukemia after treatment of small cell lung cancer, *Am. J. Med.,* 81, 962, 1986.

223. **Waksvik, H., Klepp, O., and Brogger, A.,** Chromosome analyses of nurses handling cytostatic agents, *Cancer Treat. Rep.,* 65, 607, 1981.

224. **Selevan, S. G., Lindbohm, M. L., Hornung, R. W., et al.,** A study of occupational exposure to antineoplastic drugs and fetal loss in nurses, *N. Engl. J. Med.,* 313, 1173, 1985.

225. **Kim, J. H. and Durack, D. T.,** Manifestations of human T-lymphocytic virus type I infection, *Am. J. Med.,* 84, 919, 1988.

226. **Mackowiak, P. A.,** Microbial oncogenesis, *Am. J. Med.* 82, 79, 1987.

227. **Schreiber, H., Nettesheim, P., Lijinsky, W., et al.,** Induction of lung cancer in germ free, specific-pathogen-free, and infected rats by N-nitrosoheptamethyleneimine; enhancement by respiratory infection, *J. Natl. Cancer Inst.,* 49, 1107, 1972.

228. **Craighead, J. E. and Mossman, B. T.,** The pathogenesis of asbestos-associated diseases, *N. Eng. J. Med.,* 306, 1446, 1982.

229. **Mossman, B. T. and Gee, J. B. L.,** Asbestos-related diseases, *N. Engl. J. Med.,* 320, 1721, 1989.

230. **Selikoff, I. J., Hammond, E. C., and Churg, J.,** Asbestos exposure, smoking and neoplasia, *JAMA,* 204, 104, 1968.

231. **Lemaire, I., Beaudoin, H., Masse, S., et al.,** Alveolar macrophage stimulation of lung fibroblast growth in asbestos-induced pulmonary fibrosis, *Am. J. Pathol.,* 122, 205, 1986.

232. **Case, B. W., Ip, M. P. C., Padilla, M., et al.,** Asbestos effects on superoxide production, *Environ. Res.,* 39, 299, 1986.

233. **Browne, K.,** Asbestos related malignancy and the Cairns hypothesis, *Br. J. Ind. Med.,* 48, 73, 1991.

234. **Barrett, J. C., et al.,** Multiple mechanisms for the carcinogenic effects of asbestos and other mineral fibers, *Environ. Health Perspect.,* 81, 81, 1989.

235. **Grosovsky, J. A., de Boer, J. G., de Jong, P. J., et al.,** Base substitutions, frameshifts and small deletions constitute ionizing radiation-induced point mutations in mammalian cells, *Proc. Natl. Acad. Sci. U.S.A.,* 85, 185, 1988.

236. **Garte, S. J. and Burns, F. J.,** Oncogenes and radiation carcinogenesis, *Environ. Health Perspect.,* 93, 45, 1991.

237. **Kohn, H. I. and Fry, R. J. M.,** Radiation carcinogenesis, *N. Engl. J. Med.,* 310, 504, 1984.

238. **Champlin, R. E., Kastenberg, W. E., and Gale, R. P.,** Radiation accidents and nuclear energy: medical consequences and therapy, *Ann. Intern. Med.,* 109, 730, 1988.

239. **Rowland, R. E., Stehney, A. F., and Lucas, H. F., Jr.,** Dose response relationships for female radium dial workers, *Radiat. Res.,* 76, 368, 1978.

240. **Samet, J. M., Kutvirt, D. M., Waxweiler, R. J., et al.,** Uranium mining and lung cancer in Navajo men, *N. Engl. J. Med.,* 310, 1481, 1984.

241. **Ives, J. C., Buffler, P. A., and Greenberg, S. D.,** Environmental associations and histopathologic patterns of carcinoma of the lung: the challenge and dilemma in epidemiologic studies, *Am. Rev. Respir. Dis.,* 128, 195, 1983.

242. **Vahakangas, K. H., Samet, J. M., Metcalf, R. A., et al.,** Mutations of p53 and ras genes in radon-associated lung cancer from uranium miners, *Lancet,* 339, 576, 1992.

243. **Fabrikant, J. I.,** Radon and lung cancer: the BEIR IV report, *Health Perspect.,* 59, 89, 1990.

244. **Logue, J. and Fox, J.,** Health hazards associated with elevated levels of indoor radon — Pennsylvania, *MMWR,* 34, 657, 1985.

245. National Council on Radiation Protection and Measurements, Evaluation of Occupational and Environmental Exposures to Radon and Radon Daughters, NCRP Report #78, Bethesda, MD, 1984.

246. Environmental Protection Agency, A Citizen's Guide to Radon: What It Is and What To Do About It, U.S. EPA Report OPA-86-004, Washington, D.C., 1986.

247. **Marmorstein, J.,** Lung cancer: is the increasing incidence due to radioactive polonium in cigarettes?, *S. Med. J.,* 79, 145, 1986.

248. **Shami, S. G., Thibodeau, L. A., Kennedy, A. R., et al.,** Proliferative and morphologic changes in the pulmonary epithelium of the Syrian golden hamster during carcinogenesis initiated by $^{210}$Po alpha-radiation, *Cancer Res.,* 42, 1405, 1982.

249. **Witschi, H. and Schuller, H. M.,** Diffuse and continuous cell proliferation enhances radiation-induced tumorigenesis in hamster lung, *Cancer Lett.,* 60, 193, 1991.

250. **Mole, R. H.,** Carcinogenesis by thorotrast and other sources of irradiation, especially other alpha-emitters, *Environ. Res.,* 18, 192, 1979.

251. **Boice, J. D., Monson, R. R., and Rosenstein, M.,** Cancer mortality in women after repeated fluoroscopic examinations, *J. Natl. Cancer Inst.,* 5, 115, 1981.

252. **Witt, T. R., Meng, R. L., Economou, S. G., et al.,** The approach to the irradiated thyroid, *Surg. Clin. N. Am.,* 59, 45, 1979.

253. **Schneider, A. B., Recant, W., Pinsky, S. M., et al.,** Radiation-induced thyroid carcinoma: clinical course and results of therapy in 296 patients, *Ann. Intern. Med.* 105, 405, 1986.

254. **Smith, P. G. and Doll, R.,** Mortality from cancer and all causes among British radiologists, *Br. J. Radiol.,* 54, 187, 1981.

255. **Bailar, J. C.,** Mammography: a contrary view, *Ann. Intern. Med.,* 84, 77, 1976.

256. **Health and Public Policy Committee, American College of Physicians,** The use of diagnostic tests for screening and evaluating breast lesions, *Ann. Intern. Med.,* 103, 147, 1985.

257. **Sellers, T. A., Bailey-Wilson, J. E., Elston, R. C. et al.,** Evidence for Mendelian inheritance in the pathogenesis of lung cancer, *J. Natl. Cancer Inst.,* 82, 1272, 1990.

258. **Gonzalez, F. J., Skoda, R. C., Kimura, S., et al.,** Characterization of the common genetic defect in humans deficient in debrisoquine metabolism, *Nature,* 331, 442, 1988.

259. **Ayesh, R., Idle, J. R., Ritchie, J. C., et al.,** Metabolic oxidation phenotypes as markers for susceptibility to lung cancer, *Nature,* 312, 169, 1984.

260. **Caporoso, N., Hayes, R. B., Dosemici, M., et al.,** Lung cancer risk, occupational exposure and the debrisoquine metabolic phenotype, *Cancer Res.,* 49, 3675, 1989.

261. **Caporoso, N. E., Tucker, M. A., Hoover, R. N., et al.,** Lung cancer and the debrisoquine phenotype, *J. Natl. Cancer Inst.,* 82, 1264, 1990.

262. **Kaisary, A., Smith, P., Jaczq, E., et al.,** Genetic predisposition to bladder cancer: ability to hydroxylate debrisoquine and mephenytoin as risk factors, *Cancer Res.,* 47, 5488, 1987.

263. **Idle, J. R., Mahgoub, A., Sloan, T. P., et al.,** Some observations on the oxidation phenotype status of Nigerian patients presenting with cancer, *Cancer Lett.,* 11, 331, 1981.

264. **Horsmans, Y., Desager, J. P., and Harvengt, C.,** Is there a link between debrisoquine oxidation phenotype and lung cancer susceptibility? *Biomed. Pharmacother.,* 45, 359, 1991.

265. **Kawajiri, K., Nakachi, K., Imai, K., et al.,** Identification of genetically high risk individuals to lung cancer by DNA polymorphisms of the cytochrome P450IA1 gene, *FEBS Lett.,* 263, 131, 1990.

266.  **Kellermann, G., Shaw, C. R., and Luyten-Kellerman, M.,** Aryl hydrocarbon hydroxylase inducibility and bronchogenic carcinoma, *N. Engl. J. Med.,* 289, 934, 1973.

267.  **Seidegard, J., Pero, R. W., Miller, D. G., et al.,** A glutathione transferase in human leukocytes as a marker for the susceptibility to lung cancer, *Carcinogenesis,* 7, 751, 1986.

268.  **Seidegard, J., Pero, R. W., Markowitz, M. M., et al.,** Isoenzyme(s) of glutathione transferase (class Mu) as a marker for the susceptibility to lung cancer; a followup study, *Carcinogenesis,* 11, 33, 1990.

269.  **Rudiger, H. W., et al.,** Reduced $O^6$-methylguanine repair in fibroblast cultures from patients with lung cancer, *Cancer Res.,* 49, 5623, 1989.

270.  **Perera, F. P. and Weinstein, I. B.,** Molecular epidemiology and carcinogen-DNA adduct detection: new approaches to studies of human cancer causation, *J. Chron. Dis.,* 35, 581, 1982.

271.  **Ames, B. N., Durston, W. E., Yamasaki, E., et al.,** Carcinogens are mutagens: a simple test system for combining liver homogenates for activation and bacteria for testing, *Proc. Natl. Acad. Sci. U.S.A.,* 70, 2281, 1973.

272.  **Zeiger, E., Haseman, J. K., Shelby, M. D., et al.,** A further evaluation of four in vitro genetic toxicity tests for predicting rodent carcinogenicity, *Environ. Mol. Mutagen.,* Suppl. 16, 1, 1990.

273.  **Ashby, J. and Tennant, R. W.,** Definitive relationships among chemical structure, carcinogenicity and mutagenicity for 301 chemicals tested by the U.S. NTP, *Mutat. Res.,* 257, 229, 1991.

274.  **Tomatis, L. and Bartsch, H.,** The contribution of experimental studies to risk assessment of carcinogenic agents in humans, *Exp. Pathol.,* 40, 251, 1990.

275.  **Mason, J. M., Lagenbach, R., Shelby, M. D., et al.,** Ability of short-term tests to predict carcinogenesis in rodents, *Annu. Rev. Pharmacol. Toxicol.,* 30, 149, 1990.

276.  **Slaga, T. J.,** Overview of tumor promotion in animals, *Environ. Health Perspect.,* 50, 3, 1983.

277.  **Chabra, R. S., Huff, J. E., Schwetz, B. S., et al.,** An overview of the prechronic and chronic toxicity/carcinogenicity experimental study designs and criteria used by the National Toxicology Program, *Environ. Health Perspect.,* 86, 313, 1990.

278.  **Ames, B. N., Magaw, R., and Gold, L. S.,** Ranking possible carcinogenic hazards, *Science,* 236, 271, 1987.

279.  **Ames, B. N. and Gold, L. S.,** Too many rodent carcinogens: mitogenesis increases mutagenesis, *Science,* 249, 970, 1990.

280.  **Ames, B. N. and Gold, L. S.,** Pesticides, risk and apple sauce, *Science,* 244, 755, 1989.

281.  **Cohen, S. M. and Ellwein, L. B.,** Cell proliferation in carcinogenesis, *Science,* 249, 1007, 1990.

282.  **Cohen, S. M. and Ellwein, L. B.,** Genetic errors, cell proliferation and carcinogenesis, *Cancer Res.,* 51, 6493, 1991.

283.  **Carr, C. J. and Kolbye, A. C.,** A critique of the use of the maximum tolerated dose in bioassays to assess cancer risks from chemicals, *Regul. Toxicol. Pharmacol.,* 14, 78, 1991.

284.  **Swenberg, J. A. and Maronpot, R. R.,** Chemically induced cell proliferation as a criterion in selecting doses for long term bioassays, *Prog. Clin. Biol. Res.,* 369, 245, 1991.

285.  **Infante, P. F.,** Prevention versus chemophobia: a defence of rodent carcinogenicity tests, *Lancet,* 337, 538, 1991.

286. **Weinstein, I. B.,** Mitogenesis is only one factor in carcinogenesis, *Science,* 251, 387, 1991.

287. **Hoel, D. G., Haseman, J. K., Hogan, M. D., et al.,** The impact of toxicity on carcinogenicity studies: implications for risk assessment, *Carcinogenesis,* 9, 2045, 1988.

288. **Schwetz, B. A. and Yang, R. S. H.,** in *Approaches Used by the U.S. National Toxicology Program in Assessing the Toxicity of Chemical Mixtures,* Vainio, H., Sorsa, M., McMichael, A. J., Eds., IARC, Lyon, 1990.

289. **Sporn, M. B.,** Carcinogenesis and cancer: different perspectives on the same disease, *Cancer Res.,* 51, 6215, 1991.

290. **Bond, J. H.,** Screening for colorectal cancer: need for controlled trials (editorial), *Ann. Intern. Med.,* 113, 337, 1990.

291. **Selby, J. V., Friedman, G. D., Quesenberry, C. P., Jr., et al.,** Effect of fecal occult blood testing on mortality from colorectal cancer, *Ann. Intern. Med.,* 118, 1, 1993.

292. **Sidransky, D., Tokino, T., Hamilton, S. R., et al.,** Identification of ras oncogene mutations in the stool of patients with curable colorectal tumors, *Science,* 256, 102, 1992.

293. **Birrer, M. J. and Brown, P. H.,** Application of molecular genetics to the early diagnosis and screening of lung cancer, *Cancer Res.,* Suppl. 52, 2658s, 1992.

294. **Butta, A., MacLennan, K., Flanders, K. C., et al.,** Induction of transforming growth factor beta$_1$ in human breast cancer *in vivo* following tamoxifen treatment, *Cancer Res.,* 52, 4261, 1992.

295. **Nishikura, K. and Murray, J. M.,** Antisense RNA of proto-oncogene c-fos blocks renewed growth of quiescent cells, *Mol. Cell. Biol.,* 7, 639, 1987.

296. **Rivera, R. T., Pasion, S. G., Wong, D. T. W., et al.,** Loss of tumorigenic potential by human lung tumor cells in the presence of antisense RNA specific to the ectopically synthesized alpha subunit of human chorionic gonadotropin, *J. Cell. Biol.,* 108, 2423, 1989.

297. **Feramisco, J. R., Clark, R., Wong, G., et al.,** Transient reversion of ras oncogene-induced cell transformation by antibodies specific for amino acid 12 of ras protein, *Nature,* 314, 639, 1985.

298. **Motokura, T. and Arnold, A.,** Cyclins and Oncogenesis. *Biochim. Biophys. Acta,* 1155, 63, 1993.

299. **Pezzella, F., Turley, H., Kuzu, I., et al.,** bcl-2 Protein in Non-Small-Cell Lung Cancer. *New Eng. J. Med.,* 329, 690, 1993.

300. **Ikegaki, N., Katsumata, M., Minna, J., et al.,** Expression of bcl-2 in small cell lung carcinoma cells. *Cancer Res., 54, 6, 1994.*

# Chapter 4

# The Pathology of Environmental Lung Disease

*Arnold M. Schwartz, M.D., Ph.D.*

## CONTENTS

## INTRODUCTION

During the course of normal breathing, the lung is exposed to a variety of ambient mixed gases and fumes or suspended particulate matter. Essentially, the entire airway epithelium of the lung, from proximal bronchial airways to the alveolar sacs, like the skin, comes in contact with the external environment. The factors that determine the reactions and injuries to airways and lung parenchyma depend on the biophysical chemistry of the material inhaled and the status of the host. Analogous to pharmacological determinants, the physical and chemical properties of the inhaled material, its antigenic qualities, and the dose and duration of exposure will define the nature and characteristics of the lung pathology. The responsiveness of the host lung, previous or concurrent pulmonary disease, and underlying systemic conditions will also impact on environmental or occupational lung disease. Confounding conditions of cigarette smoking or mixed occupational exposures may complicate and aggravate the pulmonary pathology.

Some of the lung exposures may lead to clinical and pathologic conditions that are transient, reversible, and nonprogressive. Other exposures produce immediate injuries, which may on their own or with prolonged exposure progress to permanent damage and/or diffuse interstitial fibrosis. Pulmonary pathology involving the progression of diffuse alveolar damage or the organization of intra-alveolar or interstitial exudates may lead to interstitial fibrosis. Incorporation of dusts with localized fibrous scars or the formation of granulomas and their organization and coalescence may also produce diffuse pulmonary fibrosis.

This chapter serves as a review of the pathologic findings in the major types of environmental or occupational lung disease.[1-5] Success in the diagnosis of these diseases rests on the classical collaboration of medical history and examination, radiologic findings, pathologic features, and physicochemical analysis.

# TOBACCO SMOKE

Inhaled cigarette smoke contains a variety of organic and inorganic solids or particulates and gases.[6,7] During inhalation, the gases easily mix with air, travel along airways, and penetrate deeply into pulmonary parenchyma. Particulate matter larger than tenths of millimeters is trapped in major airways and expectorated by the mucociliary defense system. Those particulates that are in the range of microns deposit around bronchi and bifurcations of terminal and respiratory bronchioles, with some solid material reaching the pulmonary alveolar tissue. Because a large volume of particulate matter is in this size range, it is estimated that approximately 50% of solid smoke is deposited in the lung.[6] Those particulates not removed by the mucociliary system will be deposited within the interstitial matrix, engulfed by macrophages, or transported by lymphatics. Passive or "secondhand" smoke may deposit approximately 1% of the particulate load of a regular smoker.[8] Other components of cigarette fumes consist of complicated organic compounds, in particular aromatic hydrocarbons, that are carcinogenic in laboratory animals and in the Ames test. The most noxious inhalant gas is carbon monoxide, due to its stronger affinity to hemoglobin than that of oxygen. Carbon monoxide replaces heme-bound oxygen and reduces net hemoglobin-carrying capacity of oxygen. The extent of one's smoking exposure can be assessed by the blood concentration of carboxyhemoglobin and the alveolar concentration of carbon monoxide. Smoker's polycythemia is an adaptive attempt of the body to compensate for the effective decrease in red blood cell function.

The complex components of tobacco smoke cause injury to large and small airways and pulmonary parenchyma.[9–13] The larger airways undergo hypersecretion and lose their "defensive" function, the smaller airways develop airflow resistance, and the parenchyma undergoes destruction and emphysematous change. The large tracheal and bronchial airways respond to cigarette smoke by metaplasia of the mucosal epithelium, hyperplasia of the bronchial mucus glands, and increased secretions of an inflamed thickened mucus. The extent of mucous gland hyperplasia has been quantified in the form of the Reid index—namely, the ratio of the length of mucosal gland thickness relative to the length of the bronchial mucosa internal to the cartilage.[14] Normal mucosal tissue has a Reid index of 0.4 or less, indicating that 40% or less of the mucosa is occupied by glands. In association with the reactive mucosa, a mild to moderate chronic inflammatory infiltrate is often present. Some individuals with considerable exposure develop squamous metaplasia of their respiratory ciliated epithelium. The smoker's chronic bronchitis is a clinical entity characterized by a productive cough that results from the irritation to the airway mucosa and the secondary inflamed viscous secretions.[15] Individuals with chronic bronchitis characteristically have an increased Reid index relative to normal controls. The loss of airway protective and defensive mechanisms in the tracheo-bronchial tree, render these airways more susceptible to infections.

The small airways, consisting of the distal bronchioles of the lung, are the low-resistance airflow channels. As a result of mucosal injury and inflammation, these airways develop greater airflow resistance due to mucosal edema, congestion, and increased mucus secretion with airway obstruction.[16] In certain cases, the bronchiolar pathology associated with smoking manifests a more inflammatory and clinically restrictive picture. Respiratory bronchiolitis[17–20] is a clinicopathologic entity found in adult smokers in their fourth decade, who have had 30 to 40 pack-years of smoking (Figure 1). These smokers present with dyspnea and cough and show a mixed obstructive-restrictive pattern on pulmonary function tests. Their chest X-rays may be normal but more commonly display a reticulonodular interstitial pattern. These cases may come to the attention of the pathologist due to their clinical and radiographic

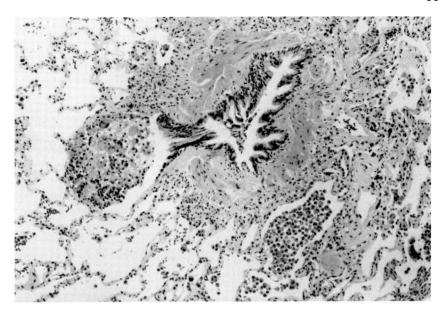

**Figure 1** Respiratory bronchiolitis demonstrating bronchiolar inflammation in a prominent intra-alveolar duct and alveolar macrophage exudate. Note the foamy cytoplasmic appearance of the macrophages and the multinucleated giant cells. (Courtesy of Dr. S. Hammar, H&E, Magnification × 40.)

similarity with desquamative interstitial pneumonitis or usual interstitial pneumonitis. The histopathologic features of respiratory bronchiolitis are seen in the bronchial lumen, mucosa, and peribronchial tissue. The airway contains abundant granular and pigmented macrophages that stain positively for iron and PAS (periodic acid-Schiff reagent) material. The staining is due to the accumulation of particulate matter phagocytosed by lysosomes. These macrophages cluster around the bronchiole and extend toward the alveolar duct and lobule. The bronchiolar mucosa shows goblet cell metaplasia and/or squamous metaplasia; prominent hyperplastic type II pneumocytes line the distal bronchiole and alveolar ducts. The bronchiole wall undergoes smooth muscle hyperplasia and peribronchiolar fibrosis, with an admixture of infiltrating mononuclear inflammatory cells. A key difference between this entity and asbestosis or other pneumoconioses is the relative absence of alveolar duct fibrosis and interstitial fibrosis in the case of respiratory bronchiolitis. The consequences of luminal desquamation, mucosal inflammation, and bronchiolar fibrosis affect the lung physiologically to increase its airflow resistance and decrease its diffusion capacity. Though some individuals with respiratory bronchiolitis have been treated with steroids, cessation of smoking often leads to a favorable outcome.

The destruction of pulmonary parenchyma, in particular the alveolar duct and lobular units, with the resultant simplification of lung tissue and reduced alveolar surface area, is termed emphysema.[14] The air spaces beyond the terminal bronchioles become enlarged due to the destruction of alveolar walls. When this damage is preferentially localized on the respiratory bronchiole, the consequence is centrilobular emphysema; destruction of the alveolar duct-acinar unit is termed panacinar emphysema. Centrilobular emphysema is characteristically seen in chronic smokers and tends to be distributed to the upper lung fields. An inflammatory bronchiolitis with anthracotic pigment, bronchiolar metaplasia and fibrosis, and scattered pigmented macrophages, features of respiratory bronchiolitis, are often seen with smoker's emphysema.

Though smoking-induced emphysema or chronic bronchitis may lead to severe pulmonary compromise, lung cancers are probably the most feared results of smoking-related lung diseases. Most authorities have documented, based on epidemiological and experimental models in animals, that smoking is causative of lung cancer.[21-23] The American Cancer Society estimates that in 1991, 21% of cancer deaths in women and 45% of cancer deaths in men were attributable to smoking.[24] The tumors found in chronic smokers are bronchogenic; they begin as bronchial epithelial dysplasia or small mucosal tumors that may erode into the bronchus or infiltrate into the adjacent lung. The four major types of cancers are all of epithelial origin and are designated as adenocarcinoma, squamous cell carcinoma, small cell carcinoma, and large cell undifferentiated carcinoma. Respectively, these tumors show histopathologically malignant features involving glands, squamous epithelium, neuroendocrine cells, and undifferentiated cells. Oncologists separate the small cell carcinomas from the other types due to their greater responsiveness to chemotherapy. Yet, whatever the tumor type, the overall 5-year survival of lung cancer combining all stages is poor.

## BRONCHIOLITIS OBLITERANS

Bronchiolitis obliterans is a clinicopathologic syndrome representing a type of small airway disease.[5,25,26] It results from a variety of specific inhalants, drug reactions, or infections or may be associated with systemic disease states. Inhalant toxins or fumes causing bronchiolitis obliterans include noxious gases, such as chlorine gas, hydrochloric acid, nitrogen dioxide, sulfur dioxide, and the smoke mixture associated with fires. Infectious etiologies include viruses, chlamydia, mycoplasma, and bacterial species. Bronchiolitis obliterans may also be present with collagen-vascular syndromes, such as rheumatoid arthritis. Presenting complaints are dyspnea and cough that may be related to a "viral-like" illness or may have developed somewhat suddenly as an acute respiratory insufficiency. The duration of pulmonary symptoms together with constitutional complaints of fever and weight loss may be present for a few months, in contrast with idiopathic pulmonary fibrosis, whose symptoms have a more insidious origin and may evolve over a year. The chest X-ray may also be confused with idiopathic pulmonary fibrosis, since bibasilar reticulonodular infiltrates may be disclosed.

Histopathologically, features of bronchiolitis obliterans are characteristic[5,25] (Figure 2). There is damage to the peripheral bronchioles with mucosal necrosis; an inflammatory obstructing process is present, which is centered on the bronchiole, with extension along the alveolar duct and acinus (Figure 3). The early phase of the disease has an inflammatory exudate filling the bronchiole lumen; bronchoalveolar lavage cellular contents are primarily lymphocytes and granulocytes. A greater component of lymphocytes tends to predominate in the lavage fluid of bronchiolitis obliterans, whereas neutrophils tend to predominate in idiopathic pulmonary fibrosis.

As the disease progresses, the active inflammatory exudate and the erosion of the bronchiolar epithelium begin to organize into an intraluminal inflamed granulation tissue polyp. The bronchiolar injury and the organizing lumenal mass produce an obstructive ventilatory defect on pulmonary function tests. The combination of loose organizing fibrous granulation tissue and the inflammatory cell infiltrate may ramify from the respiratory bronchiole to include filling of alveolar ducts and acinar airspaces. A final scarring phase may ultimately occur that produces a fibrous contracted stenotic lumen and indicates the end-stage of obliterative bronchiolitis. If the process

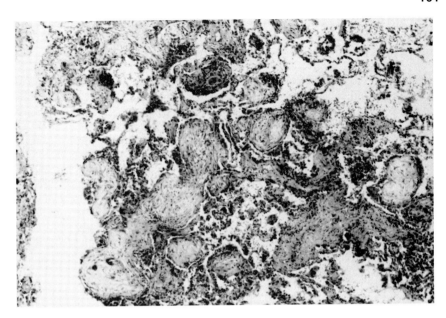

**Figure 2** Bronchiolitis obliterans organizing pneumonia showing characteristic intra-alveolar inflammatory fibrous plugs, desquamated pneumoncytes and alveolar macrophages, and thickened inflamed interstitium. (H&E, Magnification × 40.)

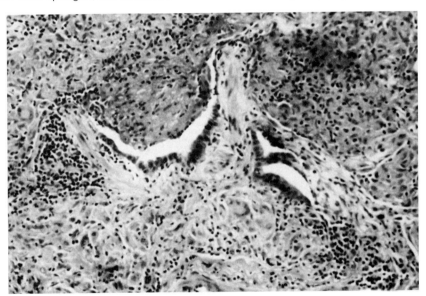

**Figure 3** Bronchiolitis obliterans with inflammatory fibrous polypoid fragment extending into partially obliterated bronchiole. (H&E, Magnification × 200.)

is self-limited or is treated with steroids to reduce the inflammatory fibroblastic process, the bronchiolar pathology may abate; repair and resolution rather than fibrous scarring may ensue. As the inflammatory fibroblastic polypoid plugs extend into the alveolar airspaces, the alveolar septa show a reactive interstitial pneumonitis.

The involvement of pulmonary parenchyma in the inflammatory process produces a restrictive ventilatory defect. As the injury of bronchiolitis obliterans proceeds to and involves more pulmonary parenchymal tissue, the pathologist may be challenged to distinguish this entity from the more rapidly progressive usual interstitial pneumonitis (UIP).[27] This differential diagnosis has both treatment and prognostic significance since bronchiolitis obliterans tends to be more steroid-responsive and less likely to have significant interstitial fibrosis with end-stage honeycomb lung as usual interstitial pneumonitis. Katzenstein has pointed out that bronchiolitis obliterans tends to be more patchy and bronchocentric, with the predominant lesion being more of an airspace pathology in contrast to the more diffusely random interstitial abnormality that is identified in UIP. The distal alveolar spaces show young fibroblasts in an edematous stroma with alveolar macrophages in bronchiolitis obliterans, whereas usual interstitial pneumonitis has a mature interstitial collagen and intra-alveolar macrophages are infrequent.

## DIFFUSE ALVEOLAR DAMAGE

If the inhalant injury is localized on the more distal alveolar airspaces and damage occurs to the alveolar epithelial-endothelial cellular units, then the pathologic result is diffuse alveolar damage (DAD).[5,28,29] A variety of agents and disease states may produce the pathologic insult that is etiologic for diffuse alveolar damage. However, because the histologic and cellular response is limited and defined, the resultant tissue pattern is essentially nonspecific, even though the insult may be as dissimilar as sepsis- or toxic inhalant-induced. Some of the inhalant or drug injuries that have been reported to be causative of bronchiolitis obliterans have also been reported to be etiologic for diffuse alveolar damage. Thus, nitrogen oxides, sulfur dioxide, and noxious fumes from fires may cause a diffuse pulmonary injury whose predominant lesion begins in alveolar airspaces. Other reported causes of diffuse alveolar damage are infections, drug reactions, shock states, including sepsis, pancreatitis, and severe trauma. Radiation or chemotherapeutic toxic injury are also well-known causes of DAD and may histopathologically be suggested by particular cytologic features. Because the categorical causes of DAD are so varied and their individual examples so numerous, usual histopathologic features of this disease process are not diagnostic for a specific etiology. In some cases, several factors may be additive or synergistic in production of DAD. Damage to the alveolar endothelial-epithelial cellular unit initiates an injury and inflammatory response characterized by increased membrane permeability, proteinaceous and cellular exudation, and an inflammatory cellular response. Clinically, these pathologic sequences give rise to noncardiogenic pulmonary edema, refractory hypoxemia, and decreased compliance of the lung. The individual complains of dyspnea, and the chest X-ray shows diffuse pulmonary infiltrates. Histologically, the microscopic findings of DAD are characteristic, however, as discussed above, not pathognomonic of any specific cause (Figure 4). The disease process evolves through a series of distinct phases. Damage to the alveolar endothelial-epithelial cellular unit leads to increasing permeability, which produces an intra-alveolar edema; hyaline membranes along alveolar ducts result from the marked proteinaceous exudate. Lost type I alveolar pneumocytes are replaced by prominent hyperplastic type II alveolar cells. The pulmonary interstitium also manifests an edematous response and the presence of a mixed inflammatory cell infiltrate. This

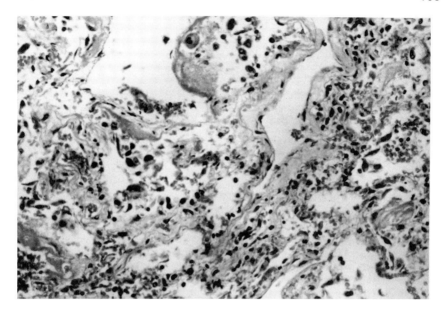

**Figure 4**  Diffuse alveolar damage, having the clinical presentation of adult (acute) respiratory distress syndrome, with hyaline membranes, sloughed alveolar lining pneumoncytes, and a widened edematous interstitium with a lymphocytic infiltrate. (H&E, Magnification × 400.)

early inflammatory and reactive cellular response with alveolar hyaline membranes is termed the exudative stage of DAD.[5] The disease may resolve at this point or may progress to the proliferative stage of DAD. In this latter stage, there is an organization of the inflammatory response, with proliferation of both inflammatory cells and fibroblasts. The organizing fibroblastic tissue is present within the interstitium and focally within alveolar spaces. Alveolar airways are now lined by prominent alveolar type II cells, often with considerable cytologic atypia. The atypia can reach extremes that cytologically are reminiscent of malignancy in certain cases of chemotherapeutic drug-induced DAD. During the organizational phase, histologic features of edema and hyaline membranes, so typical of the early phase, are minimally present. Instead, the interstitial inflammatory and organizing pattern is more suggestive of usual interstitial pneumonitis with fibrosis.

## IMMUNOLOGIC LUNG DISEASE

Mechanisms of immunologically-mediated lung disorders have been divided into four major types.[5,30] The type I disorder, or anaphylactic type, is mediated by the immunoglobulin antibody, IgE, and its characteristic inflammatory cell, the eosinophil. Asthmatic bronchial disease is a prototypical example of the type I disorder, brought about by the release of vasoactive amines and other cytokines and their action on cellular secretion, vascular permeability, and smooth muscle contraction. As a result of IgE reaction with specific antigens (allergens), mast cells and basophils degranulate and release a series of primary vasoactive and chemotactic productions and activate, through the arachidonic acid pathway, the release of such secondary mediators as leukotrienes and prostaglandins.

Goodpasture's syndrome is an example of the type II immune-mediated disorder whereby immunoglobulins IgG and IgM are formed that are directed at particular cellular antigens. These antibodies bind to antigens on target cells, initiate a variety of inflammatory cascades, and initiate cell death by complement-mediated lysis or antibody-dependent cellular cytotoxicity.

Type III hypersensitivity causes immune-mediated injury by the production of antigen-antibody complexes that deposit in tissue and result in tissue damage. Whether the antigens deposit on tissue and react with circulating antibodies or pre-formed immune (antigen-antibody) complexes deposit on tissue, the complex causes the activation of immune mediators, such as complement and stimulation of the in-flammatory response. Vasculitis is a prototypical example of type III disorders; the damaged vascular wall shows positive immunofluorescence for immunoglobulins and complement proteins.

Cell-mediated hypersensitivity, or type IV disorders, result from activation of T-helper lymphocytes that have been specifically sensitized to a particular antigen. Immune activation of the lymphocytes causes the secretion of lymphokines, such as macrophage chemotactic factors and interleukins. These immune mediators amplify the inflammatory response by the activation and proliferation of other lymphocytes and macrophages. The characteristic example of this type of immunological response is the granulomas found in tuberculosis mediated by sensitized lymphocytes.

## ASTHMA AND REACTIVE AIRWAY DISEASE

Asthma is a disease of the larger airways of the tracheo-bronchial tree, characterized by an inflammatory process that results in hyperreactive and reversible airway con-striction and obstruction. The inflammatory process may be initiated by exposure to an allergen, respiratory tract infection, occupational aerosol, or environmental con-dition or stimulus, such as exercise, cigarette smoke, or cold air. Asthma has been divided into ''extrinsic'' type, due to a specific allergen, or ''intrinsic'' type, in which the sensitizing agent is unknown. The pathogenesis of asthma may be immune-mediated, with the key immunoglobulin being IgE and the key inflammatory cells being mast cells and eosinophils.[31–34] Other types of asthma may be inflammatory but not necessarily immune-mediated. Nevertheless, all forms of asthma or hyper-reactive (hyperresponsive) airway disease are associated pathologically with parox-ysmal airway smooth muscle spasm and constriction, airway inflammation and edema, and mucus hypersecretion.

In the classical type of allergic or atopic asthma, the individual, often with a family history of asthma or atopy, is sensitized to a particular type of allergen, such as a dust, pollen, or animal protein. Following a latent period and the generation of IgE antibodies, reexposure to the sensitizing allergen permits allergen to bind to IgE-coated mast cells and induce mast cell degranulation, with release of a variety of active mediators, such as histamine, prostaglandins, and leukotrienes. The individual who is challenged with the particular allergen proceeds through a type I hypersen-sitivity response, characterized functionally by an immediate fall in the $FEV_1$. During the periods between asthmatic attacks, these atopic individuals continue to have hy-perreactive airways. Bronchoconstriction can be elicited by provocative tests with histamine or methacoline.

The released mediators cause increased vascular permeability, leading to mucosal edema.[33,34] The degranulation of mast cells releases a group of preformed vasoactive substances, including histamine, proteases, heparin, and chemotactic factors. Through the action of phospholipase and the biosynthetic arachidonic pathway, both prostaglandins and leukotrienes are generated. The histamine release and production of prostaglandins and leukotrienes cause both immediate and prolonged bronchocon-striction. The leukotrienes induce eosinophil chemotaxis and also function to increase vascular permeability and stimulate the production of mucus. Both the mucosal swelling and smooth muscle contraction cause constriction and narrowing of the airway lumen. The elaboration of mucus secretion within the airway lumen further obstructs airway flow. The mucus hypersecretion may take the form of airway plugs, with a cellular composition of desquamated epithelial cells, eosinophils, and other leukocytes (Figure 5). Charcot-Leydin crystals representing precipitated material from eosinophil granules are also identified within the viscous mucus. Histopathologically, the bronchioles will show lumenal mucus and a proteinaceous exudative plug, goblet cell metaplasia of the lumenal epithelium, thickened basement membrane, mucosal and peribronchiolar vascular congestion, hypertrophy of the smooth muscular wall, and inflammatory cells throughout the mucosa, bronchiole wall, and peribronchiolar tissue (Figure 6). As discussed above, the key cells identified will be the eosinophils and mast cells. Both of these cells have IgE receptors for the Fc component of the immunoglobulin. Activation of these cells and release of their products is directly associated with acute bronchoconstriction and the clinical features of asthma.

Both histopathologic and bronchoalveolar lavage evaluation indicate that other inflammatory cells may play a role in asthma.[34] Macrophages and lymphocytes also have IgE receptors, though of lower affinity than the receptors on mast cells or basophils. Asthmatics have higher IgE-coated macrophages than normal individuals, and cells of the mononuclear series are capable of releasing mediators such as thromboxane, platelet-activating factor, leukotrienes, interleukins, and tumor necrosis factor. These factors, along with the lysosomal contents, may enhance bronchiole epithelial injury and may be involved in chronic asthmatic conditions and prolonged bronchial hyperreactivity. B lymphocytes are the synthetic cell for IgE, T lymphocytes regulate the synthetic activity of B lymphocytes and also secrete interleukins, which modulate the activity of eosinophils. Epithelial cells are often the target of injury in asthma; desquamation and shedding of epithelial cells can be appreciated both on histological sections and in bronchoalveolar lavage. Loss of epithelial cells with exposure of underlying basement membrane and stroma increases tissue permeability and augments allergen and mediator stimulation of bronchial hyperactivity. Epithelial damage may also decrease the elaboration of a smooth muscle relaxation factor and increase the release of inflammatory and chemotactic factors.

In occupational asthma, the pathogenesis may take the form of type I hypersensitivity with a specific work-place-related allergen.[35-38] Asthmatic bronchoconstriction with possible associated rhinitis and conjunctivitis may proceed acutely with reexposure, or a delayed immune-mediated response may occur. Certain dyes and metal salts cause an immediate asthma and are associated with high titer IgE. Other types of occupational asthma may be nonimmune-mediated and consequently may have negative RAST (radioallergosorbent test), indicating the inability to discover an antibody of the IgE class. Histopathologic examination of the airway or bronchial lavage may not show mucosal eosinophilia.

Chemical substances, such as isocyanates, may act as irritants to the tracheobronchial tree and elicit a hyperreactive airway disease in susceptible individuals. Asthmatic conditions following significant exposure to irritating vapors, fumes, or smoke, have been termed reactive airways dysfunction syndrome (RADS) and occur following a short-term irritant exposure.[39,40] This type of occupational asthma may be nonallergic (i.e., not IgE-mediated), lacks a latency or sensitization period, occurs in nonatopic individuals, and leads to nonspecific airway hyperresponsiveness. The symptoms of cough, wheezing, and dyspnea, characteristic of asthma, quickly follow the exposure, and may persist for several months. These exposed individuals test positively with methacoline challenge to their airways. Similar to extrinsic

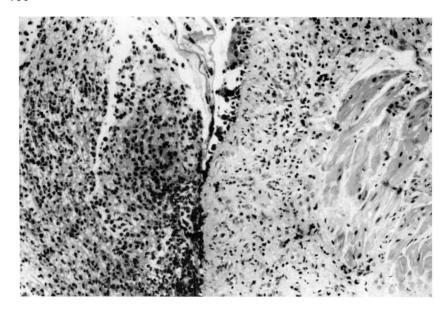

**Figure 5** Allergic asthmatic bronchitis with an eosinophilic lumenal exudate overlying an inflamed edematous mucosa. Note that a portion of the mucosa has sloughed desquamated epithelium. (H&E, Magnification × 100.)

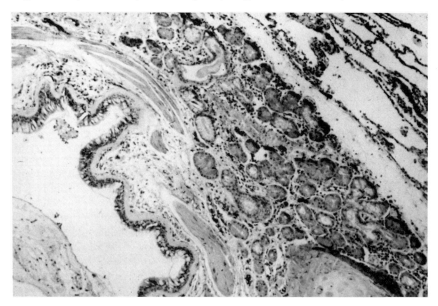

**Figure 6** Asthmatic bronchitis showing the features of mucosal goblet cell metaplasia of the surface epithelium, inflammatory exudate, mucosal inflammation, and mucous gland hyperplasia. (H&E, Magnification × 100.)

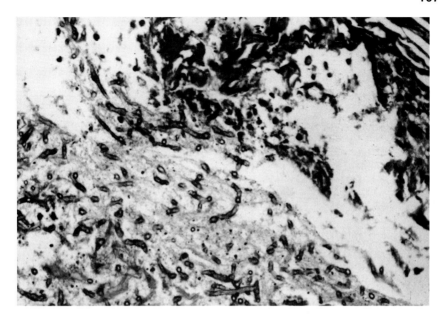

**Figure 7**  Allergic bronchopulmonary aspergillosis with bronchiolar exudate containing proteinaceous debris, desquamated bronchiolar epithelial cells, and prominent aspergillous septated hyphal forms. (H&E, Magnification × 200.)

asthma, there is inflammation of the airway, with mucosal epithelial injury, mucosal edema and vascular reactivity, and a mucosal inflammatory infiltrate. Different from extrinsic asthma is the absence of eosinophils and mast cells. Though these individuals often have no preexisting respiratory complaints, as a result of their exposure, there develops a state of airway hyperreactivity with persistence of respiratory symptoms.

Some individuals who are chronic asthmatics may develop a hypersensitivity to the fungus aspergillus and develop the chronic pulmonary syndrome of allergic bronchopulmonary aspergillosis (Figure 7). The diagnostic criteria of this chronic disease include a preexisting asthmatic condition, blood eosinophilia with increased serum IgE levels, and skin reactivity to aspergillus antigen.[41,42] The course of the disease may be characterized by pulmonary infiltrates, eosinophilic pneumonia, bronchiectasis, and mucoid impaction of the bronchi. Some authors have proposed a staging system, in which patients have acute episodes with remission and exacerbation.[41] In the acute period, the individual is symptomatic, with fever, bronchitis, and asthma; the remission period is associated with decreased symptoms, a decrement in eosinophilia and IgE levels, and resolution of the pulmonary infiltrates. With exacerbation of allergic bronchopulmonary aspergillosis, the individual enters a chronic asthmatic state that is corticosteroid dependent. Repeated airway and parenchymal injury leads to pulmonary fibrosis, with dyspnea as a prominent symptom. The pathogenesis of this disease combines type I hypersensitivity characteristic of extrinsic asthma and type III immune complex hypersensitivity.

## HYPERSENSITIVITY PNEUMONITIS

Hypersensitivity pneumonitis or extrinsic allergic alveolitis represents an immunologic lung disease resulting from inhalation of biologic or organic material.[4,5,30,43–45]

This immune-mediated pneumonitis differs from other types of pneumoconiosis, where an inorganic fiber or dust is inhaled. The sensitization brought about by vegetable or animal protein material induces an immune complex and granulomatous lung disease.

The immunopathogenesis of hypersensitivity pneumonitis appears to involve both type III and type IV immunologic mechanisms.[30] The presence of inhaled antigen in association with preformed antibody causes formation and deposition of immune complex in the pulmonary parenchyma and along vascular channels. The findings of vasculitis, though uncommon, and complement-mediated cellular damage support type III immune reactions. Serologically, the presence of skin test positivity following a few hours after antigen exposure, increased total serum immunoglobulins, and precipitating antibodies directed against the inciting antigen also suggest type III mechanisms. In addition, the demonstration of granulomas within the interstitium is indicative of type IV cell-mediated immune processes.

The types of biologic materials inhaled in hypersensitivity pneumonitis consist of thermophilic actinomycetes, molds, and animal protein antigens. The diseases are most commonly occupationally or environmentally derived and carry colorful and imaginative names. Farmer's lung, bagassosis, and mushroom worker's disease result from an idiosyncratic immune response to molds associated with hay, sugar cane, and compost, respectively. A variety of animal proteins are responsible for the pneumonitides, such as bird fancier's disease, pigeon handler's disease, and pituitary snuff-taker's lung.

The clinical findings of dyspnea, pulmonary infiltrates, restrictive disease, and nonspecific constitutional complaints can be traced to the morphologic pattern of immune-mediated inflammatory lung disease. The inhalant nature of the disease tends to center the inflammatory process around respiratory bronchioles. Surrounding these airways and extending into the interstitium is a lymphocytic and mononuclear inflammatory infiltrate, with a scattered contribution of plasma cells (Figure 8). The overall features of the process are those of a chronic interstitial pneumonitis, with widening of the interstitial areas with inflammatory cells, noncaseating granuloma, and possibly interstitial fibrosis[4,5,44-46] (Figure 9). The interstitial granuloma consists of Langhans-type multinucleated giant cells, epithelioid histiocytes, and a peripheral cuff of lymphocytes. Intraalveolar edema or exudative reaction consisting of foamy macrophages may be present; some cases may show a bronchiolitis obliterans-organizing pneumonia pattern, with aggregates of inflammatory and fibrous tissue plugs extending into bronchioles and ramifying into alveolar channels.

## SILICOSIS

Individuals exposed to the crystalline forms of silica, of which quartz is the best and most typical example, may develop a variety of pulmonary diseases denoted as silicosis.[47-49] The exposure may occur in the mining, manufacturing, or utilization of silica products. Temporally, it may occur after months or a few years, as in sandblasting, with brief exposures to high-content silica particles, or more commonly following many years of prolonged exposure, leading to a form of chronic nodular silicosis.

Acute silicosis or silicoproteinosis is a less common form of lung injury to silica exposure; it causes dyspnea and may progress to deteriorating pulmonary insufficiency in affected individuals. The histopathology of acute silicosis resembles the entity of alveolar proteinosis, in which an alveolar expansion results from an intraalveolar accumulation of granular eosinophilic lipoprotein material.[50-54] Though the chest X-ray shows interstitial-alveolar infiltrates, the underlying basic alveolar and

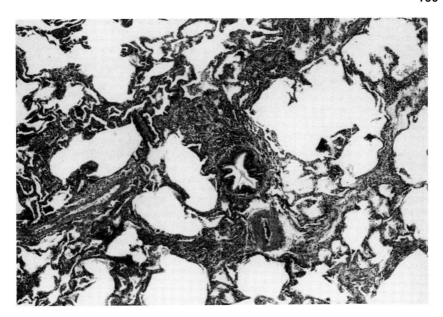

**Figure 8** Hypersensitivity pneumonitis or extrinsic allergic alveolitis with peri-bronchiolar inflammation and diffuse interstitial lymphoplasmatic infiltrate. (Courtesy of Dr. Samuel Hammar, H&E, Magnification × 40.)

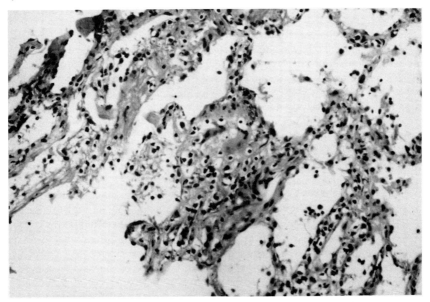

**Figure 9** Hypersensitivity pneumonitis demonstrating interstitial granuloma composed of a multinucleated giant cell and a histiocytic and lymphocytic infiltrate. (H&E, Magnification × 400.)

septal architecture is preserved. Intermixed with this intraalveolar granular material, which stains positively with PAS (periodic acid Schiff base reagent), are desquamated alveolar macrophages and type II pneumocytes. The granular exudate appears, ultrastructurally, to contain lamellar and osmiophilic bodies containing phospholipoprotein material resembling lung surfactant. The pathogenesis of acute silicoproteinosis is not completely understood; it may result from overproduction of alveolar phospholipoproteins, with imbalance between production and removal, by defective degradation and alveolar clearance by alveolar macrophages, or by excessive proliferation and breakdown of alveolar type II pneumocytes.

Exposure to silica over many years produces a spectrum of pulmonary manifestations of silicosis.[1-4,47] These reactions to crystalline silica range from asymptomatic silicotic nodules to severely debilitating progressive massive fibrosis that may be complicated with superimposed tuberculosis. The silicotic nodules are peribronchial, with interstitial accumulations of hyalinized collagen arranged in a lamellar concentric pattern and surrounded by a thick capsule (Figure 10). Grossly or at low microscopic power, the silicotic nodule appears as a whorled ball of collagen that approaches 0.5 cm in diameter. If examined chemically, these nodules contain crystalline silica and sheet silicates, which are refractile and polarizable. Though these silicotic nodules develop after decades of exposure and may be quite scattered throughout the lung, the remaining pulmonary parenchyma is essentially intact, and consequently there may be no or mild symptoms in early stages of the disease. The nodules (Figure 11) may continue to grow, obstruct bronchioles and vessels, and coalesce to form conglomerate nodules measuring several centimeters. With further fusion of the nodules, the collagenous scarring process assumes the size of bosselated tumor masses, with contraction and compromise of parenchymal tissue. This condition, termed progressive massive fibrosis, appears radiologically and pathologically as multiple well-demarcated densities predominantly on the upper lung. The adjacent lung is retracted and shows emphysematous changes.[55] If the exposure is to silica plus another mineral dust, such as the nonasbestos silicates of mica or kaolin, the resulting pulmonary fibrosis is termed mixed dust fibrosis. The nodules in mixed dust fibrosis tend to be more stellate than round and have the nonsilica component toward the periphery of the fibrotic nodule. A complication of chronic silicosis is the presence of mycobacterial tuberculosis. Epidemiologically, individuals with chronic silicosis are more susceptible to acquiring tuberculosis, and both diseases have a synergistic effect on morbidity and mortality. Enlarging nodules or cavitary necrosis within nodules is a feature of tuberculous infection; granulomatous inflammation may also be present.

The pathogenesis of silicosis is related to the toxic effect that the crystal has on alveolar macrophages.[47,48] Silica causes lysosomal membrane damage and death of the macrophage, with release of fibrogenic factors such as interleukins, platelet-derived growth factor, and fibronectin. The factors stimulate increased collagen production by fibroblasts. Lysis of macrophages also releases phagocytosed silica crystals, to be acquired by native macrophages. The macrophage damage by silica particles also compromises macrophage function in limiting mycobacterial infection. The injured macrophage may additionally be important in acute silicoproteinosis in preventing its ability to degrade secreted surfactant material and thereby lead to its accumulation.

## SILICATE PNEUMOCONIOSIS

The exposure to nonasbestos silicates also results in peribronchial scars, fibrotic nodules, and massive pulmonary fibrosis.[1-4,56-58] The mineral classification of silicates

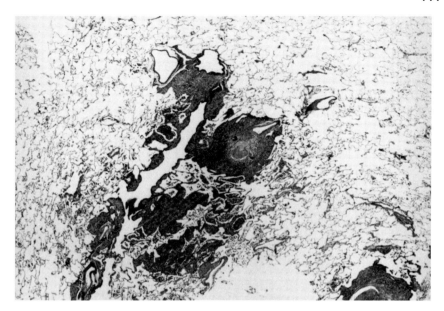

**Figure 10** Silicosis with features of mixed dust pneumoconiosis demonstrating peribronchiolar fibroinflammatory macule and silicotic nodule. (Courtesy of Dr. Samuel Hammar, H&E, Magnification × 40.)

**Figure 11** Silicosis with central hyalinized collagen and periphery of fibroinflammatory process with pigment that replaces normal lung parenchyma. (Courtesy of Dr. Samuel Hammar, H&E, Magnification × 40.)

is based on their chemical composition of silicon dioxide plus other metal cations such as calcium, magnesium, iron, or aluminum. The mineral silicates are represented by a number of different groups according to their chemical and physical properties. The best known entities are talc, mica, slate, and kaolin. Usually, exposure to silicates during industrial manufacturing combines exposure to other fibrogenic minerals such as silica (quartz) or asbestos. The resulting disease, therefore, shows many of the pathologic findings present as a result of exposure to the contaminating mineral ore. The identification of the particular species is based on X-ray spectroscopy or X-ray diffraction of the involved tissue and must be correlated with the environmental or occupational dust particle.

The earliest and most limited lesion associated with silicates is the peribronchial macule, composed of macrophages with ingested particulates and minimal fibrous reaction. The pigmentation and refractibility of the dust macule will depend on the nature of the mineral silicate. Occasionally, foreign-body-type granulomas may develop and precede the evolution of the more extensive forms of massive fibrosis and diffuse interstitial fibrosis. The formation of nodules and the coalescence of multiple nodules to form massive fibrosis are analogous to those identified in silicosis. Differentiating features among the variety of silicates include the presence of multiple refractile and polarizable particulates, pigmentation, and ferruginous bodies characteristic of the particular silicate. In a single lung, one may identify a spectrum of silicate fibrosing lesions, ranging from the microscopic peribronchial macules to the more extensive fusion of nodules forming massive fibrosis and diffuse interstitial fibrosis.

In the case of talc pneumoconiosis, the distribution and the nature of pulmonary fibrosis depends on the route of entry. Inhalation talc will take the form of silicate pneumoconiosis already discussed. In contrast, the lung from recreational intravenous drug users, especially those involving opiates processed with talc powder, will show intravascular and interstitial talc.[59,60] The intravenous route of exposure will lead to thrombosis and vascular changes indicative of pulmonary hypertension. Foreign body giant cell granulomas containing talc particles will be discovered within vascular walls and interstitium (Figure 12). The presence of intravascular and interstitial talc granuloma, pulmonary hypertensive vascular changes, and diffuse interstitial fibrosis distinguishes the intravascular exposure to talc from the inhalational talc exposure.

## COAL WORKER'S PNEUMOCONIOSIS

The exposure to coal dust consists of inhalation of amorphous carbon, silica, and silicates.[1–4,61–63] The types of coal differ in the relative proportion and chemical composition of these ingredients. Though pure carbon particles are not fibrogenic, the inclusion of silica and silicates creates fibrotic responses analogous to those seen in silicosis and in silicate pneumoconiosis.[64–67] The presence of carbon induces a macrophage peribronchial response with phagocytosis of dust particles. The combined proliferation of macrophages and reticulin and collagen production extend beyond the peribronchial distribution and involve perivascular, pulmonary interstitial, and subpleural zones. Enlarging dust macules and the accumulation of collagenous fibrosis lead to the formation of dust nodules. The presence of coal-dust-induced macules and nodules is regarded as simple coal worker's pneumoconiosis. Grossly, these black-pigmented areas appear as scattered several mm (macules) to 1 to 2 cm (nodules) patches, with associated adjacent emphysematous change. Though these lungs have impressive pathological features, there is little pulmonary compromise associated with these lesions.

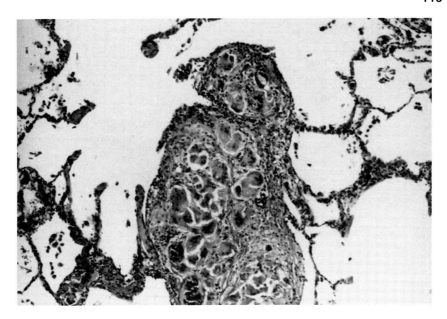

**Figure 12** Intravenous drug abuse talcosis showing interstitial intravascular multinucleated giant cells containing birefringent talc particles. (H&E, Magnification × 200.)

As in the case of chronic silicosis and silicate pneumoconiosis, the fusion of larger nodules produces massive fibrosis, and diffuse interstitial fibrosis appearing as coalescent black collagenous regions of replaced lung tissue (Figure 13). The scarring and retraction of lung tissue lead to areas of adjacent emphysema. The clinical consequence of these fibrous and destructive lesions, denoted complicated coal worker's pneumoconiosis, is more pronounced with pulmonary function tests indicative of obstructive and restrictive defects. The collagenous nodules of progressive massive fibrosis may show central cavitation due to obliteration of vascular structure. Secondary ischemic necrosis or, as noted with silicosis, superimposed tuberculosis may complicate the fibrosing pulmonary disease. Mycobacterial tuberculosis infections may also show classical tuberculoid granuloma; however, this feature is not a constant finding.

## ASBESTOS-RELATED LUNG DISEASE

Asbestos is a fibrous silicate mineral that achieved a remarkable number of commercial applications due to its favorable properties of heat, flame, and chemical resistance and its adaptability to be used as a textile material.[4,68–70] Commercial asbestos silicates may be divided into two major physicochemical groups: serpentines and amphiboles.[4,68–70] The serpentines, of which chrysotile is the major member, are curved and coiled fibers that tend to deposit more proximally along airways, fragment into smaller particles, and more easily clear from lung tissue. The amphiboles, of which amosite and crocidolite are the major minerals, are straight fibers with high aspect ratios (length to width or cross-sectional diameter) that tend to distribute more peripherally in lung tissue and are more likely, on a fiber burden basis, to be causative of asbestos-associated diseases than the serpentines. In general, fibers that are

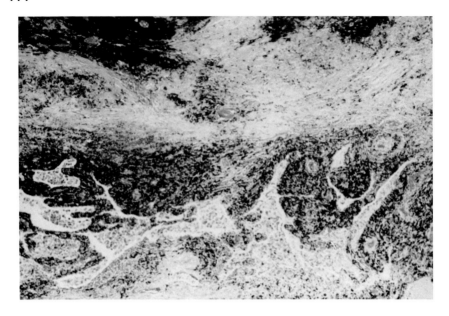

**Figure 13** Coal worker's pneumoconiosis with marked scarring fibrosis and extensive pigment. (H&E, Magnification × 200.)

retained longer are associated with a greater probability of more significant diseases. These diseases include asbestosis, lung cancer, and mesothelioma.[1–4,69–72] Prolonged asbestos exposure may also lead to pleural reactions consisting of benign asbestos-related pleural effusions, diffuse pleural thickening or fibrosis, and parietal or diaphragmatic plaques.

When asbestos fibers are inhaled, the shorter and thinner fibers will tend to deposit more peripherally within the lung tissue.[4,73–75] Very long fibers will deposit proximally and lodge in airways with elimination by mucociliary clearance. Shorter fibers will be captured at bronchiolar bifurcations, alveolar ducts, openings of alveolar sacs, or within pulmonary parenchyma. Some fibers may be captured by alveolar fluid with proximal transport and elimination by alveolar clearance mechanisms. Fibers may travel toward the visceral pleura or may be transported to the pleura through lymphatics and other mechanisms that are still unclear. Lymphatic channels may also transport the asbestos fibers to the regional draining bronchial and hilar lymph nodes. Since smaller fibers more readily clear from the lung and longer fibers with greater aspect ratios more deeply penetrate pulmonary tissue, one theory proposes that fibers >5 μm in length and <1 μm in cross-sectional diameter are those that are retained the longest and, consequently, are associated with asbestos-related pulmonary fibrosis or malignancy.[76]

When asbestos fibers are localized at bronchiolar and alveolar duct bifurcations or are deposited within pulmonary interstitium, macrophages are recruited to phagocytose the particles.[73,74] These macrophages collect at sites of fiber deposition, and reactive macrophages often desquamate within alveolar spaces. Macrophage-derived multinucleated giant cells may appear within the interstitial and alveolar spaces and may be present in bronchoalveolar lavage fluid. In addition to macrophage recruitment and activation, myofibroblasts, fibroblasts, and lymphocytes accumulate and begin the process of cytokine release and collagen production. Fibrosis occurs in the region of the respiratory bronchioles, alveolar ducts, and septa. Alveolar epithelial cells are damaged by asbestos fibers and by the release of toxic substances elaborated by macrophages, and their loss stimulates the proliferation of type II alveolar lining cells. The early lesion of asbestosis consists of bronchiolar and alveolar duct fibrosis in a setting of macrophage activation, an interstitial and alveolar inflammatory infiltrate, reactive epithelial changes, and the presence of asbestos fibers and asbestos bodies.

As the disease progresses, the fibrosing process extends centrifugally from the initial site of injury and involves pulmonary parenchyma in a patchy distribution (Figure 14). An early accompaniment of asbestos-related lung disease is visceral pleural (interlobar fissural) thickening.[77] Fissural thickening is a sensitive marker of asbestos exposure and may be radiographically present in pathologically established pulmonary asbestosis, despite a normal chest radiograph. The established disease state of asbestosis is evidenced by diffuse and bilateral pulmonary interstitial fibrosis, predominantly lower lobes, with scattered asbestos bodies. The interstitial fibrosis and scarring causes contraction of the lung, thickened and obliterated alveolar structures, and the resultant reticulonodular pattern on chest radiograph and restrictive disease on pulmonary function tests. Pulmonary asbestosis occurs following a long latency of 1 to 2 decades, and lung tissue in asbestotics has a significant asbestos burden relative to that in the general population.[70] A grading scheme marking the progression of the fibrogenic process was proposed by the Pneumoconiosis Committee of the College of American Pathologists and consists of a ''severity'' and ''extent'' grade.[69] Severity refers to the level of fibrosis involving the distal airways and parenchyma, with features of contiguous fibrosis and restructured lung being the most severe grade; the extent of the process refers to the amount of lung tissue involved. Though it is generally accepted that interstitial fibrosis plus asbestos bodies are diagnostic for pulmonary asbestosis, there remains some controversy in recognizing the earliest or mildest form of the disease process.[77-86] Peribronchiolar and alveolar duct fibrosis with pigment deposition may represent a low severity grade of asbestosis or may indicate a nonspecific mineral dust airway disease. As a result of the bronchiolar and alveolar duct fibrosis, small airways disease may demonstrate obstructive findings on pulmonary function tests.

Epidemiologically, workers in asbestos-related industries tend to be cigarette smokers, creating an interactive and confounding effect of these two exposures on lung function and pathology. Though there are considerable controversies, it appears clear that asbestos does not cause emphysema and that cigarette smoke does not cause hyaline pleural plaques. A majority of reviews and studies of the effect of cigarette smoking and asbestos-exposure support a positive interaction between the two inhalants.[87-93] Whether the focus is directed at radiographic small irregular opacities or pulmonary fibrosis, the cohort of asbestos-exposed individuals who smoke show a greater percentage of radiographic or pathologic abnormalities than the non-smoking cohort. Though the positive effect may be additive or multiplicative, pathogenetically, smoking may enhance pulmonary asbestosis by increasing the effective dose of asbestos fibers due to impaired clearance mechanisms.[94] Similarly, asbestos may increase the effective dose of cigarette smoke by adsorption of toxic materials and gases on asbestos fibers.

Within the lung of an exposed individual, only a minority of asbestos fibers are coated with an iron-containing proteinaceous layer, producing a type of ferruginous body termed an asbestos body[95,96] (Figure 15). The asbestos body contains a thin translucent asbestos fiber, as the needlelike core of the foreign body surrounded by beaded iron material and frequently resembling a ''dumbbell'' structure (Figure 16). One may distinguish by light microscopy a variety of minerals by the features of their particular ferruginous bodies; however, only those iron-coated bodies formed

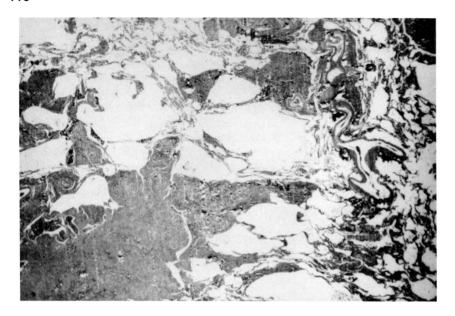

**Figure 14** Combined asbestosis and nonasbestos silicate pneumoconiosis showing coalescent interstitial fibrosis producing a fibrous scar. (H&E, Magnification × 20.)

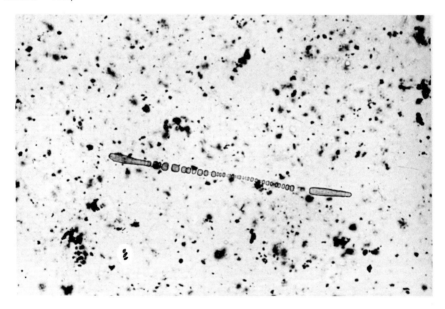

**Figure 15** Asbestos body derived from ashing lung tissue with bleach and filtering product on grid. (Unstained, Magnification × 400.)

with an asbestos core are to be labeled as asbestos bodies. The iron coating of asbestos bodies is initiated by macrophges and is presumably an attempt to limit the cytotoxicity of the asbestos fiber. The extent of exposure or the fiber burden present in an individual's lung may be assessed microscopically by counting the number of asbestos bodies in a given weight of lung; the uncoated fibers are not identifiable in tissue sections at the microscopic level.[70] Counting asbestos bodies using light microscopy is a relatively simple yet insensitive method of assessing asbestos exposure. Staining the tissue with iron stains (Prussian blue stain) enhances the identification of asbestos bodies. Lung tissue may be chemically disintegrated and ashed, and counting the filtered residual asbestos bodies and fibers provides a more sensitive light microscopic technique[70,97] (Figure 15). Ultrastructural methods, including transmission and scanning electron microscopy, have been used as research tools to more accurately quantitate the lung asbestos burden. These methods have demonstrated that, in comparison to the nonexposed control population, individuals with asbestosis have markedly excessive asbestos fiber contents.[98,99] Studies have shown that the lung asbestos fiber burden correlates with the nature of the lung pathology; namely, benign pleural disease or pleural plaques are associated with lower lung fiber content than that identified in lungs with pulmonary asbestosis,[100] and, in general, the severity of interstitial fibrosis increases monotonically with an increase in lung fiber content.[4]

Benign asbestos-related pleural disease consists of benign asbestos effusions, diffuse pleural thickening, and hyaline plaques.[101] Benign asbestos pleural effusions are exudates that may appear within a few years of exposure and thereby represent the first tissue response to asbestos inhalation. The finding of asbestos effusions preceding pleural thickening or plaques suggests that asbestos effusions represent a focus of injury from which other asbestos-related pleural diseases may evolve. Diffuse pleural thickening, representing visceral pleural fibrosis, may also occur in the absence of antecedent pleural effusion. Pleural plaques are parietal or diaphragmatic acellular collagenous tissue deposits that appear as discontinuous islands and may calcify[1-4] (Figure 17). Both of these reactions occur following a latency period of at least a decade and are associated with an asbestos fiber burden that is 10 to 100 times lower than that present in established pulmonary asbestosis. Pleural fibrosis and pleural plaques are not premalignant and signify asbestos exposure and not pulmonary asbestosis. Commonly, pleural plaques produce no symptoms, whereas infrequently, diffuse pleural thickening may encase considerable lung tissue and have a restrictive component. If a significant amount of pulmonary parenchyma is trapped within a pleural fibrous reaction, the radiographic feature of rounded atelectasis is created. Pathologically, there is visceral pleural fibrosis surrounding collapsed lung tissue.

In addition to diffuse interstitial fibrosis, an ominous consequence of prolonged asbestos exposure is the development of lung cancer. Similar to pulmonary asbestosis, asbestos-related lung cancer usually occurs with a long latency of two or more decades and often in the setting of pulmonary asbestosis. It is not understood whether asbestosis clinically precedes asbestos-related lung cancer or is pathogenetic for its initiation. A minority of lung cancer cases with asbestos exposure and identifiable asbestos bodies on tissue examination fail to show established asbestosis.[70,71] Historically, most asbestos workers tend to be cigarette smokers, and though smoking is etiologically causative of lung cancer, the combined inhalation of cigarette smoke and asbestos exposure seems to have a multiplicative risk for the development of cancer.[102,103] Thus, smokers, relative to nonasbestos-exposed nonsmokers, have an 11-fold relative risk of dying from lung cancer; asbestos-exposed smokers have a multifold increase in relative risk. Of the variety of bronchogenic cancers, small cell and non-small cell carcinomas, all occur with roughly the same frequency seen

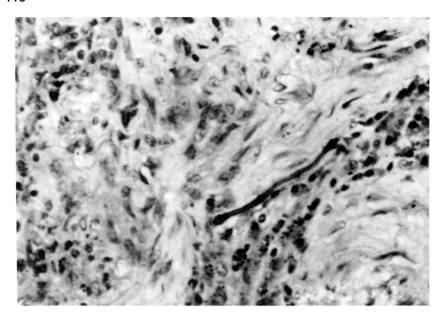

**Figure 16** Asbestos body identified within diffuse interstitial fibrosis of the lung (asbestosis). Note the beaded poles and the linear translucent core of the coated fiber. (H&E, Magnification × 400.)

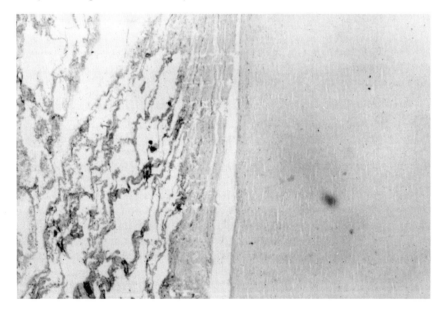

**Figure 17** Pleural plaque with linear collagenous sparsely cellular tissue adherent to underlying lung tissue. (H&E, Magnification × 100.)

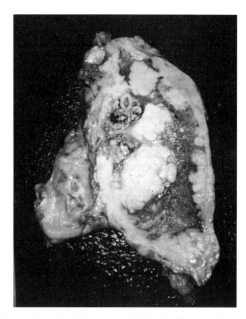

**Figure 18** Diffuse pleural malignant mesothelioma encasing the
periphery of the lung and invading
in a nodular pattern the pulmonary
parenchyma.

in nonasbestos-exposed cigarette smokers. The implication of this finding is that asbestos exposure increases the risk but not the type of bronchogenic carcinomas.

Diffuse malignant mesothelioma is a pleural- (and peritoneal-) based malignancy that has a strong epidemiological association with occupational or environmental asbestos exposure.[4,104–106] Environmental exposure has typically been in the form of household contact of asbestos worker's clothing or residence adjacent to asbestos factories. Malignant mesothelioma has been reported to occur sporadically in the general population, but its incidence, in the absence of asbestos exposure, is unusual.[107] Comparable to asbestos-related lung cancer, malignant mesothelioma occurs following a prolonged latency period, but in contrast, its development is not enhanced with cigarette smoking and frequently occurs with lower levels of asbestos fiber burden and in the absence of asbestosis.[108] As with other asbestos-related lung disease, the amphiboles are more effective in producing tumors than are the serpentines; and controversy exists whether chrysotile (or its impurities and contaminants) are capable of causing malignant mesothelioma.[109]

Diffuse malignant mesothelioma usually presents with chest pain, dyspnea, pleural effusion, and a typical chest X-ray or CT scan. Though gross examination of malignant mesothelioma demonstrates a characteristic thick, hard visceral pleural mass, with a gelatinous surface, surrounding the lung with direct nodular invasion into the parenchyma (Figure 18), the light microscopic diagnosis may prove more difficult. Malignant mesotheliomas must be distinguished from other primary pleural-based tumors and from pleural metastases. The histopathologic finding of combined epithelial tublopapillary component with a spindle cell sarcomatous component is the most distinctive pattern (Figure 19). However, malignant mesotheliomas may show (or tumor sampling may reveal) a monophasic tumor resembling a nondistinctive sarcoma or a purely epithelial tumor with glandular-like differentiation resembling an adenocarcinoma. An additional dilemma is the misinterpretation of reactive hyperplastic mesothelial cells for a malignant mesothelioma. The differential diagnosis and interpretive difficulty are compounded when only a sampling of tissue is obtained in a fine needle aspiration or needle biopsy. The pathological diagnosis of malignant

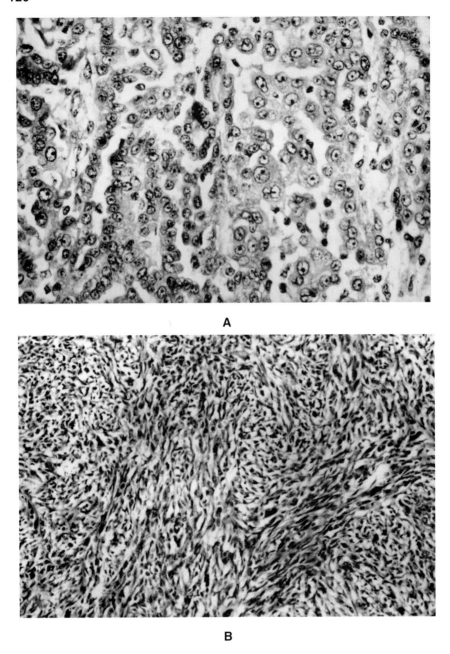

**Figure 19** Diffuse malignant mesothelioma demonstrating biphasic components:
**A.** epithelial cell component, **B.** spindle cell (sarcomatous) component. (H&E, Magnification × 400 and × 200.)

mesothelioma rests on a panel of supportive findings from histochemical, immuno-histochemical, and ultrastructural analyses.[4,70,106] Malignant mesotheliomas have hy-aluronic acid, which is positive in particular mucin stains, but lack abundant epithelial mucins characteristic of adenocarcinomas. Immunohistochemical demonstration of keratin and epithelial membrane antigen will tend to be positive in both mesothelioma and adenocarcinoma; however, malignant mesotheliomas will lack carcinoembryonic antigen, in contrast to most endodermal adenocarcinomas. Ultrastructurally, meso-theliomas have numerous thin, elongated surface microvilli, different from the shorter, blunter microvilli identified in usual adenocarcinomas (Figure 20).

## METAL LUNG DISEASE

The exposure to metals and their chemical products, such as metallic salts and metal oxides, produces a complicated array of pulmonary reactions and lung injuries.[4,110,111] The exposure may be in the form of fine particulate metallic fragments or compounds or their associated fumes or gases. Industrial procedures involving metals, such as mining, refining, grinding, and welding, may be performed at extremely high tem-peratures, and consequently gases, such as ozone, carbon monoxide, and oxides of nitrogen and sulfur, are produced.[112] These by-products are frequently more toxic to the lung than the metal exposure and may complicate or aggravate the pulmonary inhalant damage. It is well accepted that the lung responds to the variety of these inhalants in a specified and limited manner. Agents and inhalants may be toxic at the level of bronchioles or, more distally, at alveolar ducts and alveolar sacs. The resul-tant pulmonary disease may show the clinical and histopathological injuries of bron-chiolitis obliterans or diffuse alveolar damage. Though the initial insult is usually of rapid onset and may be associated with considerable morbidity, these exposures may also lead to a chronic course of interstitial fibrosis and end-stage lung disease. Sep-aration of metal-induced pulmonary fibrosis from other types of pneumoconiosis is based on an adequate occupational, enviromental, and clinical history and exposure record, distribution of pulmonary disease and histopathologic features, and chemical analysis of lung tissue.

A less damaging form of metallic inhalation injury on lung tissue is the dust macule, which is a localized fibroinflammatory response centered around small bron-chovascular structures. These microscopic collagenous scars often have considerable metallic pigment deposition. Metals, such as iron, tin, and titanium, are relatively inert and nonfibrogenic; consequently, prolonged exposure leads to accumulated macules without the progression to interstitial fibrosis. In contrast, hard metals, such as fused tungsten carbide and cobalt, and soft metals, such as beryllium, can produce a variety of pulmonary diseases leading to interstitial fibrosis. Hard metal disease has an interesting histologic appearance, resembling desquamative and giant cell inter-stitial pneumonitis.[4,111,113] Sections of the lung from this pneumoconiosis show a prominent intra-alveolar macrophage exudate with abundant multinucleated giant cells in a setting of an interstitial fibrosis with interstitial cellular reaction (Figure 21). The refractile or birefringent metallic particulates may be identified within the giant cells. The interstitial and alveolar pneumonitis leads to interstitial fibrosis. Ber-ylliosis may occur with short significant exposures and lead to diffuse alveolar damage or, in the chronic form, may present as an interstitial pneumonitis that may progress to interstitial fibrosis.[114] Chronic berylliosis is also characterized by a nod-ular noncaseous granulomatous reaction similar to sarcoidosis. Berylliosis may be distinguished from sarcoidosis by an appropriate occupational and exposure history and by identifying beryllium in lung tissue or urine.[111] Berylliosis may also be di-agnosed by immunologic studies on bronchoalveolar derived lymphocytes.[115]

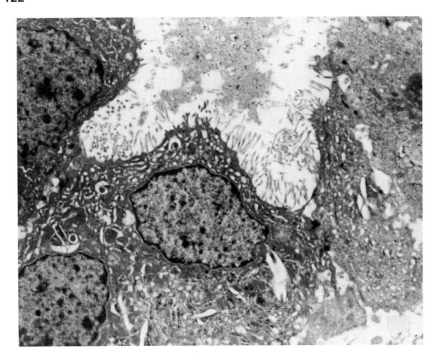

**Figure 20**  Electron micrograph of the epithelial cell component of diffuse malignant mesothelioma showing the multiple elongated apical microvillus processes. (Courtesy of Dr. J. M. Orenstein, Uranyl acetate-lead citrate, Magnification × 7000.)

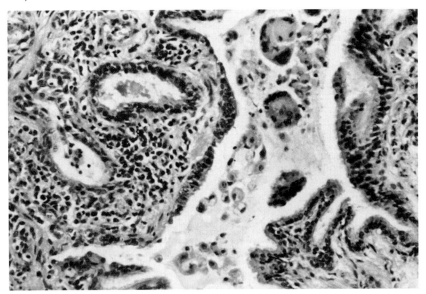

**Figure 21**  Hard metal disease with interstitial pneumonitis and prominent desquamative component of alveolar macrophages and multinucleated giant cells. (H&E, Magnification × 100.)

# REFERENCES

1. **Parkes, W. R.,** *Occupational Lung Disorders,* 2nd ed., Butterworths, London, 1982.
2. **Morgan, W. K. C. and Seaton, A.,** *Occupational Lung Diseases,* 2nd ed., W. B. Saunders, Philadelphia, 1984.
3. **Coles, J. E. and Steel, J.,** *Work-related Lung Disorders,* Blackwell Scientific, Oxford, 1987.
4. **Churg, A. and Green, F. H. Y.,** *Pathology of Occupational Lung Disease,* Igaku-Shoin, New York, 1988.
5. **Katzenstein, A. A. and Askin, F. B.,** *Surgical Pathology of Non-neoplastic Lung Disease,* W. B. Saunders, Philadelphia, 1990.
6. **Phalen, R. F.,** *Inhalation Studies: Foundations and Techniques,* CRC Press, Boca Raton, 1985, 7.
7. **Harley, R. A.,** Tobacco, in *Pulmonary Pathology,* Dall, D. H. and Hammer, S. P., Eds., Springer-Verlag, New York, 1988.
8. **Hiller, F. C., McCusker, K. T., Mazumder, M. K., et al.,** Deposition of side-stream cigarette smoke in the human respiratory tract, *Am. Rev. Respir. Dis.,* 125, 406, 1982.
9. **Auerbach, O., Stout, A. P., Hammond, E. C., et al.,** Smoking habits and age in relation to pulmonary changes, *N. Engl. J. Med.,* 269, 1045, 1963.
10. **Auerbach, O., Garfinkle, L., and Hammond, E. C.,** Relation of smoking and age to findings in lung parenchyma: A microscopic study, *Chest,* 65, 29, 1974.
11. **Niewoehner, D., Kleinerman, J., and Rice, D.,** Pathologic changes in the peripheral airways of young smokers, *N. Engl. J. Med.,* 291, 755, 1974.
12. **Cosio, M., Hale, K., and Niewoehner, D.,** Morphologic and morphometric effects of prolonged cigarette smoking on the small airways, *Am. Rev. Respir. Dis.,* 122, 265, 1980.
13. **Wright, J., Lawson, L., Pare, P., et al.,** Morphology of peripheral airways in current smokers and ex-smokers, *Am. Rev. Respir. Dis.,* 127, 474, 1983.
14. **Reid, L.,** *The Pathology of Emphysema,* Year Book Medical Publishers, Chicago, 1967.
15. **Medical Research Council,** Definition and classification of chronic bronchitis for clinical and epidemiological purposes, *Lancet,* 1, 775, 1965.
16. **Churg, A. and Wright, J. L.,** Small airways disease and mineral dust exposure, *Pathol. Ann.,* 18(2), 233, 1983.
17. **Myers, J. L., Veal, C. F., Shun, M. S., and Katzenstein, A. A.,** Respiratory bronchiolitis causing interstitial lung disease, *Am. Rev. Respir. Dis.,* 135, 880, 1987.
18. **Brody, A. and Craighead, J.,** Cytoplasmic inclusions in pulmonary macrophages of cigarette smokers, *Lab. Invest.,* 32, 125, 1975.
19. **Yousem, S. A., Colby, T. V., and Gaensler, E. A.,** Respiratory bronchiolitis and its relationship to DIP, *Mayo Clin. Proc.,* 64, 1373, 1989.
20. **Yousem, S. A., Colby, T. V., and Gaensler, E. A.,** Respiratory bronchiolitis-associated interstitial lung disease and its relationship to desquamative interstitial pneumonia, *Mayo Clin. Proc.,* 64, 1380, 1989
21. Office of Smoking and Health, Public Health Service, U.S. Department of Health and Human Services: The Health Consequences of Smoking: Cancer, DHHS Publ. No. (PHS) 82-50179, Washington, D.C., 1982.
22. **Doll, R. and Peto, R.,** Cigarette smoking and bronchial carcinoma: dose and time relationships among regular smokers and lifelong non-smokers, *J. Epidemiol. Community Health,* 32, 303, 1978.

23. **Peto, R., Lopez, A. D., Boreham, J., et al.,** Mortality from tobacco in developed countries: indirect estimation from national vital statistics, *Lancet,* 339, 1268, 1992.

24. **Shopland, D. R., Eyre, H. J., and Pechacek, T. F.,** Smoking attributable cancer mortality in 1991, *J. Nat. Cancer Inst.,* 83, 1142, 1991.

25. **Yousem, S. A.,** Small airways disease, *Pathol. Ann.,* 26 (part 2), 109, 1991.

26. **Epler, G. R., Colby, T. V., McCloud, T. C., et al.,** Bronchiolitis obliterans organizing pneumonia, *N. Engl. J. Med.,* 312, 152, 1985.

27. **Katzenstein, A. A., Myers, J. L., Prophet, W. D., et al.,** Bronchiolitis obliterans and usual interstitial pneumonia, *Am. J. Surg. Pathol.,* 10, 373, 1986.

28. **Rinaldo, J. E. and Rogers, R. M.,** Adult respiratory distress syndrome, *N. Engl. J. Med.,* 306, 900, 1982.

29. **Murray, J. F., Matthay, M. A., Luce, J. M., et al.,** An expanded definition of the adult respiratory distress syndrome, *Am. Rev. Respir. Dis.,* 138, 720, 1988.

30. **Schatz, M., Patterson, R., and Fink, J.,** Immunological lung disease, *N. Engl. J. Med.,* 300, 1310, 1979.

31. **Flint, K. C., Leung, K. B., Hudspith, B. N., et al.,** Bronchoalveolar mast cells in extrinsic asthma: a mechanism for the initiation of antigen specific bronchoconstriction, *Br. Med. J.,* 291, 923, 1985.

32. **Gleich, G. J., Frigas, E., Loegering, D. A., et al.,** Cytotoxic properties of the eosinophil major basic protein, *J. Immunol.,* 123, 2925, 1979.

33. **Barnes, P. J., Chung, K. F., and Page, C. P.,** Inflammatory mediators and asthma, *Pharmacol. Rev.,* 40, 49, 1988.

34. **Town, G. I. and Holgate, S. T.,** The role of inflammation in airways disease, in *Mediators of Pulmonary Inflammation,* Bray, M. A. and Anderson, W. H., Eds., Marcel Dekker, New York, 1991.

35. **Brooks, S. M.,** The evaluation of occupational asthma in the laboratory and workplace, *J. Allergy Clin. Immunol.,* 70, 56, 1982.

36. **Chan-Yeung, M. and Lam, S.,** Occupational asthma—state of the art, *Am. Rev. Respir. Dis.,* 133, 686, 1986.

37. **Yeung, M. and Grzybowski, S.,** Prognosis in occupational asthma, *Thorax,* 40, 241, 1985.

38. **Cullen, M. R., Cherniak, M. A., and Rosenstock, L.,** Occupational medicine, *N. Engl. J. Med.,* 322, 594, 1990.

39. **Tarlo, S. M. and Broder, I.,** Irritant-induced occupational asthma, *Chest,* 96, 297, 1989.

40. **Brooks, S. M., and Weiss, M. A., Bernstein, I. L.,** Reactive airways dysfunction syndrome (RADS). Persistent airway hyperreactivity after high level irritant exposure, *Chest,* 88, 376, 1985.

41. **Mendelson, E. B., Fisher, M. R., Mintzer, R. A., et al.,** Roentgenographic and clinical staging of allergic bronchopulmonary aspergillosis, *Chest,* 87, 334, 1985.

42. **Bosken, C. H., Myers, J. L., Greenberger, P. A., et al.,** Pathologic features of allergic bronchopulmonary aspergillosis. *Am. J. Surg. Pathol.,* 12, 216, 1988.

43. **Kawanami, O., Bassett, F., Barrios, R., et al.,** Hypersensitivity pneumonitis in man, *Am. J. Pathol.,* 110, 275, 1983.

44. **Hogg, J. C.,** Extrinsic immunologic lung disease, in *Pathology of the Lung,* Thurlbelk, W. M., Ed., Thorne Med. Publ., New York, 1988.

45. **Hammar, S. P.,** Extrinsic allergic alveolitis, in *Pulmonary Pathology,* Dall, D. H. and Hammar, S. P., Eds., Springer-Verlag, New York, 1988.

46. **Reyees, C. N., Wenzel, F. J., Lawton, B. R., et al.,** The pulmonary pathology of farmer's lung disease, *Chest,* 81, 142, 1982.

47. **Craighead, J. E., Kleinerman, J., et al.,** Diseases associated with exposure to silica and nonfibrous silicate minerals, *Arch. Pathol. Lab. Med.,* 112, 673, 1988.

48. **Ziskind, M., Jones, R. H., and Weil, H.,** State of the art: silicosis, *Am. Rev. Respir. Dis.,* 113, 643, 1976.

49. **Nagelschmidt, G.,** The relation between lung dust and lung pathology in pneumoconiosis, *Br. J. Ind. Med.,* 17, 247, 1960.

50. **Beuchner, H. A. and Ansari, A.,** Acute silico-proteinosis: a new pathologic variant of acute silicosis in sandblasters, characterized by histologic features resembling alveolar proteinosis, *Dis. Chest,* 55, 274, 1969.

51. **Suratt, P. M., Winn, W. C., Jr., Brody, A. R., et al.,** Acute silicosis in tombstone sandblasters, *Am. Rev. Respir. Dis.,* 115, 521, 1977.

52. **Xipell, J. M., Ham, K. N., Price, C. G., et al.,** Acute silicoproteinosis, *Thorax,* 32, 104, 1977.

53. **Heppleston, A. G.,** Silicotic fibrogenesis: a concept of pulmonary fibrosis, *Ann. Occup. Hyg.,* 26, 449, 1982.

54. **Prakash, U., Barham, S. S., Carpenter, H. A., et al.,** Pulmonary alveolar proteinosis: experience with 34 cases and a review, *Mayo Clin. Proc.,* 62, 499, 1987.

55. **Hnizdo, E., Sluis-Cremer, G. K., and Abramowotz, J. A.,** Emphysema type in relation to silica dust exposure in South African gold miners, *Am. Rev. Respir. Dis.,* 143, 1241, 1991.

56. **Vallyathan, N. V. and Craighead, J. E.,** Pulmonary pathology on workers exposed to nonasbestiform talc, *Hum. Pathol.,* 12, 28, 1981.

57. **Vallyathan, N. V.,** Talc pneumoconiosis, *Respir. Ther.,* 10, 34, 1980.

58. **Craighead, J. E., Emerson, R. J., and Stanley, D. E.,** Slateworker's pneumoconiosis, *Hum. Pathol.,* 23, 1098, 1992.

59. **Tomashefski, J. F. and Hirsch, C. S.,** The pulmonary vascular lesions of intravenous drug abuse, *Hum. Pathol.,* 11, 133, 1980.

60. **Crouch, E. and Churg, A.,** Progressive massive fibrosis of the lung secondary to intravenous injection of talc, *Am. J. Clin. Pathol.,* 80, 520, 1983.

61. **Green, F. and Laquer, W. A.,** Coal worker's pneumoconiosis, *Pathol. Annu.,* 15, 333, 1980.

62. **Davis, J., Chapman, J., Collings, P., et al.,** Variations in the histological patterns of the lesions of lung dust content, *Am. Rev. Respir. Dis.,* 128, 118, 1983.

63. **Soutar, C. A.,** Update on lung diseases in coal miners, *Br. J. Ind. Med.,* 44, 145, 1987.

64. **Sheenan, D. H., Washington, J. S., and Thomas, D. F.,** Factors predisposing to the development of progressive massive fibrosis in coal miners, *Br. J. Ind. Med.,* 38, 321, 1981.

65. **Kleinerman, J., Green, F., Harley, R. A., et al.,** Pathology standards for coal worker's pneumoconiosis, *Arch. Pathol. Lab. Med.,* 103, 375, 1979.

66. **Pratt, P. C.,** Role of silica in progressive massive fibrosis, *Arch. Environ. Health,* 16, 734, 1968.

67. **Ruckley, V. A., Fernie, J., Chapman, J. S., et al.,** Comparisons of radiographic appearances with associated pathology and lung dust content in a group of coal workers, *Br. J. Ind. Med.,* 41, 459, 1984.

68. **Becklake, M. R.,** Asbestos-related diseases of the lung and other organs: their epidemiology and implications for clinical practice, *Am. Rev. Respir. Dis.,* 114, 187, 1976.

69. **Craighead, J. E., Abraham, J. L., Churg, A., et al.,** The pathology of asbestos-associated diseases of the lungs and pleural cavities: diagnostic criteria and proposed grading schema, *Arch. Pathol. Lab. Med.,* 106, 544, 1982.

70. **Roggli, V. L., Greenberg, S. D., and Pratt, P. C.,** *Pathology of Asbestos-Associated Diseases,* Little, Brown, Boston, 1992.

71. **Mark, E. J. and Shin, D. H.,** Asbestos and the histogenesis of lung cancer, *Semin. Diag. Pathol.* 9, 110, 1992.

72. **Mossman, B. T. and Gee, J. B. L.,** Asbestos-related diseases, *N. Engl. J. Med.,* 320, 1721, 1989.

73. **Kagan, E.,** Current issues regarding the pathobiology of asbestos, *J. Thorac. Imag.,* 3, 1, 1988.

74. **Rom, W. N., Travis, W. D., and Brody, A. R.,** Cellular and molecular basis of the asbestos-related diseases, *Am. Rev. Respir. Dis.,* 143, 409, 1991.

75. **Pinkerton, K. E., Plopper, C. G., Mercer, R. R., et al.,** Airway branching patterns influence asbestos fiber location and the extent of tissue injury in the pulmonary parenchyma, *Lab. Invest.,* 55, 688, 1986.

76. **Stanton, M. F., Layard, M., Tegeris, A., et al.,** Relation of particle dimension to carcinogenicity in amphibole asbestosis and other fibrous minerals, *J. Natl. Cancer Inst.,* 67, 965, 1981.

77. **Rockoff, S. D., Kagan, E., Schwartz, A., et al.,** Visceral pleural thickening in asbestos exposure: the occurrence and implications of thickened interlobular fissures, *J. Thorac. Imag.,* 2, 58, 1987.

78. **Murphy, R. L., Becklake, M. R., Brooks, S. M., et al.,** The diagnosis of nonmalignant diseases related to asbestos, *Am. Rev. Respir. Dis.,* 134, 363, 1986.

79. **Churg, A. and Golden, J. G.,** Current problems in the pathology of asbestos-related diseases, *Pathol. Ann.,* 17 (pt. 2), 33, 1982.

80. **Churg, A. and Wright, J. L.,** Small airways disease in persons exposed to non-asbestos mineral dusts, *Hum. Pathol.,* 14, 688, 1983.

81. **Wright, J. L. and Churg, A.,** Morphology of small airways disease induced by asbestos exposure, *Hum. Pathol.,* 15, 68, 1984.

82. **Churg, A., Wright, J. L., Wiggs, B., et al.,** Small airways disease and mineral dust exposure, *Am. Rev. Respir. Dis.,* 131, 139, 1985.

83. **Craighead, J. E. and Mossman, B. T.,** The pathogenesis of asbestos-associated diseases, *N. Engl. J. Med.,* 306, 1446, 1982.

84. **Roggli, V. L.,** Pathology of human asbestosis: a critical review, *Adv. Pathol.,* 2, 31, 1989.

85. **Rockoff, S. D. and Schwartz, A. M.,** Noninvasive detection of early asbestos-related disease of the thorax, *Ann. N.Y. Acad. Sci.,* 643, 121, 1991.

86. **Schwartz, A., Rockoff, S. D., Christianni, D., et al.,** A clinical diagnostic model for assessment of asbestosis, *J. Thorac. Imag.,* 3, 29, 1988.

87. **Weiss, W.,** Cigarette smoke, asbestos, and small irregular opacities, *Am. Rev. Respir. Dis.,* 130, 293, 1984.

88. **Lilis, R., Selikoff, I. J., Lerman, Y., et al.,** Asbestosis: interstitial pulmonary fibrosis and pleural fibrosis in a cohort of asbestos insulation workers, influence of cigarette smoke, *Am. J. Ind. Med.,* 10, 459, 1986.

89. **Kilburn, K. H., Lilis, R., Anderson, H. A., et al.,** Interaction of asbestos, age, and cigarette smoking in producing radiographic evidence of diffuse pulmonary fibrosis, *Am. J. Med.,* 80, 377, 1986.

90. **Blanc, P. D. and Gamsu, G.,** The effect of cigarette smoking on the detection of small radiographic opacities in inorganic dust diseases, *J. Thorac. Imag.,* 3, 51, 1988.

91. **Hnizdo, E. and Sluis-Cremer, G. K.,** Effect of tobacco smoking on the presence of asbestosis at postmortem and on the reading of irregular opacities on roentgenograms in asbestos-exposed workers, *Am. Rev. Respir. Dis.,* 138, 1207, 1988.

92. **Hobson, J., Gilks, B., Wright, J., et al.,** Direct enhancement by cigarette smoke of asbestos fiber penetration and asbestos-induced epithelial proliferation in rat tracheal explants, *J. Natl. Cancer Inst.,* 80, 518, 1988.

93. **Blanc, P.,** Cigarette smoking, asbestos, and parenchymal opacities revisited, *Ann. N.Y. Acad. Sci.,* 643, 133, 1991.

94. **Cohen, D., Arai, S. F., and Brain, J. D.,** Smoking impairs long-term dust clearance from the lung, *Science,* 204, 514, 1979.
95. **Churg, A. and Warnock, M. L.,** Asbestos fibers in the general population, *Am. Rev. Respir. Dis.,* 122, 669, 1980.
96. **Churg, A. and Warnock, M. L.,** Asbestos and other ferruginous bodies, *Am. J. Pathol.,* 102, 447, 1981.
97. **Roggli, V. L. and Pratt, P. C.,** Number of asbestos bodies on iron-stained tissue sections in relation to asbestos body counts in lung tissue digests, *Hum. Pathol.,* 14, 355, 1983.
98. **Churg, A.,** Asbestos fiber content of the lungs in patients with and without asbestos airways disease, *Am. Rev. Respir. Dis.,* 127, 470, 1983.
99. **Churg, A. and Warnock, M. L.,** Correlation of quantitative asbestos body counts and occupation in urban patients, *Arch. Pathol. Lab. Med.,* 101, 629, 1977.
100. **Churg, A.,** Asbestos fibers and pleural plaques in a general autopsy population, *Am. J. Pathol.,* 109, 88, 1982.
101. **Wain, S. L., Roggli, V. L., and Foster, W. L.,** Parietal pleural plaques, asbestos bodies, and neoplasia, *Chest,* 86, 707, 1984.
102. **Selikoff, I. J., Hammond, E. C., and Churg, J.,** Asbestos exposure, cigarette smoking, and neoplasia, *J. Am. Med. Assoc.,* 204, 104, 1968.
103. **Hammond, E. C., Selikoff, I. J., and Seidman, H.,** Asbestos exposure, cigarette smoking, and death rates, *Ann. N.Y. Acad. Sci.,* 330, 473, 1979.
104. **Craighead, J. L.,** Current pathogenesis concepts of diffuse malignant mesothelioma, *Hum. Pathol.,* 18, 544, 1987.
105. **Kannerstein, M., Churg, J., and McCaughey, W. T. E.,** Asbestos and mesothelioma: a review, *Pathol. Ann.,* 13 (pt. 1), 81, 1978.
106. **Roggli, V. L., Kolbeck, J., Sanfillipo, F., et al.,** Pathology of human mesothelioma: etiologic and diagnostic considerations, *Pathol. Ann.,* 22 (pt. 2), 91, 1987.
107. **Peterson, J. T., Greenberg, S. D., and Butler, P. A.,** Non-asbestos-related malignant mesothelioma, *Cancer,* 54, 951, 1984.
108. **Roggli, V. L., McGavran, M. H., Subach, J., et al.,** Pulmonary asbestos body counts and electron probe analysis of asbestos body cores in patients with mesothelioma, *Cancer,* 50, 2423, 1982.
109. **Churg, A.,** Chrysotile, tremolite, and malignant mesothelioma in man, *Chest,* 93, 621, 1988.
110. **Wright, J. L. and Churg, A.,** Diseases caused by metals and related compounds, fumes, and gases, in *Pathology of Occupational Lung Disease,* Igaku-Shoin, New York, 1988.
111. **Roggli, V. L. and Shelburne, J. D.,** Mineral pneumoconiosis, in *Pulmonary Pathology,* Dall, D. H. and Hammer, S. P., Eds., Springer-Verlag, New York, 1988.
112. **Sferlazza, S. J. and Beckett, W. S.,** The respiratory health of welders, *Am. Rev. Respir. Dis.,* 143, 1134, 1991.
113. **Ohori, N. P., Sciurba, F. C., Owens, G. R., et al.,** Giant cell interstitial pneumonia and hard metal pneumoconiosis, *Am. J. Surg. Pathol.,* 13, 581, 1989.
114. **Colby, T. V.,** Berylliosis, in *Pathology of Occupational Lung Diseases,* Igaku-Shoin, New York, 1988.
115. **Newman, L. S., Reiss, K., King, T. E., et al.,** Pathologic and immunologic alterations in early stages of beryllium disease, *Am. Rev. Respir. Dis.,* 139, 1479, 1989.

*Chapter 5*

# Causality Assessment: Causal Inference in Toxicology

*Sorell L. Schwartz, Ph.D., Nancy J. Balter, Ph.D., and Philip Witorsch, M.D.*

## CONTENTS

## CAUSALITY DEFINED

The enterprise that we term "causality assessment" is one of the most intricate of logical endeavors. Its conduct is frequently characterized by an underestimation of the difficulty in design, process, and achievement of the assessment goal, and an overestimation of the validity of the assessment per se. Causality is often thought of in relationship to an event; i.e., given observation of an event, **E**, which of the infinite number of antecedent events can be identified such that its (their) absence precludes the observation of **E**. But the causality concept is much broader; it constitutes the underpinnings of scientific thought and logic. Consider the following from Popper.[1]

> To give a *causal explanation* of an event means to deduce a statement which describes it, using as premises of the deduction one or more *universal laws,* together with certain singular statements, the *initial conditions . . .*
>
> We have thus two different kinds of statements, both of which are necessary ingredients of a complete causal explanation. They are (1) *universal statements,* i.e., hypotheses of the character of natural law, and (2) *singular statements,* which apply to the specific event in question . . . call[ed] 'initial conditions.' It is from universal statements in conjunction with initial conditions that we *deduce* the singular statement. . . . We call this statement a specific or singular prediction. *[Emphasis his.]*

Popper includes in the term, "prediction," statements about the past. He also notes that "initial conditions" characterize what is commonly called the "cause" of the event in question, and the "prediction" depicts what is commonly called the "effect" (although Popper declines to use either of the terms "cause" and "effect").

Popper's construct can be contained within the framework of one suggested by Lane[2] (though not strictly adhered to here) that asks three questions. They are:

(1) "Can it?"

Based on experimental animal data and/or a series of cases, where exposure to chemical or physical agent, **M**, has occurred and events of the type, **R**, have been observed, is there sufficient cumulative evidence to support the implication that **M** can cause **R**, even if the evidence in each individual case is not strong enough for there to be a reliable causal inference?

(2) "Did it?"

In a particular case, did exposure **M** cause **R** to occur? That is, if exposure to **M** had not occurred, would **R** have happened anyway?

(3) "Will it?"

What is the chance that **R** will occur to the next person exposed to **M**, and how does this compare to the chance that **R** will happen to the same individual who is not exposed to **M**?

In a general sense, the answer to "Can it?" can be considered a universal statement, i.e., a recognized biological property of **M**. "Will it?" and "Did it?" are, obviously, expressions of a singular statement or prediction, i.e., an expression of causal inference under the observed (initial) conditions.

This framework provides some of the logical background for appreciating and considering the principles and problems inherent in arriving at a causal inference. Not the least of these is understanding what question is asked. There is an abundance of examples where this seemingly elementary condition is neglected. In both the regulatory and the civil litigation domains, a risk estimate or injury causation decision has been made with "scientific and medical certainty," while the underlying biological property of the "causal" agent in question remained uncertain. See, for example, Marshall's commentary on daminozide (Alar).[3]

It is currently incontrovertible that there are no means by which a theory can be shown to be true with certainty. If there is no resolution to this conflict, then it follows that there can be no proven cause for empirical observations, an epistemological dead end. Rigid adherence to that idea would make short work of the scientific enterprise. In his distinguished treatise, *Causality in Modern Science,* Bunge[4] points out that a scientific explanation does not have to entail absolute certainty, but that the use of the word *scientific* implies certain qualities:

[N]ot every explanation affording what is loosely called intellectual satisfaction is scientific; only general, meaningful, and verifiable ideas are involved in genuine scientific explanation.

The balance of this discussion is directed at approaches to achieving "genuine scientific explanation" in arriving at a causal inference. It will also be evident that, especially with respect to the "Did it?" and "Will it?" questions, this goal is subject to revision.

## THE UNIVERSAL STATEMENT: "CAN IT?"

Is there sufficient scientific evidence from prior experience to support the hypothesis that exposure to chemical or physical agent **M** can cause a response in the form of a health disturbance, **R**, in humans. Two types of data will be considered, that from epidemiological studies, and that from experimental animals.

### EPIDEMIOLOGIC STUDIES

Cohort and case–control studies provide the bulk of the epidemiological data from which the relationship between a defined exposure and a measured response is assessed. Briefly, in cohort (prospective) studies, subjects are classified according to their exposure status. Ideally, incidence of the measured response in the exposed cohort is compared to the incidence of the response in a cohort with no known exposure. In case–control (retrospective) studies, subjects are classified on the basis of whether or not they have (or had) the measured response, and are then compared on whether or not they were exposed to **M**. Results are then expressed in one form or another that describes a rate ratio (or relative risk), **RR**:

$$Rate\ Ratio = \frac{Incidence\ of\ R\ in\ M\text{-}Exposed\ Group}{Incidence\ of\ R\ in\ M\text{-}Unexposed\ Group}$$

An **RR** significantly $>1$ implies an association between **M** and **R**. This, of course, cannot, in and of itself, be interpreted as demonstrating causation. Achievement of statistical significance does not constitute proof of a hypothesis. The fundamental premise of testing the null hypothesis is that the *null hypothesis is true.* The finding of a "statistically significant difference" is a basis for hypothesis refutation, in this case refuting the null hypothesis. The finding *does not represent a test of any alternative hypothesis,* particularly a universal statement. True, it may lead to hypothesis generation, but that is decidedly different from arriving at an explicit answer to the "Can it?" problem. Similarly, if statistical significance does not result, hypothesis rejection is not necessarily appropriate. The *power* of the study, i.e., the probability that a difference of a specified magnitude that actually exists will be observed in the study, is a necessary consideration. If a given epidemiologic study would be insensitive to the detection of the specified, real difference in the parameter of interest, then the failure of the study to identify a difference could be meaningless.

One particular way to examine the causal assessment value of the **RR** is the *attributable risk,* i.e., the fraction of **R** in the entire population that is attributable to **M**, assuming that the fraction of the population exposed to **M** $(P_e)$ is also known:

$$Attributable\ Risk = \frac{P_e(RR-1)}{P_e(RR-1)+1}$$

A high attributable risk is not an obligatory criterion for causality, but a high attributable risk can have a strong influence on interpreting the association. Furthermore, as is discussed in greater detail in the next section, an attributable risk $>0.5$ does not necessarily mean, in a specific individual case of **R** in an individual exposed to **M**, that **M** caused **R**.

## Sources of Bias
Epidemiologic studies are always subject to the influence of variables other than those that the study is designed to measure. Bias can introduce error into a study, resulting in an observed **RR** (**ORR**) that is different from the true **RR** (**TRR**). The effect of the bias can be either to cause the **ORR** to be closer to or farther from 1 (i.e., the study can be biased such that it would be less likely to detect a true difference, or such that it would be more likely to find a difference that does not, in fact, exist). There are a number of potential sources of bias in an epidemiologic study.

*Selection bias.* Selection bias occurs when the criteria for entry into the cohort make the study population nonrepresentative of the general population of interest. One example is self-selection bias. A simple survey of a group of former students attending a college reunion may reveal that the "success indices" (e.g., income, education, position) for the class are significantly higher than for the general population of college graduates from similar socioeconomic backgrounds. However, there is a good possibility that graduates who attend a reunion are more likely to have higher success indices than those who do not attend, so the group attending the reunion may not be an unbiased sample of the class.

*Observation (information) bias.* This source of bias can occur in either a cohort or case–control study when the methods of data collection are not comparable among the various groups. In a cohort study, this could occur, for example, if the outcome of interest was looked for more aggressively in the exposed group than in the comparison group. In a case–control study, bias could be introduced, for example, in the collection of exposure data if cases were queried more extensively about their exposure experience than controls. This is the reason for double-blind controlled clinical trials. That is, if the observer knows the identification of the treatment and placebo group members, then the possibility exists that the questioning of the treatment group members regarding adverse reactions will be more persistent than that of the placebo group.

*Misclassification bias.* Misclassification refers to the incorrect classification of a study subject with respect to exposure or disease. It can be the result of selection or observation bias and can be either nondifferential (random) or differential. Random misclassification means that the misclassification affects the groups on both sides of the comparison. For example, in an occupational cohort study, where exposure is assumed by virtue of the employment, it is likely that some members of the occupational cohort are misclassified with respect to exposure. For example, someone may be assigned to a manufacturing area where exposure is presumed to occur, but the individual could be a clerical worker in an office adjacent to the actual physical space where exposure occurs. Since this misclassification is as likely to occur in cohort members with the disease of interest as in those without the disease of interest, the misclassification is said to be "random." Random misclassification will have the effect of biasing the study, such that the **ORR** will be closer to 1 than the **TRR**— that is, toward a conclusion that there is no difference between the exposed and control groups. Misclassification can also be differential, affecting the compared groups differently. For example, in an occupational cohort study, there will be misclassification of the cause of death in both the exposed and comparison groups. However, if the cause-specific mortality experience of the nonexposed group is based on death certificate-based statistics and if that of the exposed group is based on a review of medical records, the misclassification can be differential, affecting one group more than the other.

*Confounding.* In measuring the association between **M** and **R**, a confounding factor is something that is associated with the exposure and can independently be the cause of the disease. For example, assume an epidemiologic study is designed to examine the effect of gasoline fumes on cancer mortality, and it finds a statistically significant increased **RR** for lung cancer among gas station attendants. Smoking must be considered as a potential confounder, and if the incidence of smoking among gas station attendants is greater than that in the general population, the elevated **RR** could represent an effect of smoking and be independent of any effect of gasoline fumes. If confounding factors are recognized and appropriate data collected in the study, their effect on the outcome of the study can be controlled for in the analysis of data. The problem is that all of the possible confounding factors may not be recognized, and even when a confounding factor is suspected, the data necessary to adjust for it in the analysis may not be available.

As noted, epidemiologic studies can demonstrate an association between **M** and **R**, but association does not necessarily justify a universal statement of causality. Often, specified criteria are used to define causality. This is a source of significant controversy.[5] Several articulations of these criteria have been published. For example, the following is a modified version of the Henle-Koch-Evans' postulates:[6]

1. The prevalence rate of the disease should be significantly higher in those exposed to the hypothesized cause than in controls not so exposed.

2. Exposure to the hypothesized cause should be more frequent among those with the disease than in controls without the disease, when all other risk factors are held constant.
3. Incidence of the disease should be significantly higher in those exposed to the cause than in those not so exposed, as shown by prospective studies.
4. A spectrum of host responses should follow exposure to the hypothesized agent along a logical biological gradient from moderate to severe, depending upon dose.
5. The dose necessary to cause any specific effect should be normally distributed for the population.
6. Temporally, the disease should follow exposure to the hypothetical causative agent over a time period that is normally distributed for the population.
7. Experimental reproduction of the disease should occur more frequently in animals or man appropriately exposed to the hypothesized cause than in those not so exposed.
8. Elimination or modification of the hypothesized cause should decrease the incidence of the disease.
9. All of the relationships and findings should make biological and epidemiological sense.

The problem with such criteria is not with the assistance they provide in evaluating associations; it is that they tend to become the sine qua non of causal inference. That is an abuse of such guidelines.

This leads to another conflict, i.e., deterministic (objective and deductive) vs. probabilistic (subjective and inductive) approaches to causal inference. The influential 20th-century scientific philosopher Karl Popper, who was quoted above, is generally recognized as the archetypical proponent of the determinism of deductive logic and as an equally strong critic of induction and arriving at a level of belief by probabilistic methods.[1] It was Buck[7] who introduced the Popperian philosophy to epidemiology (in the sense that it was Popper's philosophy). The justification that Popper gives for his logical approach to science is quite detailed and complex. Buck's summary is an adequate one for present purposes, i.e., the pitfalls of assuming that association is the equivalent of causation:

> The traditional view of science has been that induction, the formation of a hypothesis based upon observation, is the mainstay of the scientific process. This has always presented a logical difficulty because a hypothesis so derived is forever vulnerable to denial by the first observation that proves an exception. If by induction we believe that all swans are white, our belief can be overthrown by the appearance of a black swan. [*The swan example is Popper's not Buck's.*] Popper has rejected the traditional view and regards induction as a dispensable concept. He believes that all observations are made to test some hypothesis already in mind, the derivation of the hypothesis being a work of the imagination. One uses deductive reasoning to make predictions from the hypothesis and thus to state what it prohibits. Scientific discovery is based solely upon a hypothetico-deductive process and *advances by disproof rather than by proof.* [*Emphasis added.*]

In other words, there are an infinite number of hypotheses that can explain any set of empirical observations. Tests designed to verify a single hypothesis, no matter how far-reaching and corroborative, cannot adequately enrich the hypothesis (i.e., reduce the uncertainty to some indistinct margin) in the absence of considering, testing, and rejecting alternative hypotheses, i.e., attempts at refutation. In fact, since refutation has so much more potent an influence on evaluating a hypothesis than does verification (corroboration), hypothesis testing should be asymmetrical, favoring attempts at refutation.

In reality, an argument can be made that causality decisions must be examined from two points of view—pure scientific theory and uncertainty judgments needed for public health purposes. In most basic scientific endeavors, a causal inference has neither a decision-making quality nor requirement. That is, hypotheses can be pursued and tested until such time as they are refuted. This is the general rule of scientific research. However, interpretation of data for public health purposes presents a decision-making requirement, a requirement for an uncertainty judgment. Such judgments generally demand an element of induction, including some expression of a *degree of belief* in the uncertainty judgment. It is this necessity to arrive at a degree of belief that is considered the underlying weakness of induction, and to strict Popperians, it is unacceptable. In the realm of public health decisions, however, there is *often* very little choice. Still, when assessments are heavily biased toward verification with little recognition given to the testing strength of refutation, irrespective of the quality of that verification, the paucity of alternate-hypothesis consideration weighs heavily on the analytical value of the causal inference.

## EXPERIMENTAL ANIMAL STUDIES

Animal studies involving experimental exposure to **M** under controlled conditions can take a variety of forms. The exposure can be single or repeated, acute or chronic, and by any one of a number of routes of exposure, including injection, inhalation, or dermal, and oral. The quantity and quality of available animal data will vary greatly from one chemical to another. The more the data and the greater the similarity among various animal species in the response to **M**, the greater the confidence in extrapolating the effects seen in animals to humans.

Clearly, experiments with animals present the advantage of controlled hypothesis testing. That hypothesis applies to the animal studied, the route and length of exposure, and the exposure level studied. The question of extrapolation to humans still prevails.

### Interspecies Extrapolation

There is always some uncertainty in extrapolating the effects caused by **M** in animals to predict effects in man. Issues to be addressed concerning the appropriateness of extrapolation include species differences in the absorption, distribution or elimination of **M** and species differences in susceptibility to the effects of **M**. Some toxicological endpoints in animals are notably unreliable as predictors of human toxicity. That is, the toxicity associated with **M** in animals may not be seen in humans exposed to equivalent doses of **M**, and in some cases, human toxicity associated with **M** may not be predicted by animal studies. A important example of this is teratogenicity.[8]

### Route of Exposure Extrapolation

The data available from animal studies very often do not involve a comparable exposure to that which occurred in the case being evaluated. Extrapolation can be based on the ability to calculate equivalent systemic doses of **M** associated with the various exposure routes. This is a well-known phenomenon that can cause dosage differences for drugs in humans. For example, when morphine is taken orally, it, like all materials absorbed from the GI tract (except for some rectal absorption), must make a first pass through the liver before entering the general circulation. It is extensively metabolized by the liver, such that a major hepatic extraction of the dose occurs before it becomes systemic. This "first-pass" effect does not occur following parenteral administration. Thus, a higher oral dose of morphine is necessary to obtain a systemic blood level equivalent to that obtained after a parenteral dose.

### Dose Extrapolation

The concept of threshold is a key principle of toxicology; that is, for any **R** caused by **M**, there will be a dose of **M** below which **R** is not seen. Animal studies may establish thresholds for specific effects. It must be appreciated, however, that identification of the threshold reflects the sensitivity of the animal studies in detecting an effect that is occurring, which in turn depends upon the sensitivity of the endpoint measurement and the number of animals tested in the study. The issue of low dose extrapolation becomes particularly important in dealing with carcinogenic effects. Two general processes appear to be involved in chemical carcinogenesis: a mutation in the DNA of a cell, causing it to be transformed into a cell with the potential of becoming a tumor cell, and stimulation of the proliferation of cells containing the mutated DNA. There has been a prevailing view that, unlike other manifestations of toxicity, the carcinogenic effect of a chemical has no threshold. This view is based on a working hypothesis that a single mutation in DNA, caused by a single molecule of **M**, can initiate the sequence of events that ultimately results in tumor formation. This no-threshold hypothesis for chemical carcinogenesis is based more on theory than on scientific evidence and does not take into account the existence in most cells of active DNA repair mechanisms. It is also not applicable to a chemical that causes tumors in animals by a nongenotoxic mechanism (i.e., that does not affect DNA), such as phenobarbital in mouse liver.[9] However, since an assumption of no-threshold represents a highly conservative approach to low-dose extrapolation, it has been generally applied in the area of regulatory toxicology. This is discussed in more detail in the section on answering the "Will it?" question.

## "DID IT?": A POSTERIORI CAUSAL INFERENCE IN TOXIC INJURY

During the course of or following an occupational or environmental exposure of an individual to a specific level of **M**, an abnormality **R** is observed. Did the exposure cause the effect? Stated another way, would the effect have occurred in this individual in the absence of the exposure? This question of a posteriori probability (the probability that an event *was* caused by a specified preceding occurrence) is what usually comes to mind with respect to the issue of causality. Medically and legally, it is often considered an issue of diagnosis. It is, in fact, a situation that often takes the physician outside of the usual heuristic components of diagnosis. The use of the term "a posteriori probability" deserves emphasis. Causal inference for a clinical situation of the type described is an uncertainty judgment, and it relies on the probability assigned by the evaluator; i.e., a causality *decision* rests on the strength of a causal *inference*. Accordingly, a point raised earlier is reiterated: i.e., an hypothesis cannot be used to explain a set of empirical observations with absolute certainty. The following points apply to the *clinical situation under consideration:*

1. For decision-making purposes, causal inference must be approached from a probabilistic (inductive) standpoint, not a deterministic (deductive) one.
2. Acceptance that a cause-and-effect relationship has been proven reflects a level of confidence that a causal inference is correct. The problem is twofold: (a) how to measure that confidence; (b) agreement on what level of confidence constitutes proof.*

---

*It is important to recognize that the probability assigned to a causal inference needs only to be above 0.5 for the inference to be that causation is more likely than not. It should not be confused with probability computed for testing the null hypothesis, which ordinarily is rejected only if the probability of chance is <0.01 to 0.05. The latter refers to the acceptability of data; the former refers to its evaluation.

3. If scientific, medical, legal, social, political, economic, etc. decisions are to be made, then it must be assumed that it is possible to arrive at a level of confidence that a causal inference is correct.

4. If it is possible to arrive at a level of confidence that a causal inference is correct, then it must be possible to arrive at a level of confidence that a causal inference is incorrect. It follows, then, that there is little philosophical difference between *proving a positive* and *proving a negative,* and the often stated maxim, that it is scientifically impossible to prove a negative, is trite.

Two general approaches to the problem will be discussed. They are Bayesian and algorithmic. The essential element of any causality assessment, however, is the data from which that assessment is being made.

Causal inference as applied to single individuals utilizes two types of data: the exposure data, i.e. data regarding concentration, duration, and routes of exposure, etc., for the exposure(s) in question, and the clinical data. The clinical data is derived from history, physical examination, laboratory studies, and medical records. This aspect of data collection is discussed in detail elsewhere in this book.

## THE DATA: EXPOSURE ASSESSMENT

Exposure assessment is a key component of causal analysis. The parameters defining the individual's exposure to **M**, including route(s), dose, and time-course, need to be quantified to the extent possible, based upon the available data. Depending upon the health disturbance being evaluated, total dose or peak exposure can be the more significant exposure variable. There are a number of approaches to exposure assessment. In situations where the only available information is anecdotal, the types of acute effects reported, or details such as whether or not a characteristic odor was detected, can provide a basis for estimating the exposure concentration. Industrial hygiene or other monitoring data (concentrations in water, food, air) provide a more quantitative estimate of exposure. Such data, used in conjunction with assumptions or information concerning the individual's daily activity patterns, dietary habits, water ingestion/use patterns, etc., allow estimation of the dose of **M** received by the exposed individual over time.

Biomonitoring is another approach to exposure assessment; it can be directed at monitoring exposure/dose or exposure/effect. Biomonitoring for exposure/dose involves measurement of the concentration of **M** or a metabolite of **M** in the individual's breath or in body fluids, usually blood or urine. The most significant limitation associated with this approach is that, even after relatively high exposures, **M** will be detected for only a short period, usually measured in days, following the exposure. Some chemicals accumulate in hair or nails, and analysis of the concentration of **M** in these substrates can reflect exposures that occurred over the previous several months. Similarly, chemicals (or their metabolites) that are highly fat-soluble will accumulate in adipose tissue, and analysis of the concentration of **M** in tissue obtained from fat biopsies can reflect exposures to **M** that occurred over the previous years to, in some cases, decades. Under specific circumstances, biomonitoring can provide a quantitative estimation of the dose of **M** received by the individual, although more commonly, the biomonitoring provides only qualitative or semiquantitative exposure estimates.

Biomonitoring can also be used as a measure of exposure/effect if the endpoint measured is sensitive to the effect of **M**. For example, the exposure/effect of a pesticide that inhibits the activity of acetylcholinesterase can be monitored by measuring the activity of that enzyme in the exposed individual. Biomonitoring for exposure/effect of genotoxic substances has involved the enumeration of white blood cells that exhibit evidence of chromosomal damage, usually either chromosomal aberrations or sister chromatid exchanges. Even in circumstances where epidemiologic studies have demonstrated that, as a cohort, individuals exposed to **M** have a greater percentage of cells with evidence of chromosomal damage than an appropriately matched control group, the findings in a single individual are rarely conclusive with respect to the relationship between the finding and exposure to **M**. This is because exposure to genotoxic substances is ubiquitous, and the endpoints of chromosomal damage being measured are usually not unique to **M**.

Quantifying adduct formation is a relatively new approach to biomonitoring that takes advantage of the finding that many chemically reactive substances, including genotoxic carcinogens, will react with proteins or DNA to form stable adducts. These adducts can be measured for several months following exposure and could potentially serve as biomarkers of exposure/dose as well as exposure/effect. Although in some cases the adducts formed are chemical-specific, in most cases the adducts could reflect exposure to one of a number of chemicals. In this situation, measurement of protein or DNA adducts in an individual will have only limited applicability in the assessment of that individual's exposure to a specific chemical.

## APPLICATION OF BAYES' THEOREM

It will be recalled that *attributable risk* is that fraction of the disease in the entire population that is attributable to **M** and that is determined by the relative risk (**RR**) and the fraction of the population exposed (**$P_e$**) to the **M** under consideration.

$$Attributable\ Risk = \frac{P_e(RR-1)}{P_e(RR-1)+1}$$

For example, assume that the relative risk of lung cancer in all U.S. cigarette smokers is 12, and that the fraction of smokers in the U.S. population is 0.3. Therefore, the fraction of lung cancer in the U.S. that can be considered as attributable to cigarette smoking is:

$$\frac{0.3(12-1)}{0.3(12-1)+1} = 0.77 = 77\%$$

One interpretation might be that if the attributable risk of a health disturbance from exposure to a substance is $> 0.5$, then it can be concluded that it is more likely than not that the suspected agent caused the effect. Such an assumption is wrong. Each case has its *individual characteristics* (initial conditions). For example, what was the histological type of lung cancer? If the case was adenocarcinoma, there is a lesser likelihood of causality. Was the exposure (number of packs per day for a specified number of years) consistent with the database relating cigarette smoking and lung cancer, from which the relative risk was computed? One method for evaluating case-specific characteristics is the application of Bayes' theorem.

Consider two events **A** and B occurring together. What is the probability, **P**, that the *occurrence of A is conditional on the occurrence of B.* That is, what is **P(A |B)**? (The symbol " |" is read as "conditional on.") This can be shown to be:

$$P(A\mid B) = \frac{P(A\&B)}{P(B)}$$

where **P(A&B)** is the probability of **A** and **B** occurring together and **P(B)** is the unconditional probability of the occurrence of **B**. Similarly, the probability that the *occurrence of **B** is conditional on the occurrence of **A*** can be expressed as:

$$P(B \mid A) = \frac{P(A\&B)}{P(A)}$$

and

$$P(A\&B) = P(A) \cdot P(B \mid A)$$

and, substituting,

$$P(A \mid B) = \frac{P(A) \cdot P(B \mid A)}{P(B)}$$

This is Bayes' theorem. It can be utilized in hypothesis testing. For example, let **Hy** be the hypothesis and let **Obs** be the observation that will confirm or not confirm the hypothesis. Accordingly,

$$P(Hy \mid Obs) = \frac{P(Hy) \cdot P(Obs \mid Hy)}{P(Obs)}$$

**P(Hy | Obs)** is referred to as the *posterior probability*. It is the probability of the validity of the hypothesis conditional upon the new observation. The *prior probability* of **Hy**, i.e., **P(Hy)** is the probability that the hypothesis is correct prior to considering the new observation. If **P(Hy | Obs)** is > **P(Hy)**, then **Obs** confirms **Hy**. **P(Obs | Hy)** is an expression of *conditional probability*, i.e. the *likelihood* that **Obs** would be observed if **Hy** were correct. **P(Obs)** is the unconditional probability of making the observation, **Obs**, i.e., the probability of **Obs** for any reason at all.

Bayes' theorem can be applied to the evaluation of a clinical test with respect to diagnosis. For example, based on the initial description of symptoms and the initial observation of a patient, it is hypothesized that the patient has Disease Entity **Q** (**DEQ**); i.e., **Hy**: Patient has **DEQ**. Furthermore, based on previous experience and the clinical literature, it is estimated that there is a 40% probability that the diagnosis is correct; i.e., **P(Hy)** = 0.4. Clinical Test **W** (**CTW**) results are obtained for the patient, and it is known that these results are seen in 60% of the patients diagnosed with **DEQ**. The conditional probability, i.e., **P(Obs | Hy)**, is, therefore 0.6. Is the probability of **Hy** now any greater after the observation of the **CTW** results; i.e. is **P(Hy | Obs)** > **P(Hy)**? Or, stated another way, does **Obs (CTW)** confirm **Hy** (the presence of **DEQ**)? The answer is *not* as intuitively evident as it might seem. Consider two situations: (1) 25% of the general population will give the **CTW** results obtained; (2) 75% of the general population will give the **CTW** results obtained. Accordingly:

$$P(Hy \mid Obs) = \frac{P(Hy) \cdot P(Obs \mid Hy)}{P(Obs)} = \frac{0.4 \cdot 0.6}{0.25} = 0.96$$

and:

$$P(Hy \mid Obs) = \frac{P(Hy) \cdot P(Obs \mid Hy)}{P(Obs)} = \frac{0.4 \cdot 0.6}{0.75} = 0.32$$

respectively, for the two situations. In the first case, **CTW** strongly confirmed **DEQ**. In the second situation, the **CTW** result, though present in the majority of patients with **DEQ**, is still less characteristic of the **DEQ** patients than it is of the general population, irrespective of **DEQ**.

Accordingly, the question can be asked as to the probability that **Obs** verifies that **Hy** *is not correct*, represented as ~**Hy**. Deriving as above:

$$P(\sim Hy \mid Obs) = \frac{P(\sim Hy) \cdot P(Obs \mid \sim Hy)}{P(Obs)}$$

Dividing the previous equation by this one, the *odds form* of Bayes' theorem is obtained:

$$\frac{P(Hy \mid Obs)}{P(\sim Hy \mid Obs)} = \frac{P(Hy)}{P(\sim Hy)} \cdot \frac{P(Obs \mid Hy)}{P(Obs \mid \sim Hy)}$$

| Posterior Odds | Prior Odds | Likelihood Ratio |
|:---:|:---:|:---:|
| **(PtO)** | **(PrO)** | **(LR)** |

The equation can be written in shorthand form as

$$PtO = PrO \text{ o } LR$$

and it states that:

> *the odds the **Obs** confirms that **Hy** is correct*
> is equal to
> *the odds that **Hy** is correct prior to **Obs** being made*
> times
> *the ratio of likelihood (probability) that **Obs** would be observed of **Hy** is correct to the likelihood that **Obs** would be observed if **Hy** is not correct.*

The means of applying this to causality assessment is adapted from the Bayesian approach to causality assessment with respect to adverse drug reactions given by Lane.[10] Accordingly,

**M**→**R** is a statement of **M** causing **R**
**M**×**R** is a statement of **M** not causing **R**
**B** represents background information about the relationship of **M** to **R**, e.g., dose–response information, factors that enhance or reduce the toxicity of **M**.

**C** represents the case-specific details, including the patient's state of health, the presence or absence of risk factors that affect the toxicity of **M**, the quality of the collected data (e.g., the qualitative and quantitative reliability of exposure information), and other exposures or factors that can increase the likelihood of **R**, irrespective of **M**.

Bayes' theorem is then represented as:

$$\underbrace{\frac{P(M \times R,B \mid C)}{P(M \times R,B \mid \bar{C})}}_{\text{PtO}} = \underbrace{\frac{P(M \times R,B)}{P(M \times R,B)}}_{\text{PrO}} \cdot \underbrace{\frac{P(C \mid M \times R,B)}{P(\bar{C} \mid M \times R,B)}}_{\text{LR}}$$

This is read as:

*The odds that **M** caused **R** based on the available background information and under the conditions specific to the case*
is equal to
*the ratio of incidence of **R** among those exposed to **M** under similar back-ground conditions to the incidence of **R** among those not so exposed*
times
*the ratio of the probability that if **M** caused **R** in this case its details would resemble what was observed to the probability that those case details would have been observed if **M** did not cause **R** in this case.*

Consider, for example, a worker with no previous history of asthma who begins training as a stainless steel welder. He develops what is confirmed by all tests to be asthma. A review of the literature reveals a recognizable incidence of IgE-mediated asthma following sensitization to the chromium and nickel components of stainless steel welding rod. Based on the background information and the conditions known to date, the probability of the occupational welding being the cause of the asthma was estimated to be 80%, i.e., prior odds of 4:1. Further evaluation of the case revealed that 2 years after being removed from the work environment, the worker's asthma progressed; dechallenge had no effect on the course of the disease. There was no evidence of any other source of a similar exposure. It was determined that there was only a 10% probability, i.e., a likelihood ratio of 1:9, that lack of dechallenge effect would be observed if the causal hypothesis were true. Consequently,

$$PtO = PrO \cdot LR = \frac{4}{1} \cdot \frac{1}{9} = \frac{4}{9}$$

That is, the probability of the occupational welding being the cause of the asthma has fallen from 80 to 31% (4:9).

A decomposition of the analysis reveals the intricacies and the information required to make a causality assessment in an individual case. It is obviously not adequate to conclude that because **M** can cause **R** under some circumstance or another, and a subject experienced **R** following exposure to **M**, a causal inference is justified. A careful examination of what is required for the prior odds discloses the importance of knowing the details of the background conditions under which **M** has been observed to cause **R**, and, as well, the conditions under which **R**-like responses have been observed. This raises a causality assessment error that is not all that uncommon. A patient develops **R** after exposure to **M**, a known cause of **R**. All known alternatives have been eliminated, leaving **M** as a singular cause, and it is considered as such.

The error often made in such situations is not to consider *idiopathic* as a viable alternative. When a clinical manifestation is commonly of idiopathic origin, that fact can and should have as much of an impact on the denominator of the prior odds estimate as any identifiable etiological factor.

Returning to the previous example, the further evaluation of evidence implied < 1:1 odds that the welding exposure was the cause of the asthma. However, the presumption was made here that the asthma was atopic. If, on the other hand, the persistent condition was actually a reflection of reactive airways dysfunction syndrome (RADS), then persistence would not be as unlikely. On the other hand, it would be necessary to evaluate exposure data (actual or surrogate) in order to determine the physical and/or chemical burden. This creates a new likelihood ratio to be tested. It can be seen that there can be many likelihood ratios encompassing many forms of evidence. *The overall likelihood ratio is the product of the individual ratios.* That is,

$$PtO = PrO \ o \ LR_1 \ o \ LR_2 \ o \ ... \ LR_n$$

where **n** is the number of evidence categories tested. A careful review of the likelihood ratio components, such as patient occupational, family, social, and medical histories and exposure data—including the accuracy of exposure assessment, information on exposure to other causes of **R**, etc.—will show the strong influence that case-specific data should have on the causality assessment.

Obtaining quantitative information for estimates of posterior odds and likelihood ratios can and usually does require an extensive effort at obtaining relevant information, and it can involve a good deal of speculation on how to use that information. That, in turn, can make the output speculative. That does not detract from the utility of the approach, as long, of course, as the final estimate is kept in appropriate perspective. As an alternative, or in the absence of adequate data, an algorithmic approach can be useful. As will be seen, the proper algorithm can be philosophically similar to Bayesian analysis.

## CAUSAL INFERENCE BY THE USE OF ALGORITHMS

The use of algorithms in evaluating scientific data evokes mixed views. In the course of scientific endeavor, and the proffering of universal statements, algorithms have little knowledge value. But, as discussed above, causality decisions are very often made under conditions where vigorous and repeated hypothesis testing is not practical. These conditions comprise the realm of uncertainty judgments and represent the major circumstances under which decisions are made.

Algorithms provide a tool for making such decisions. They are the analogues of the probability estimations made by Bayesian analysis. In fact, a good algorithm will have identifiable Bayesian components and could, in fact, be used to organize information ultimately used for a Bayesian analysis. Such is the situation for the algorithm developed by Kramer et al.[11] for adverse drug reactions. This algorithm is extremely detailed and takes into account much of the information that would go into the estimation of prior odds and the multitude of components comprising the likelihood ratio in a Bayesian analysis. The algorithm is readily adaptable for use with toxic exposures, in general. Such an adaptation, which is quite true to the original, is shown in Appendix A. The "scoring system" is intended to provide some form of relative value to particular pieces of information. The same group[12] also classified the scores into ordinal categories. Our adaptation of those categories is as follows:

| | |
|---|---|
| +6 | DEFINITE |
| +4, +5 | PROBABLE |

0 TO +3          POSSIBLE
<0              UNLIKELY

These categories are useful in that PROBABLE can be interpreted as a >50% probability of causation, and DEFINITE can be considered to represent a high enough probability to justify actions with high competitive risks. The UNLIKELY category is the operative equivalent of NEGATIVE.

## CAVEAT EMPTOR

No numerical or algorithmic method used to assist causal inference, regardless of how sophisticated, has any value in the absence of the evaluator's own critical, analytical judgment.

## "WILL IT?": THE ISSUE OF A PRIORI CAUSAL INFERENCE

Causality is generally considered from the a posteriori viewpoint. From the discussion at the outset, however, it should be apparent that causality can be considered from an a priori viewpoint. That is, given a set of initial conditions, will those conditions cause an effect in an individual or group of individuals? More precisely, what is the probability of a specific response? There are two readily apparent applications of quantitative risk assessment (**QRA**), i.e., applicable to populations and applicable to an individual. On the surface, these appear identical. If the probability of a specific response to the initial conditions for a population is 1:1,000,000, then it should follow that the probability of the response in a particular individual is also 1:1,000,000. However, the information required and how it is used can be quite different. **QRA** applied to populations is generally a regulatory activity pursued in the interest of a prudent public health policy. The applicable data are the following: **M** *can* cause **R** (the universal statement) and the available background data, i.e., dose–response and demographic considerations. With such information, assuming its appropriateness, of course, an a priori probability of **M** causing **R** can be estimated. *Presumably,* individual (case-specific) information that would raise or lower the population probability estimate would null to within any particular unimodal distribution, and therefore, a prediction of overall probable causes of response in the population can be made. However, assessing risk for a particular individual, i.e., a factory worker within the occupational setting, requires the utilization of data on the individual characteristics of the subject. Thus, an individual with a history of RADS might be considered at higher risk from an exposure level of a demonstrable airway irritant than the group from which the a priori population estimate was made. These characteristics must then be used to adjust the a priori population estimate to tailor an assessment for the individual. In this respect, an a priori assessment of risk probability in an individual can be made by a process similar to that used for a posteriori causal inference ("Did it?") discussed in the previous section. Accordingly, the balance of this section will concern the matter of **QRA** applied to the population. It is a matter of significant scientific and socio-political controversy.

Clearly, based on both the definitions presented at the outset and on common sense, the universal statement of causality ("Can it?") must be answered in the affirmative before the "Will it?" question can be addressed. Regulatory standards, which define allowable concentrations of **M** in such media as outdoor air, food, or water, or in occupational environments, are based, in part, on risk assessments. Regulatory agencies routinely develop risk assessments based on a presumption that the answer to the ''Can it?'' question is affirmative. This presumption is generally based on worst-case scenarios or prudent public health policy considerations. This usually means the qualitative and quantitative extrapolation from animal data. Only in rare circumstances is it based on epidemiological evidence. Because of the worst-case (or ''weight-of-evidence'') basis of prudent public health policy (''Why take chances?'') risk assessments, regulatory standards cannot necessarily be used in individual causal inferences.

For noncarcinogenic effects, the regulatory standard establishes a reference dose (**RfD**), based upon application of a safety factor to the threshold dose established in experimental or human studies. If the threshold is based upon the results of animal studies, the threshold for the most sensitive animal species is used, and a safety factor of 100 to 1000 is applied, depending upon the quality and/or quantity of data. If the threshold is based upon data in humans, then a smaller safety factor, usually 10, is applied, since the uncertainties of extrapolation from animals to humans have been eliminated. In this case, the safety factor is intended to account for members of the population who are more sensitive to the adverse effects of **M**. By definition, if the exposure at issue results in estimated daily doses of **M** that are below the **RfD**, the risk of noncarcinogenic health effects is minimal. However, the converse is not necessarily true. Because of the conservative assumptions used in deriving the **RfD**, chronic daily doses exceeding the **RfD** will not necessarily be associated with adverse health effects. Assessing the risks of such exposures must take into the account the dose and duration of **M**, the mechanism of action of **M**, and the qualitative and quantitative plausibility, based on prior experience with **M**, that **R** will occur.

Regulatory standards for carcinogenic effects generally involve extrapolations based on rodent bioassays and incorporate an assumption of no threshold. This necessitates a different approach in the risk assessment. Dose–response data from chronic rodent bioassays, usually involving a response that reflects a minimum of a 1:10 increase in cancer risk, are fit into a mathematical model, in order to extrapolate to doses where the predicted response is on the order of an increased lifetime cancer risk of 1:100,000 or 1:1,000,000. A number of mathematical models for this extrapolation have been used; the current mathematical model of choice is the linearized multistage model. Essentially, the bioassay data are fit to a multiexponential equation, which has the following form:

$$P(d) = 1 - \exp[-(q_0 + q_1 d + q_2 d^2 + ... + q_k/d^k)]$$

where **P(d)** is cancer risk at dose, **d**, (and **P(0)** is the background risk), **q** is the exponential rate function reflecting the slope of the section of the dose-risk curve, and $q_i > 0$ and $i = 0, 1, ... k$.

Low dose extrapolation is done through $q_1$. Specifically, the 95% upper confidence limit on $q_1$, designated $q_1*$, is used to assess the 95% upper confidence limit on excess risk and the 95% lower confidence limit on the dose producing that excess risk. This is an estimate that effectively provides the highest risk-to-dose relationship. Theoretically, the dose-specific excess cancer, **A(d)**, is equal to $[P(d) - P(0)]/[1 - P(0)]$. However, *at low doses*, $A(d) \approx q_1* \times d$, and this equation can be used to estimate risk presented by any particular dose. The calculated risks are generally well below the 1-in-3 to 1-in-4 risk of cancer that applies to the population as a whole, independent of any specific exposure. However, it must be recognized that, because the risk assessment process is based on highly conservative and, in some cases, poorly supported assumptions, reliable estimates of *real* human risk cannot be assured by current carcinogenicity risk assessment methods. What this method can be and is used for is to provide reference values for the establishment and consistency of regulatory guidelines.

## REFERENCES

1. **Popper, K. R.,** *The Logic of Scientific Discovery,* English Translation, Basic Books, New York, 1975.
2. **Lane, D. A.,** A probabilist's view of causality assessment, *Drug Inf. J.,* 18, 323, 1984.
3. **Marshall, E.,** A is for apple, Alar, and ... alarmist? *Science,* 254, 20, 1991.
4. **Bunge, M.,** *Causality in Modern Science,* 3rd ed., Dover Publications, New York, 1979.
5. **Rothman, K. J.,** *Modern Epidemiology,* Little, Brown, Boston, 1986, 16.
6. **Evans, A. S.,** Causation and disease: the Henle-Koch postulates revisited, *Yale J. Biol. Med.,* 49, 175, 1976.
7. **Buck, C.,** Popper's philosophy for epidemiologists, *Int. J. Epidemiol.,* 4, 159, 1975.
8. **Hogan, M. D. and Hoel, D. G.,** Extrapolation to man, in *Principles and Methods of Toxicology,* 2nd ed., Hayes, A. W., Ed., Raven Press, New York, 1989, 879.
9. **Newberne, P. M., Suphakarn, V., Punyarit, P., and de Camargo, J.,** Nongenotoxic mouse liver carcinogens, in *Nongenotoxic Mechanisms in Carcinogenesis,* Banbury Report 25, Butterworth, B. E. and Slaga, T. J., Eds., Cold Spring Harbor Laboratory, Cold Spring Harbor, ME, 1987, 165.
10. **Lane, D. A.,** The Bayesian approach to causality assessment: an introduction, *Drug Inf. J.,* 20, 455, 1986.
11. **Kramer, M. S., Leventhal, J. M., Hutchinson, T. A., and Feinstein, A. R.,** An algorithm for the operational assessment of adverse drug reactions: I. background, description, and instructions for use, *JAMA,* 242, 623, 1979.
12. **Hutchinson, T. A., Leventhal, J. M., Kramer, M. S., Karch, F. E., Lipman, A. G., and Feinstein, A. R.,** An algorithm for the operational assessment of adverse drug reactions: II. demonstration of reproducibility and validity, *JAMA,* 242, 633, 1979.

# APPENDIX A

# CAUSATION ALGORITHM

146

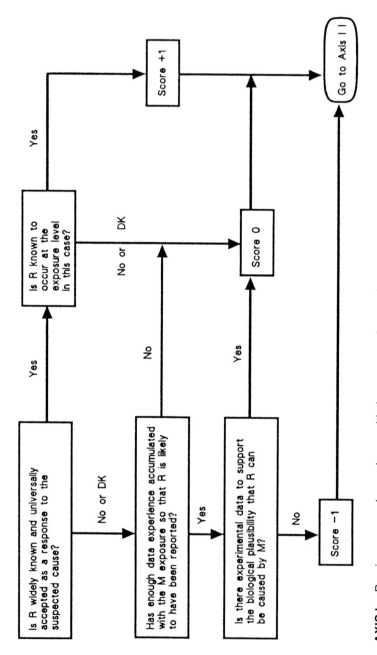

**AXIS I.** Previous general experience with the suspected causative exposure (**M**).

147

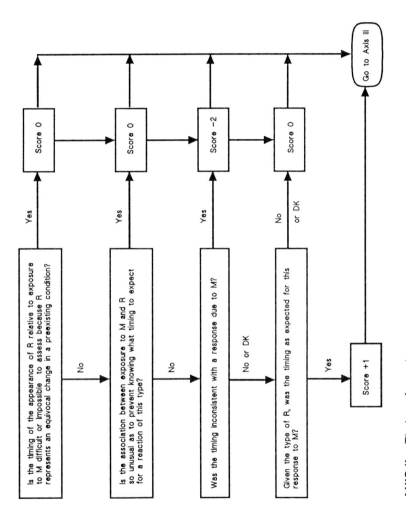

**AXIS II.** Timing of events.

148

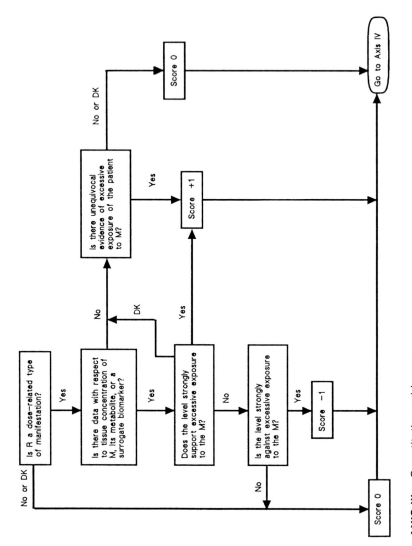

**AXIS III.** Quantitative evidence.

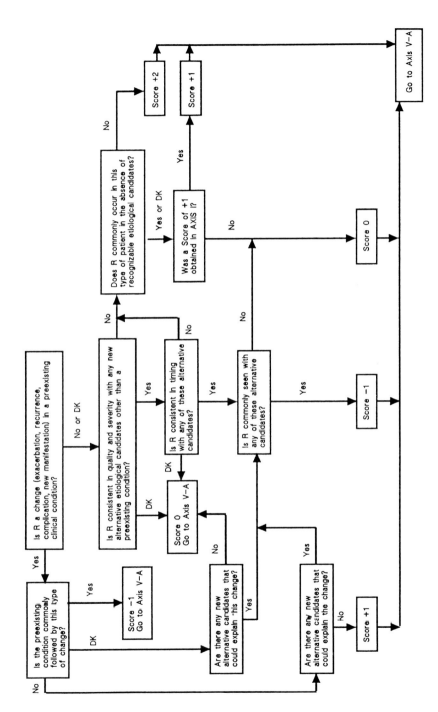

**AXIS IV.** Alternative etiologic candidates.

150

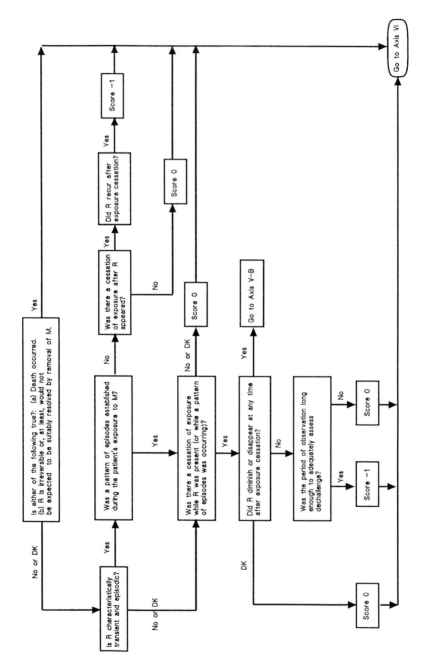

**AXIS V–A.** Initial dechallenge assessment.

151

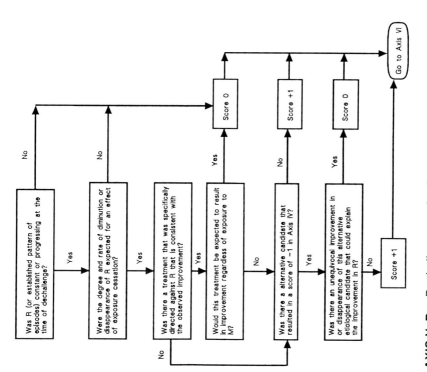

**AXIS V–B.** Dechallenge evaluation.

152

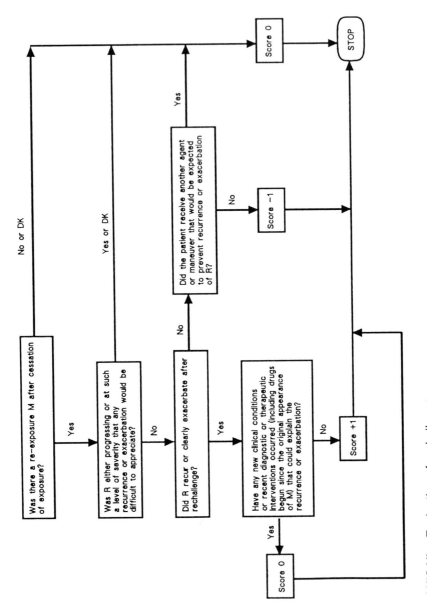

**AXIS VI.** Evaluation of rechallenge.

# Chapter 6

# Clinical Evaluation of the Individual Patient: Diagnosis and Differential Diagnosis

*Philip Witorsch, M.D. and Sorell L. Schwartz, Ph.D.*

## CONTENTS

## INTRODUCTION

The clinical evaluation of the patient who might have a disorder related to exposure to a toxic substance is not significantly different from that of a patient who might have such a disorder related to some other cause. Both situations require attention to detail and careful consideration of all reasonable differential diagnostic possibilities. It is indeed unfortunate that some physicians feel so intimidated by issues involving possible toxic exposure and chemical injury that when they address such issues in individual patients, they appear to suspend critical reasoning and sound clinical judgment. Instead, they either jump to unwarranted diagnostic conclusions and inappropriate causal inferences or, alternatively, throw up their hands in frustration and defeat and fail to address the problem at all. Both of these undesirable extremes can be avoided.

There are two types of data that are utilized in causal inference as applied to possible toxicologic disorders in individuals, clinical data and exposure data (Table 1). Both are necessary for causal inference as applied to individuals in such disorders, much the same as both clinical data and microbiological data are utilized for causal inference in possible infectious diseases. Exposure data and the utilization of both types of data in causal inference are discussed in detail elsewhere in this book. In this chapter, we will discuss various aspects of the clinical data.

The sources of the clinical data in possible toxicologic disorders are basically the same as those in nontoxicologic disorders: history, both as obtained from the patient and as may be obtained from review of medical records and other sources, physical examination, and pertinent laboratory studies. These are discussed in detail below.

The question of a disorder that might be related to exposure to an inhaled toxin can arise in one of two contexts. The first is that of the patient who presents to the physician with symptoms and/or clinical findings but who may not volunteer (or at least not emphasize) a history of possible toxicologic exposure. In such a circumstance, the most likely inappropriate outcome will be the underdiagnosis of a toxicologic

**Table 1  Types of Data Utilized in Causal Inference as Applied to a Possible Toxicologic Disorder in an Individual**

I. Clinical data
   1. Complete medical history obtained from patient
   2. Detailed toxicologic (occupational and environmental) history obtained from patient
   3. Review of prior medical records
   4. Information from family members and co-workers
   5. Complete physical examination
   6. Routine laboratory studies
   7. Special diagnostic studies

II. Exposure data
   1. Concentration/dose of exposure
   2. Duration of exposure
   3. Route(s) of exposure
   4. Other conditions of exposure
   5. Any modifying factors

disorder, i.e., the failure to make such a diagnosis due to failure of the physician to consider such a possibility, when it is, in fact, the basis of the patient's problem. It is thus necessary to maintain a high index of suspicion for toxicologic disorders and to obtain a thorough and appropriate occupational and environmental history, in order to avoid this significant error of omission.

The second type of presentation is that of the individual who has had (or believes he or she has had) an exposure to a potentially toxic substance and who has health complaints that he or she attributes to that exposure. Sometimes (or alternatively), there may also be concerns about possible subclinical disease and/or future disease related to the exposure. This situation is commonly encountered in relation to toxic tort litigation and workers' compensation claims. In such cases, the most likely inappropriate outcome results from assuming a causal relationship between the alleged exposure and the patient's health complaints, without considering other reasonable differential diagnostic possibilities or even evaluating the plausibility of the assumed causal relationship from both an exposure and a clinical point of view. That is, the most likely error is one of commission, with overdiagnosis of toxicologic disorders. Needless to say, it is essential that this extreme be avoided as well.

In determining whether or not a symptom or a physical, physiological, biochemical, or histopathological change in an individual is related to the exposure of that individual to a chemical or physical agent, mere temporal eligibility (i.e., the occurrence of the symptom or change after the exposure) is not adequate for causal inference. There are a relatively limited number of ways in which the body responds to various insults, whether these are the result of an intrinsic disease or an extrinsic agent. Thus, diverse intrinsic disease states and extrinsic agents can result in similar, and at times virtually identical, pathophysiological events, often resulting in similar or even identical symptoms, as well as physical and laboratory findings.

It is important to keep in mind that many symptoms are, consequently, nonspecific in character and compatible with a multitude of medical conditions and diagnoses. In this regard, one only need consider the nonspecificity of such respiratory symptoms as cough, dyspnea, and chest pain, with their long list of associated differential diagnoses. In the case of an individual manifesting health complaints following exposure to a chemical or physical agent, possible causes of the health complaints include (1) the agent under suspicion, (2) other chemical or physical agents, (3) a combination of chemical and/or physical agents, (4) adverse reactions to therapeutic drugs, (5) intrinsic disease states, (6) infectious agents, (7) traumatic injuries, and (8) psychogenic disorders (Table 2). The evaluation of these possible causes involves the process of differential diagnosis, by which the various alternative possible causes of the patient's symptoms are considered and evaluated. This differential diagnostic process is not substantially different in situations involving possible toxic exposure than it is in circumstances where there may be no toxicologic considerations. It includes history obtained from the patient, available medical records, and other sources, complete and thorough physical examination, appropriate routine laboratory testing, and in some cases, special diagnostic procedures (Table 1). Such clinical evaluation provides a clinical database, which, together with the corresponding exposure database obtained by exposure assessment, serves as the basis for causal analysis.

## HISTORY OBTAINED FROM THE PATIENT, MEDICAL RECORDS, AND OTHER SOURCES

In obtaining clinical data relevant to causal analysis in an individual with symptoms that might be related to exposure to a chemical or physical agent, we usually begin with the history obtained from the patient, much as we do in any other clinical evaluation. Where there is a possibility of toxicologic injury, the history must include not only the usual chief complaint, present illness, review of systems, past history, habits, social history, family history, childhood illnesses, operations, allergies, and medications taken, but a detailed and complete toxicologic (i.e., occupational and environmental) history as well (Table 1). A number of standard forms that can be filled out by or administered to the patient have been used to obtain such toxicologic information, and their use may have advantages with respect to efficiency and reducing the possibility of overlooking areas of relevance.[1] Nevertheless, it must be emphasized that such forms are only a start, and it is necessary to use the information obtained from them as points of departure for asking the patient pertinent follow-up questions, rather than considering the information obtained on such forms as definitive and complete. In some circumstances, administration of a toxicologic history questionnaire and follow-up questioning by the examiner may require as long as 2 to 3 hours to obtain a complete toxicologic profile on the patient, but it is time well spent. In many situations, it is appropriate and, in fact, preferable to have the toxicologic history obtained by an experienced and knowledgeable toxicologist, who, from the points of view of education, training, background, experience, and expertise, is often the most qualified examiner to most effectively pursue relevant follow-up questioning with regard to potential toxicologic exposure and injury.

A thorough history facilitates full elucidation of all of the symptoms that the patient may have, not just those that prompted the evaluation. While the patient may present with a chief complaint of respiratory symptoms, such as cough, shortness of breath, wheezing, chest pain, etc., with or without associated constitutional symptoms, such as fever, weight loss, malaise, etc., these are relatively nonspecific complaints and are consistent with a wide range of differential diagnostic possibilities. In many cases it may, in fact, be the nonrespiratory complaints that suggest a toxicologic source for a patient's respiratory illness, where one might not otherwise be considered. In other instances, extrapulmonary complaints in a patient presenting with respiratory symptoms thought to be related to exposure to an inhaled toxicant may, in fact, suggest a more likely alternative diagnosis for the patient's respiratory symptoms than the alleged toxic exposure. In the special but increasingly common case of an individual seeking compensation by means of litigation, where the official

Table 2 **Possible Causes of Health Complaints Following
Exposure to a Chemical or Physical Agent**

1. The agent under suspicion
2. Other chemical or physical agents
3. A combination of chemical and/or physical agents
4. Adverse reaction to therapeutic drugs
5. Intrinsic disease states
6. Infectious agents
7. Traumatic injuries
8. Psychogenic disorders

"complaint" may have been drafted by the patient's attorney, a detailed history often serves to verify whether or not the patient, in fact, has the symptoms claimed, as opposed to their being the product of the attorney's imagination or misunderstanding.

A very important part of the history is not obtained from the patient but, rather, from prior medical records. Review of such records may reveal information that supports or refutes assumed temporal relationships between exposure and onset of symptoms, as well as provide data relevant to alternative diagnostic possibilities. In some instances, where the history obtained from the patient has failed to disclose a potential toxic exposure that may be relevant to the patient's symptoms, such information may be found in prior medical records. It is important that complete medical records, not just summaries or abstracts, be obtained and reviewed, lest potentially important information be overlooked. Similarly, when pulmonary function studies, X-rays, and other diagnostic studies have been done in the past, it is important to review the actual data whenever possible, rather than rely on someone else's interpretation of those data. In the case of pulmonary function studies, actual tracings and raw data should be obtained and all "runs" of any test should be reviewed, not only of those selected to be included in the final report. In the case of radiologic studies, such as X-rays, CT scans, and nuclear imaging, the original radiographs, etc., are usually of better technical quality and allow for more accurate evaluation of detail than copies.

Another potentially useful source of historical information that should not be overlooked is that of individuals other than the patient, including family members and co-workers. Information obtained from such observers may sometimes more accurately portray both the patient's symptoms and the exposure conditions than such information obtained from the patient. For example, a spouse may note snoring and episodes of apnea during sleep not noted by the patient. A co-worker may report such observations as odors, as well as details regarding the circumstances of exposure, that might not have been noted by the patient. Also, the presence or absence of symptoms among family members or co-workers similarly exposed may serve to provide more or less support for a possible relationship to the alleged exposure (keeping in mind, of course, that there are often wide individual variations in thresholds of reactivity to various chemical and physical agents and that Mass Psychogenic Illness is a well-known phenomenon associated with purported toxic exposures).[2,3]

It cannot be emphasized too strongly that it is essential to fully document information obtained during the history. This includes both "positives" and "significant negatives." Such thorough documentation is necessary for later review relative to refining a differential diagnosis and relative to problems that may come up in the future, as it is in all other medical situations. In addition, in the case of possible exposure- related disorders that may be seen in the context of, or at some future date become the subject of litigation and/or compensation proceedings, it is critical to its

**Table 3  Evaluation of a Patient Who May Have a Respiratory Disorder Related to a Toxic Exposure: Routine Laboratory Studies**

1. Complete blood count
2. Blood chemistry panel
3. Urine analysis
4. Stool occult blood
5. PA and lateral chest roentgenogram
6. 12-lead electrocardiogram
7. Pulmonary function studies
    a. Spirometry before and after aerosolized bronchodilator
    b. Lung volumes
    c. Diffusing capacity (DLCO)
    d. Resting arterial blood gases

credibility that such information be contemporaneously documented in the patient's medical record.

## PHYSICAL EXAMINATION

Careful physical examination is, of course, essential to diagnosis and to differential diagnosis in any situation, and this is no less the case when dealing with disorders possibly related to toxic exposure than it is in any other clinical circumstance.[1,4,5] The physical examination may provide objective correlates and validation of the patient's subjective complaints. In addition, physical examination may lead to the detection of previously unconsidered conditions, which could account for the patient's symptoms, or it may result in the exclusion of certain other possibilities that might otherwise have been considered. It is important to emphasize that, even though the patient may have complaints that appear to be limited to the respiratory system and the possible toxicant may be an inhaled air pollutant, it is mandatory that the physical examination be thorough and not limited to examination of the respiratory and related systems. While this usually should be the rule in any patient with respiratory complaints, even when there is no issue of possible toxicologic exposure, it is especially important in instances of possible toxic injury. Inhaled pollutants may not only injure the respiratory system, but they may be absorbed through the respiratory tract into the systemic circulation and cause injury involving a variety of other organ systems as well. For example, inhalation of arsenic may affect the skin, mucous membranes, and nervous system, in addition to the respiratory tract, while inhalation of cadmium may cause damage to the liver and kidneys, as well as to the lungs.[6,7] Alternatively, physical findings involving other than the respiratory system may suggest a condition other than the alleged exposure that may account for the patient's symptoms, such as the skin lesions of systemic lupus erythematosus, the joint changes of rheumatoid arthritis, or the ocular and nailbed manifestations of infective endocarditis.[8-10]

## LABORATORY STUDIES

Laboratory studies can be divided into ''routine'' laboratory studies and ''special'' diagnostic studies (Table 1). Certain laboratory studies are worth doing routinely as part of the initial evaluation of a patient who may have a disorder related to a toxic exposure, because they represent a reasonably cost-effective contribution to resolution of the differential diagnosis (Table 3). These are generally the same studies that are often done as a part of a standard, periodic medical evaluation and include the following:

1. complete blood count, including a differential white blood cell count and erythrocyte sedimentation rate;
2. standard, automated blood chemistry panel, which usually includes glucose, blood urea nitrogen (BUN), transaminases, lactic dehydrogenase (LDH), alkaline phosphatase, bilirubin, cholesterol, triglycerides, total protein, albumin, calcium, phosphorous, electrolytes, and, sometimes, creatinine;
3. urine analysis, including microscopic examination of urine sediment;
4. stool occult blood determination.[11,12]

These routine laboratory studies may shed light on possible perturbations of various organ systems that may have occurred as a result of the alleged exposure, as well as on various intrinsic disease states or other conditions that may alternatively account for the patient's complaints.

A patient who presents with respiratory complaints or who may have been exposed to an inhaled toxin should, additionally, routinely undergo roentgenologic examination of the chest, including PA and lateral views, as well as pulmonary function studies, including the following: (a) spirometry, both before and after aerosolized bronchodilator; (b) determination of lung volumes; (c) diffusing capacity (DLCO); and (d) resting arterial blood gases (Table 3). In addition, because of the propensity of cardiovascular disorders to present with respiratory complaints, such as shortness of breath and cough, and because pulmonary disorders may secondarily compromise the heart, a 12-lead, standard electrocardiogram should be considered routine in such patients as well.

Pulmonary function studies have an important role in differential diagnosis, prognosis, evaluation of impairment and disability, and assessment of the benefit of therapeutic interventions (Table 4).[13–15] In some cases, abnormalities on pulmonary function testing may be the only findings indicating a respiratory disorder in a patient who may have been exposed to an inhaled toxin but who has normal findings on both clinical and X-ray examination. Thus, for example, a reduced DLCO may be the only significant finding in a patient with early interstitial lung disease secondary to a pneumoconiosis or chronic hypersensitivity pneumonitis, while persistent airflow obstruction may be the only finding in an individual chronically exposed to inhaled cotton dust or who has asthma or bronchitis related to an inhaled irritant or allergen.[16–26] In other cases, the pattern of abnormality, that is, whether the patient has a restrictive, obstructive, or mixed disorder, may suggest certain conditions and make others unlikely.[14,16] For example, in a patient with hypersensitivity pneumonitis, one would expect to find a restrictive impairment, a reduced DLCO, no significant airflow obstruction and, in some patients, arterial hypoxemia without carbon dioxide retention.[20–22] In an individual with occupational or environmental asthma, one would expect to find airflow obstruction when the patient was symptomatic, evidence of significant reversibility of the obstruction, and variation in airflow, either spontaneously or as a result of therapy.[23–26] On the other hand would be a person with occupational exposure to coal dust who also smoked two packs of cigarettes per day for 50 years and who had symptoms of chronic cough and shortness of breath, no greater than category 1 changes on chest X-ray, and a marked degree of fixed airflow obstruction with hyperinflation, moderate reduction in DLCO, no restrictive impairment, and moderate hypoxemia associated with mild chronic carbon dioxide retention on pulmonary function testing. This patient is more likely to have chronic obstructive

**Table 4    Role of Pulmonary Function Testing in a Patient Who May Have a Respiratory Disorder Related to a Toxic Exposure**

1. Pulmonary function abnormalities may be the only ''positive'' findings indicative of a respiratory disorder.
2. Pattern of abnormality may suggest certain conditions and make other conditions unlikely.
3. Testing permits quantification of degree of impairment.
4. Testing may show possible indication for institution of and basis for evaluating response to therapy.

pulmonary disease (COPD) related to cigarette smoking than a pneumoconiosis secondary to occupational coal dust exposure.[13–17]

Routine pulmonary function studies also permit quantification of the degree of respiratory impairment.[13,15] This has important implications with regard to prognosis and disability evaluation. In chronic conditions, such as interstitial fibrosis or chronic airflow obstruction, there is good correlation between degree of abnormality on pulmonary function testing, extent of disease demonstrated by other methods, and prognosis.[13] In asthma, on the other hand, where marked variations in airflow are typical, the quantitative level of airflow obstruction on any one occasion or even on multiple occasions is, at best, of very limited prognostic value.[25] Similarly, while there are reasonable published criteria that allow for determination of degree of impairment and disability in chronic interstitial and fixed airflow obstructive disorders,[27] the situation is much more problematic in the case of asthma, with its associated variable airflow obstruction.[25]

Finally, the results of pulmonary function testing may serve both as an indication for the institution of therapy and as a basis upon which to evaluate the response to treatment.[13] Especially in those situations where therapy includes the use of drugs such as corticosteroids and/or immunosuppressive agents, which are often associated with significant side effects, it is important that both the indications for treatment and the response to treatment be supported by objective pulmonary function test data.

## SPECIAL DIAGNOSTIC TESTS

In addition to the foregoing fairly standard, routine laboratory studies, further special diagnostic testing may be indicated in selected instances (Table 5). Such special studies may include the following:

1. immunologic studies, including skin test and serologic studies;
2. tests relevant to the diagnosis of collagen-vascular disorders, including such parameters as rheumatoid factor and antinuclear antibodies (where such disorders are appropriately considered in the differential diagnosis);[28]
3. tests of endocrine function, including thyroid and/or adrenal function studies in patients with nonspecific constitutional complaints, such as fatigue, weight loss, etc., that could reasonably relate to such conditions;[29,30]
4. liver function studies, where there is reason to suspect liver disease as a cause of some of the patient's symptoms or, alternatively, where the toxin that the patient inhaled has the propensity to cause liver damage, e.g. carbon tetrachloride or arsenic;[6,31,32]
5. appropriate microbiological studies, including cultures and serologic studies, where disease caused by an infectious agent is a reasonable possibility;
6. toxicologic studies of blood and urine, where appropriate;

**Table 5   Evaluation of a Patient Who May Have a Respiratory Disorder Related to a Toxic Exposure: Special Diagnostic Tests**

1.  Immunologic studies (skin tests and serology)
2.  Tests relevant to "collagen-vascular" disorders
3.  Endocrine function tests
4.  Liver function studies
5.  Microbiological studies
6.  Toxicologic studies (blood, urine, etc.)
7.  Examination for parasites
8.  Neurological diagnostic studies
9.  Psychological testing
10. Specialized cardiovascular evaluative procedures
11. Special pulmonary imaging techniques
    a.  Computed tomography (CT)
    b.  High resolution CT (HRCT)
    c.  Gallium 67 scintigraphy
12. Special pulmonary function studies
    a.  Nonspecific bronchoprovocation testing (methacholine, histamine, cold air, etc.)
    b.  Specific inhalation challenge testing (antigens, other)
    c.  Cardiopulmonary exercise testing
13. Invasive diagnostic techniques
    a.  only infrequently indicated

---

7.  examination of stool, blood, urine, and other body fluids for parasites that could be the cause of the patient's complaints;
8.  neurological diagnostic studies relevant to both the central and peripheral nervous systems, where some of the patient's complaints may have a neuropathic basis or where the suspect toxicant has the potential for causing neurologic damage (e.g., arsenic);
9.  psychological testing, where there is reason to suspect a psychogenic basis for the patient's complaints.

In patients with respiratory complaints, such as dyspnea or chest pain, that remain obscure in origin, specialized cardiac evaluative procedures, including cardiac stress testing, radioisotopic techniques, and even cardiac catheterization and angiography may be appropriate. Also, in some patients, special imaging and/or pulmonary functional diagnostic tests may be indicated, as discussed below. In a few patients, an invasive procedure, such as bronchoscopy, transbronchial lung biopsy, or even open lung biopsy may be indicated.

## SPECIAL PULMONARY IMAGING TECHNIQUES

Special pulmonary imaging techniques that may be used in addition to and in conjunction with standard chest radiography include conventional computed tomography (CT), high resolution (thin cut) computed tomography (HRCT), and gallium 67 scintigraphy (Table 5).[33–36]

While conventional CT has increased sensitivity for the detection and characterization of interstitial lung disease compared to the plain chest radiograph, HRCT is even more sensitive and has been shown to correlate well with pathologic/histologic evaluations relative to the presence or absence of interstitial disease (although not for specific histopathology). HRCT is most productively used in combination with conventional CT. HRCT is especially useful for the detection of interstitial lung disease not readily apparent on the plain chest roentgenogram and may even permit better evaluation of the particular interstitial pattern, e.g., nodular vs. linear-reticular. This ability to detect and even characterize otherwise "subradiographic" interstitial disease is especially useful in individuals who may have early pneumoconioses, subacute or chronic hypersensitivity pneumonitis, and other interstitial conditions possibly related to inhaled agents. CT scanning (although not necessarily HRCT) also has a major role with regard to diagnosis and staging of tumors of the lung, which may be among the delayed consequences of chronic exposure to certain air pollutants, as well as in detection and characterization of hilar and/or mediastinal lymphadenopathy, which may be of significance relative to differential diagnosis.[33–36] For example, the presence of hilar adenopathy would be more suggestive of sarcoidosis than hypersensitivity pneumonitis in a patient in whom these two conditions were being considered.[20–22,37]

There is little to no experience with or indication for the application of magnetic resonance imaging (MRI) to the clinical evaluation of respiratory disorders possibly related to inhalation of toxic air pollutants, except possibly with respect to differential diagnosis relative to other possible causes of the patient's symptoms. MRI of the chest is most useful relative to vascular structures, which are usually not affected by these disorders, and is of little value for assessment of the lung parenchyma or airways, which are the usual targets of inhaled toxicants. MRI may also be useful for imaging the mediastinum, especially in patients in whom CT with contrast cannot be done due to intolerance for the contrast agent.[38,39]

Gallium 67 scintigraphy is useful for the diagnosis of inflammatory conditions of the lung, such as *Pneumocystis carinii* pneumonia, and the staging of inflammation in such interstitial lung diseases as sarcoidosis and idiopathic interstitial fibrosis. However, thus far, its role in diagnosis, staging, or management of pulmonary disorders related to exposure to inhaled toxicants appears to be quite limited. A few studies suggest the possibility that gallium 67 scanning may have a role in the early detection of such interstitial lung conditions as asbestosis, silicosis, and hypersensitivity pneumonitis, as well as in determining indications for and following response to treatment. However, its value in this regard remains to be confirmed by further studies.[33,40]

## SPECIAL PULMONARY FUNCTION STUDIES

The generally accepted criteria for the diagnosis of asthma include the presence of nonspecific bronchial hyperreactivity.[15,25,41] The demonstration of such reactivity to nonspecific bronchial inhalation challenge, i.e., bronchoprovocation testing using methacholine, histamine, or cold air, has an important role in the diagnosis, differential diagnosis, and management of occupational and environmental asthma (Table 5). While a "positive" nonspecific bronchial inhalation challenge test is not, in and of itself, diagnostic of asthma, a "negative" test is strong evidence against such a diagnosis. Furthermore, serial measurements of bronchial responsiveness to nonspecific bronchoprovocation testing has proven useful in documenting an association with a particular exposure circumstance (i.e., by demonstration of nonspecific bronchial hyperreactivity associated with exposure, improvement in such bronchial reactivity when away from exposure, and exacerbation of hyperreactivity when reexposed). In addition, the degree of bronchial reactivity to nonspecific bronchoprovocation testing (i.e., the quantitative result, such as the methacholine $PC_{20}$, the cumulative dose of methacholine resulting in a 20% fall in $FEV_1$) may be a useful endpoint with which to follow a patient's response to therapy, thus making repeated determinations of this parameter of value in the follow-up and management of patients with asthma, including individuals with occupational and environmental asthma.[23–25,41,42]

Specific (as opposed to nonspecific) inhalation challenge may also sometimes be useful in the diagnosis and differential diagnosis of respiratory disorders possibly related to the inhalation of toxic airborne pollutants (Table 5). Specific inhalation challenge with certain allergens and causative agents of occupational asthma has proven useful in the diagnosis of such conditions. However, such testing may be problematic due to the limited standardization of materials and techniques, as well as to potential risks to the patient. Thus, unless done in specialized centers that have the appropriate experience and use well-standardized materials and techniques, such testing should be undertaken only with great caution, if at all, and the results obtained interpreted with similar caution.[23–25,41,42]

Cardiopulmonary exercise testing is sometimes of use in the evaluation of a patient who may have a respiratory disorder possibly related to exposure to air pollutants (Table 5). Such testing may permit the exclusion of both pulmonary and cardiac limitations to exercise in patients in whom psychogenic factors may be primarily responsible for their symptoms. Exercise testing may also prove useful in differentiating between a cardiac and a pulmonary basis for unexplained dyspnea. Another important role for cardiopulmonary exercise testing relates to disability evaluation and determination (discussed in detail elsewhere in this book).[43–46]

## INVASIVE DIAGNOSTIC TECHNIQUES

Most cases of environmentally related respiratory disorders can be diagnosed using measures short of lung biopsy. In many instances, a diagnosis can be made with confidence based upon exposure information (e.g., occupational exposure to asbestos), consistent findings in the history and during physical examination (e.g., shortness of breath, nonproductive cough, and rales at the lung bases), consistent or even typical findings on chest radiography (with or without CT and/or HRCT), pulmonary function testing, and, most importantly, adequate exclusion of other reasonably plausible causes. It is, therefore, only infrequently necessary or even appropriate to obtain tissue, especially just to satisfy a legal standard of proof for a litigation or compensation proceeding.

On occasion, however, especially where the clinical, pulmonary functional, and radiologic pictures are nonspecific and relevant exposure information is questionable, resolution of a differential diagnosis and proper management of the patient may require sampling of lung tissue (Table 5). The particular technique used will depend, in part, on the specific differential diagnostic considerations that are raised by the clinical (including radiologic and laboratory) features of the case. Thus, for example, if sarcoidosis is a leading consideration (as would be the case if there were hilar adenopathy present), transbronchial lung biopsy or mediastinoscopy are high yield, relatively safe diagnostic procedures for this condition.[37] On the other hand, when there is a nonspecific picture of interstitial lung disease, without associated adenopathy and with nonspecific findings of restrictive impairment and reduced DLCO on pulmonary function testing, the differential diagnosis would include such conditions as idiopathic interstitial fibrosis, hypersensitivity pneumonitis, pneumoconiosis, etc. In such a case, the differential diagnosis often cannot be resolved by measures short of lung biopsy, and an open lung biopsy may be the procedure of choice, even though it is associated with a risk of morbidity and even mortality. Thoracoscopy (often with video enhancement) or a limited thoracotomy, while still requiring general anesthesia, may, in some cases, permit lung and/or lymph node biopsy productive of adequate tissue samples, while still avoiding some of the morbidity and mortality of full thoracotomy.[47]

## WHO SHOULD EVALUATE THE PATIENT?

The patient who has respiratory complaints that may reflect a disorder related to exposure to an inhaled toxicant often presents a difficult and at times vexing problem in differential diagnosis and cause-and-effect analysis. The situation is frequently complicated by a setting involving litigation, claims for compensation, the participation of attorneys and insurance adjusters, and issues of motivation and secondary gain. In addition, questions may develop regarding the validity and reliability of subjective complaints, and potential conflicts may arise between legal requirements and definitions of reasonable medical or scientific probability and what we often consider acceptable in the way of scientific medical proof. It is necessary, on the one hand, to consider all reasonable possible causes of the patient's complaints, including the alleged exposure. On the other hand, it is critical to avoid unfounded speculation and inappropriate causal inference. The physician who evaluates such a patient needs to have a broad background and experience in general medicine, together with significant depth of expertise and interest in both pulmonary disorders and environmental and occupational medical problems.

It is our strongly held opinion, based on more than two decades of experience with many hundred such patients, that the initial evaluation of such a patient, together with the overall management and coordination of any subsequent evaluation, should thus be in the hands of either a pulmonary internist with particular interest and expertise in environmental and occupational medical problems or an occupational medicine physician with particular interest and expertise in pulmonary disorders (Table 6). Not only must this physician have sufficient expertise in and experience with the relevant medical diagnostic issues, but he or she must also have a degree of sensitivity to the legal requirements that come into play relative to litigation and compensation proceedings, as well as the ability to remain totally objective and nonadvocative in what may be an otherwise heated adversarial situation.

In most, if not all, cases, in addition to medical evaluation by a pulmonologist/occupational-environmental medicine specialist, evaluation by a toxicologist is necessary and appropriate. This includes the obtaining of a detailed toxicologic and exposure history from the patient as well as evaluation of all available exposure-related data. As noted earlier, a knowledgeable and experienced toxicologist is usually the best qualified individual to obtain a meaningful toxicologic history and to evaluate the toxicologic data (Table 6).

Since psychogenic disorders may be plausible differential diagnostic possibilities in such situations, it is often necessary to have the patient evaluated by a psychiatrist, as well as to obtain psychological and/or neuropsychological testing by a clinical psychologist or neuropsychologist (Table 6). Whether or not consultation by specialists and/or subspecialists in other disciplines (e.g., dermatology, neurology, allergy-immunology, etc.) is indicated will depend upon what other complaints and/or other clinical and laboratory findings are noted or upon the potential effects on other organ systems that may be associated with the alleged exposure (Table 6). Thus, for example, a patient with a complaint of ''multiple chemical sensitivities and immune system dysfunction,'' allegedly resulting from exposure to a pesticide, might have symptoms that include inability to concentrate, mental confusion, headaches, irritability, pins and needles sensation and numbness in the hands and feet, cough, dyspnea, chest pain, itching of the skin, skin rash, and arthralgias, among others.

Table 6    **Who Should Evaluate the Patient Who May Have a Respiratory Disorder Related to a Toxic Exposure**

I. Basic evaluation
  1. Pulmonologist or occupational medicine specialist
     a. Basic medical evaluation and coordination of work-up
  2. Toxicologist
     a. Exposure evaluation
     b. Toxicologic history

II. Possible consultant evaluations
  1. Psychiatrist
     a. Frequently necessary
  2. Clinical psychologist or neuropsychologist
  3. Dermatologist in selected cases
  4. Neurologist in selected cases
  5. Allergist in selected cases
  6. Other consultants in selected cases

Such a patient would require evaluation initially by a pulmonologist/occupational-environmental medicine specialist and a toxicologist. Then, depending upon the results of the initial clinical and laboratory evaluation, the patient might also require evaluation by consultant(s) with expertise in one or more of the following: psychiatry, clinical or neuropsychology, allergy-immunology, neurology, cardiology, dermatology, and rheumatology.

# REFERENCES

1.  **Frank, A. L.,** The occupational and environmental history and examination, in *Environmental and Occupational Medicine,* 2nd ed., Rom, W. N. Ed., Little, Brown, Boston, 1992, 29.
2.  **Klaassen, C. D. and Eaton, D. L.,** Principles of toxicology, in *Casarett and Doull's Toxicology: The Basic Science of Poisons,* 4th ed., Amdur, M. O., Doull, J., and Klaassen, C. D., Eds., Pergamon Press, New York, 1991, 12.
3.  **Boxer, P. A., Singal, M., and Hartle, R. W.,** An epidemic of psychogenic illness in an electronics plant, *J. Occup. Med.,* 26, 381, 1984.
4.  **Rom, W. N.,** The discipline of environmental and occupational medicine, in *Environmental and Occupational Medicine,* 2nd ed., Rom, W. N., Ed., Little, Brown, Boston, 1992, 3.
5.  **Boehlicke, B. A. and Bernstein, R. S.,** Recognition and evaluation of occupational and environmental health problems, in *Environmental and Occupational Medicine,* 2nd ed., Rom, W. N., Ed., Little, Brown, Boston, 1992, 7.
6.  **Landrign, P. J.,** Arsenic, in *Environmental and Occupational Medicine,* 2nd ed., Rom, W. N., Ed., Little, Brown, Boston, 1992, 773.
7.  **Bhamra, R. K. and Costa, M.,** Trace elements—aluminum, arsenic, cadmium, mercury, and nickel, in *Environmental Toxicants: Human Exposures and Their Health Effects,* Lippman, M., Ed., Van Nostrand Reinhold, New York, 1992, 575.
8.  **Schur, P. H.,** Clinical features of SLE, in *Textbook of Rheumatology,* 4th ed., Kelley, W. N., Harris, E. D., Jr., Ruddy, S., and Sledge, C. B., Eds., W. B. Saunders, Philadelphia, 1993, 1017.
9.  **Harris, E. D., Jr.,** Clinical features of rheumatoid arthritis, in *Textbook of Rheumatology,* 4th ed., Kelley, W. N., Harris, E. D., Jr., Ruddy, S., and Sledge, C. B., Eds., W. B. Saunders, Philadelphia, 1993, 874.

10. **Scheld, W. M. and Sande, M. A.,** Endocarditis and intravascular infections, in *Principles and Practice of Infectious Diseases,* 3rd ed., Mandell, G. L., Douglas, R. G., Jr., and Bennett, J. E., Eds., Churchill Livingstone, New York, 1990, 670.

11. **Eddy, D. M.,** Ed., *Common Screening Tests,* American College of Physicians, Philadelphia, 1991.

12. **Hayward, R. S. A., Steinberg, E. P., Ford, D. E., Roizem, M. M., and Roach, K. W.,** Preventive care guidelines: 1991, *Ann. Intern. Med.,* 114, 758, 1991.

13. **Miller, A.,** Applications of pulmonary functions tests, in *Pulmonary Function Tests in Clinical and Occupational Lung Disease,* Miller, A., Ed., Grune and Stratton, Orlando, FL, 1986, 3.

14. **Miller, A.,** Patterns of impairment, in *Pulmonary Function Tests in Clinical and Occupational Lung Disease,* Miller, A., Ed., Grune and Stratton, Orlando, FL, 1986, 249.

15. **Bates, D. V.,** *Respiratory Function in Disease,* 3rd ed., W. B. Saunders, Philadelphia, 1989.

16. **Attfield, M. and Wagner, G.,** Respiratory disease in coal miners, in *Environmental and Occupational Medicine,* 2nd ed. Rom, W. N., Ed., Little, Brown, Boston, 1992, 325.

17. **Lapp, N. L. and Parker, J. E.,** Coal workers' pneumoconiosis, *Clin. Chest Med.,* 13, 243, 1992.

18. **Balaan, M. R. and Banks, D. E.,** Silicosis, in *Environmental and Occupational Medicine,* 2nd ed., Rom, W. N., Ed., Little, Brown, Boston, 1992, 344.

19. **Graham, W. G. B.,** Silicosis, *Clin. Chest Med.,* 13, 253, 1992.

20. **Fink, J. N.,** Hypersensitivity pneumonitis, in *Environmental and Occupational Medicine,* 2nd ed., Rom, W. N., Ed., Little, Brown, Boston, 1992, 367.

21. **Fink, J. N.,** Hypersensitivity pneumonitis, *Clin. Chest Med.,* 13, 303, 1992.

22. **Fink, J. N.,** Hypersensitivity pneumonitis, in *Allergy: Principles and Practice,* 4th ed., Middleton, E., Jr., Reed, C. E., Ellis, E. F., Adkinson, N. F., Jr., Yunginger, J. W., and Busse, W. W., Eds., C. V. Mosby, St. Louis, 1993, 1415.

23. **Brooks, S. M.,** Occupational and environmental asthma, in *Environmental and Occupational Medicine,* 2nd ed., Rom, W. N., Ed., Little, Brown, Boston, 1992, 393.

24. **Brooks, S. M.,** Occupational asthma, in *Bronchial Asthma: Mechanisms and Therapeutics,* 3rd ed., Weiss, E. B. and Stein, M., Eds., Little, Brown, Boston, 1993, 585.

25. **Alberts, W. N. and Brooks, S. M.,** Advances in occupational asthma, *Clin. Chest Med.,* 13, 281, 1992.

26. **Bernstein, D. I. and Bernstein, I. L.,** Occupational asthma, in *Allergy: Principles and Practice,* 4th ed., Middleton, E., Jr., Reed, C. E., Ellis, E. F., Adkinson, N. F., Jr., Yunginger, J. W., and Busse, W. W., Eds., C. V. Mosby, St. Louis, 1993, 1369.

27. **American Medical Association,** The respiratory system, in *Guides to the Evaluation of Permanent Impairment,* 3rd ed. (revised), American Medical Association, Chicago, 1990, 115.

28. **Arnold, W. J. and Ike, R. W.,** Specialized procedures in the management of patients with rheumatic diseases, in *Cecil Textbook of Medicine,* 19th ed., Wyngaarden, J. B., Smith, L. H., Jr., and Bennett, J. C., Eds., W. B. Saunders, Philadelphia, 1992, 1503.

29. **Larsen, P. R.,** The thyroid, in *Cecil Textbook of Medicine,* 19th ed., Wyngaarden, J. B., Smith, L. H., Jr., and Bennett, J. C., Eds., W. B. Saunders, Philadelphia, 1992, 1248.

30. **Tyrrell, J. B.,** Laboratory evaluation of adrenocortical function, in *Cecil Textbook of Medicine,* 19th ed., Wyngaarden, J. B., Smith, L. H., Jr., Bennett, J. C., Eds., W. B. Saunders, Philadelphia, 1992, 1279.

31. **Sherlock, S. and Dooley, J.,** *Diseases of the Liver and Biliary System,* 9th ed., Blackwell Scientific, Oxford, 1993, 17.

32. **Andrews, L. S. and Snyder, R.,** Toxic effects of solvents and vapors, in *Casarett and Doull's Toxicology: The Basic Science of Poisons,* 4th ed., Amdur, M. O., Doull, J., and Klaassen, C., Eds., Pergamon Press, New York, 1991, 681.

33. **Leonard, J. F. and Templeton, P. A.,** Pulmonary imaging techniques in the diagnosis of occupational interstitial lung disease, *Occup. Med.,* 7, 241, 1992.

34. **Muller, N. L. and Ostrow, D. N.,** High resolution computed tomography of chronic interstitial lung disease, *Clin. Chest Med.,* 12, 97, 1991.

35. **Nadich, D. P. and Garay, S. M.,** Radiographic evaluation of focal lung disease, *Clin. Chest Med.,* 12, 77, 1991.

36. **Webb, W. R. and Golden, J. A.,** Imaging strategies in the staging of lung cancer, *Clin. Chest Med.,* 12, 133, 1991.

37. **Fanburg, B. L. and Pitt, E. A.,** Sarcoidosis, in *Textbook of Respiratory Medicine,* Murray, J. F. and Nadell, J. A., Eds., W. B. Saunders, Philadelphia, 1988, 1486.

38. **Shepard, J. O. and McCloud, T. C.,** Imaging the airways: computed tomography and magnetic resonance imaging, *Clin. Chest Med.,* 12, 151, 1991.

39. **Gamsu, G. and Sostman, D.,** Magnetic resonance imaging of the thorax, *Am. Rev. Respir. Dis.,* 139, 254, 1989.

40. **Kramer, E. L. and Divgi, C. R.,** Pulmonary applications of nuclear medicine, *Clin. Chest Med.,* 12, 55, 1991.

41. National Asthma Education Program: Expert Panel Report, Guidelines for the Diagnosis and Management of Asthma, Publication No. 91-3042, U.S. Department of Health and Human Services, Bethesda, MD, 1991.

42. **Spector, S. L.,** Bronchial provocation tests, in *Bronchial Asthma: Mechanisms and Therapeutics,* 3rd ed., Weiss, E. B. and Stein, M., Eds., Little, Brown, Boston, 1993, 501.

43. **Wasserman, K., Hansen, J. E., Sue, D. Y., and Whipp, B. J.,** *Principles of Exercise Testing and Interpretation,* Lea and Febiger, Philadelphia, 1987.

44. **Barnhart, S.,** Evaluation of impairment and disability in occupational lung disease, *Occup. Med.,* 2, 227, 1987.

45. **Sherman, C. B.,** Cardiopulmonary exercise testing to assess respiratory impairment in occupational lung disease, *Occup. Med.,* 2, 243, 1987.

46. **Harber, P.,** Respiratory disability: the uncoupling of oxygen consumption and disability, *Clin. Chest Med.,* 13, 367, 1992.

47. **Zavala, D. C.,** Bronchoscopy, lung biopsy, and other procedures, in *Textbook of Respiratory Medicine,* Murray, J. F. and Nadell, J. A., Eds., W. B. Saunders, Philadelphia, 1988, 562.

Chapter 7

# Spirometry and Cardiopulmonary Exercise Testing in Impairment Evaluation

*Richard C. Bernstein, M.D. and*
*Samuel V. Spagnolo, M.D.*

## CONTENTS

## INTRODUCTION

Exposure of an individual to air pollutants or pulmonary toxins may lead to many pathological and clinical outcomes, as discussed in previous chapters. Various classification schemes and diagnostic approaches have been suggested for evaluating the degree of respiratory impairment. Most methods currently include spirometry and cardiopulmonary exercise testing along with the chest roentgenogram. Dyspnea is a nonspecific symptom that frequently develops as a consequence of these pulmonary exposures, and we have previously defined dyspnea as "breathing" that is "perceived" by an individual "to be difficult, uncomfortable, or unpleasant."[1] This increased awareness by the individual that more effort is now required to breathe can result in decreased performance of daily activities, ranging from slightly diminished maximal exercise tolerance to marked shortness of breath at rest.[2] This chapter will focus on the methodology currently used to quantitate respiratory impairment.

## BACKGROUND

The American Thoracic Society (ATS) defines impairment as ". . . a purely medical condition resulting from a functional abnormality, which may or may not be stable at the time evaluation is made and may be temporary or permanent."[3] In contrast, the American Medical Association (AMA) defines impairment as ". . . any anatomic or functional abnormality or loss."[4] This addition of anatomic loss as a criterion can be important in asbestos workers, who may have pleural thickening or interstitial changes on chest roentgenogram without evidence of any significant functional loss.

Impairment, however, is only one factor that can be used in determining disability. Again, as defined by the ATS, disability is ". . . the total effect of impairment on the patient's life. It is affected by such diverse factors as age, gender, education, economic and social environment, and energy requirements of the occupation."[3] And again in contrast, the 1984 AMA guides define disability as ". . . the inability to perform at a specific level of activity."[5]

Since these definitions of disability include factors other than impairment, such as educational, economic, and occupational considerations, it becomes a multidisciplinary task to determine the level of disability. The process may require persons knowledgeable in the fields of law, education, and economics, as well as medical sciences.[6,7,8]

Many disability claims are made under specific government or worker compensation programs, and the criteria for disability in these programs are only relevant to the program or policy from which they were generated.[9] A specific example would be the compensation to coal miners for black lung disease. Although no anatomic or functional loss or abnormality is demonstrated in many cases, the claimants may still receive benefits. The determination of impairment is much less controversial and is performed by the physician.[10]

## MOTIVATION OF DISABILITY CLAIMANTS

It should be noted that individuals who undergo evaluation for respiratory impairment as part of a disability or worker's compensation proceeding may claim a far greater amount of dyspnea than a control group with identical tests of pulmonary function.[6] This suggests that the perception of symptoms by an individual is very closely related to the reason the person sees the physician.[11] Other factors unrelated to the extent of the lung disease that may alter an individual's response to dyspnea include communication skills, socioeconomic and education backgrounds, and their level of concern regarding health.[9] Therefore, dyspnea should be considered only in combination with other clinical and physiological factors in the determination of disability. Cardiopulmonary exercise testing is unique in its ability to identify mechanisms of limitation in patients with exercise intolerance and may be very useful in identifying psychogenic dyspnea or deconditioning.[12]

## CLINICAL ASSESSMENT OF IMPAIRMENT

### HISTORY

As with any medical assessment, the physician must thoroughly explore the patient's prior medical history. This would include objective quantification of dyspnea, sputum production, cough, and tobacco use, as well as an exhaustive occupational and environmental history.[2,8]

### PHYSICAL EXAMINATION

The physical examination should be done with particular attention given to patterns of speech (how many words the patient can say between breaths), the use of accessory

Table 1    **Classification of Respiratory Impairment Based on Spirometry**

| Degree of Impairment | FEV$_1$ | FVC | DLCO | FEV$_1$/FVC |
|---|---|---|---|---|
| None | >80% predicted | >80% predicted | >80% predicted | >.75 |
| Mild | 60–79% predicted | 60–79% predicted | 60–79% predicted | .60–.74 |
| Moderate | 41–59% predicted | 51–59% predicted | 41–59% predicted | .41–.59 |
| Severe | <40% | <50% | <40% | <.40 |

From American Thoracic Society, *Am. Rev. Respir. Dis.*, 133,1205, 1986. With permission.

muscles to breathe, the quality and character of the breath sounds, the respiratory rate, the presence of clubbing of the fingers and toes, cyanosis, or objective evidence of cor pulmonale (enlargement of the right ventricle secondary to a primary respiratory disorder). Cor pulmonale is suggested clinically by evidence of right-sided heart failure, right ventricular heave, gallop and/or murmur, and ECG or chest X-ray evidence of right ventricular hypertrophy. Cor pulmonale is accepted by the ATS as indicating severe respiratory impairment, despite the result of the pulmonary function rating.[3]

## SPIROMETRIC TESTS

In determining impairment, tests of lung function are fundamental. Current guidelines recommend the following tests for primary disability evaluation: FEV$_1$, FVC, FEV$_1$/FVC, and DLCO.[3] The evaluation should only be performed while the patient is receiving optimal therapy, which might include beta-agonist, glucocorticoid and ipratropium bromide-metered dose inhalers (MDIs), oral glucocorticoids, theophyline, etc. All spirometry equipment should meet the ATS official statement on standardization for spirometry or any subsequent official revisions.[13]

Baseline information necessary for pulmonary function testing (PFT) includes height, age, weight, gender, and ethnic background. These are factors that were used to develop database population standards for predicted normal values. It should be noted whether the patient smoked within several hours prior to testing, since this can affect the results of the DLCO due to increased back pressure from carbon monoxide present in cigarette smoke.

If height cannot be measured due to spinal deformity or the inability to stand, then it may be considered equal to armspan, which is defined by the distance between the tips of the middle fingers when the arms are outstretched.[14] The relationship for adults of armspan to height is within 3%.[10]

When evaluating different ethnic groups, predicted standards for these ethnic groups should be used if proper studies exist.[10,3] It has been shown that blacks have an FVC and FEV$_1$ 15% lower than whites, adjusted for height and age.[15]

Tables 1 and 2 summarize various medical organizations' criteria for impairment, based on spirometry.[3,5] The AMA utilizes the DLCO only in the determination of severe impairment (<40% of predicted values), in contrast to the ATS. In the determination of impairment, we recommend that the FEV$_1$ value be taken from the trial that has the highest peak flow value, because this would reflect the trial in which the greatest effort was given.

Both the AMA and ATS guides use a percentage of predicted values in the classification of respiratory impairment. By contrast, the Social Security Administration

Table 2  **Classification of Respiratory Impairment Based on Spirometry**

| Degree of Impairment | FEV$_1$ | FVC | DLCO | FEV$_1$/FVC |
|---|---|---|---|---|
| Class 1<br>0%<br>None | Above the lower limit of normal for the predicted value defined by the 95% confidence interval. | | — | Above the lower limit of normal for the actual ratio as defined by the 95% confidence interval. |
| Class 2<br>10–25%<br>Mild | Below the 95% confidence interval but >60% predicted. | | — | Below the 95% confidence interval but ratio >60%. |
| Class 3<br>30–45%<br>Moderate | 40–60% predicted | 50–60% predicted | — | 40–60% actual ratio |
| Class 4<br>>50%<br>Severe | <40% predicted | <50% predicted | <40% predicted | <40% actual ratio |

From American Medical Association, *Guides to the Evaluation of Permanent Impairment,* 2nd ed., AMA, Chicago, 1984, 85. With permission.

defines impairment as a value of function below which a person would be considered unable to perform a given activity. This is called the minimum value method.[16] The biases of the different methods of impairment evaluation have been reviewed by Harber.[17]

## OTHER FACTORS

### Arterial Blood Gases
Since the partial pressure of arterial oxygen (PaO$_2$) normally falls with age, arterial hypoxemia does not indicate severe impairment unless cor pulmonale is also present.[3] In our own evaluation of normal subjects, we found the regression equation for PaO$_2$ in mmHg to be $104 - (0.49 \times$ age), with a SD of 5.5, which is in close agreement with published studies.[18] If we assign a PaO$_2$ value 2SD below the calculated value from the above equation as being the lower limit of normal, then a PaO$_2$ below $93 - (0.49 \times$ age)[mmHg] would be considered abnormal.

However, other somewhat more complex formulae are available, along with computer programs to assist in complex calculations. These equations include correction factors for age and barometric pressure (BP).[19]

$$PaO_2 = -0.279008 \times Age + 0.1134332 \times BP + 14.6324 \tag{1}$$

$$PaCO_2 = 0.409901 \times Age - (0.2264837 \times BP) + (0.00021668 \times BP)^2$$

$$- (0.0005306 \times Age \times BP) + 87.0423 \tag{2}$$

In the equations, Age = years and BP = barometric pressure. Controversy remains about methods for determining the lower limits of normal. Many laboratories prefer the 95% confidence interval (CI) method, since it is widely accepted and can be applied to all pulmonary function tests including blood gases and permits a consistent approach to the interpretation. Using Intermountain Thoracic Society guidelines, the following is suggested:[19]

$$PaO_2 = -(0.280357 \times Age) + (0.1134470 \times BP) + 3.1003 \qquad (3)$$

$$PaCO_2 = PaCO_2 \pm 6.71 \qquad (4)$$

Using the above equations, the predicted $PaO_2$ for a normal 60-year-old person at sea level (BP 760 mmHg) is 84 mmHg, and the lower limit of normal (CL) for the same individual is 72 mmHg.

Moving the same individual to an altitude of 5000 ft (1524 m) with a BP of 632 mmHg results in a predicted $PaO_2$ of 70 mmHg with a lower limit of normal (CL) of only 58 mmHg. The $PaCO_2$ would change slightly and is approximately 5 mmHg lower than the $PaCO_2$ at sea level because of the slight hyperventilation caused by the altitude induced hypoxemia.

## Asthma

Asthmatics are considered severely impaired if they have carefully documented attacks that require emergency room or hospital treatment six or more times a year and bronchospasm between acute attacks despite optimal therapy.[3] This determination should be made by a physician experienced in asthma treatment.

## CARDIOPULMONARY EXERCISE TESTING

### INTRODUCTION

Cardiopulmonary exercise (CPX) testing is an extremely important modality in impairment evaluation when spirometry does not give definitive answers about a individual's performance status. If spirometry results meet the criteria for severe disability, there is no need for further evaluation. Conversely, exercise limitation is not likely to be due to a pulmonary source when resting spirometric tests are normal or minimally reduced.[20] Despite resting spirometry results indicating either mild or no impairment, many individuals will allege difficult breathing with work activities or exertion. In this subset, CPX testing permits physiologic measurements of the functional capacities of individuals. The information obtained assists in the evaluation of subjective complaints and frequently permits the physician to determine a cause.[21–23]

### OVERVIEW

CPX testing can be done using invasive and noninvasive methods. Unfortunately even today, there are no universally accepted guidelines for testing and interpretation. Testing usually incorporates the analysis of inspired and expired gasses to determine the quantity of oxygen consumed ($VO_2$) and carbon dioxide produced ($VCO_2$). In addition, the end-tidal partial pressure of $O_2$ and $CO_2$ is measured. The rate of airflow is determined using a pneumotachometer or mass flow amenometer, and from this measurement and the respiratory rate, tidal volumes are determined. The EKG will record heart rate and monitor for dysrhythmia and ischemia. Blood pressure is periodically measured by sphygmomanometer. An arterial line is placed for easy arterial blood gas (ABG) sampling and is especially important for bicarbonate and lactate measurements.

Although the various commercially available metabolic exercise systems have slightly different set-ups, the basics are quite similar (Figure 1). A tight mouth seal is maintained to avoid leakage of air, and a nonrebreathing valve separates inspired

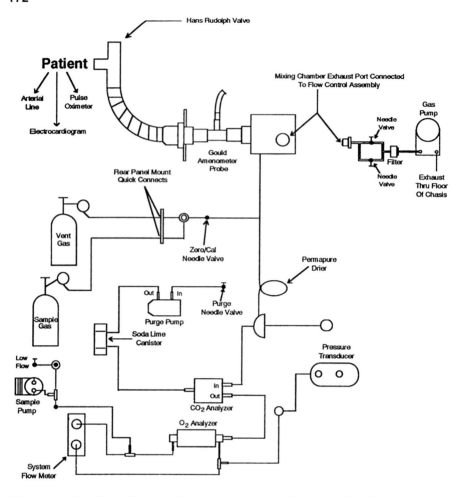

**Figure 1**  Simplified diagram of a commercially available cardiopulmonary exercise system. (From the Sensormedics 2900, adapted by permission.)

room air from exhaled air.[24] $O_2$ and $CO_2$ tensions are determined by rapid-response analyzers that are supplied from a small sampling line near the mouth. The $VO_2$ and $VCO_2$ measurements are adjusted to compensate for the small amount of additional dead space in the mouthpiece and collecting chamber.

Before testing begins, all calibrations of the CPX equipment must be performed as required by the manufacturer. Flow measurements are calibrated using a 3-l. syringe. Temperature and barometric pressures are obtained with an approved mercury thermometer and barometer. These are compared to the measurements obtained by the CPX equipment. When using a bicycle ergometer, it must be periodically evaluated for accuracy of workload. Any recommended changes in respiratory tubing or other basic maintenance must be adhered to strictly.

## ENERGY SUPPLY IN EXERCISE

Normally the human body generates adenosine triphosphate (ATP) as a high-energy bond to be used in muscular activity.[25] During nonstressful activity, ATP is generated almost exclusively from glucose or lipids. Intracellular glucose is metabolized into pyruvate, which is then converted to acetlyl-CoA, which enters the Krebs cycle.

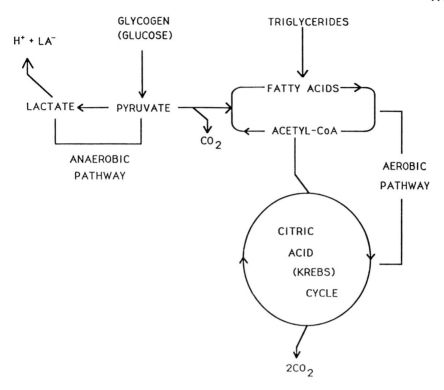

**Figure 2** The common pathways for energy metabolism of carbohydrates and lipids. Once the aerobic pathway is limited by oxygen delivery, then pyruvate is converted to lactate in the anaerobic pathway.

When lipids such as palmitate are utilized, large amounts of ATP are produced (Figure 2). The ATP produced for each fuel source are illustrated by the formula below.

*ATP PRODUCTION* (AEROBIC)

$C_6H_{12}O_6 + 6O_2 >>>> 6CO_2 + 6H_2O + 36ATP + Heat$
(glucose)

$C_{16}H_{32}O_2 + 23O_2 >>>> 16CO + 16H_2O + 130ATP + Heat$
(palmitate)

There is no limitation to the aerobic process when nutrition is normal and the body is not severely stressed. During exercise, small amounts of energy are available from energy stored in the muscle as creatine phosphate. However, energy needs are primarily met by the aerobic process, and lipid is the predominant fuel if plasma levels of free fatty acids are adequate and if muscle blood flow increases normally in response to exercise.[25] As exercise increases, so does the amount of carbohydrate utilized, and muscle glycogen stores are gradually depleted.[26]

Anaerobic metabolism is used to supplement energy production when oxygen demand exceeds oxygen delivery.[25,27,28] The "turn on" of pyruvate oxidation of reduced nicotinamide adenine dinucleotide [NADH + H+] occurs without immediate use of oxygen and is thus termed "anaerobic." During this process, increased lactate production occurs, and as lactate dissociates within the cell, hydrogen ions

(H+) produced are then buffered by bicarbonate. This sudden drop in bicarbonate is an important marker of anaerobic metabolism and results in the development of a metabolic acidosis.

At high work rates, both aerobic and anaerobic mechanisms share in energy production, although the latter mechanism provides an increasing proportion of energy as the work rate increases.

The point where aerobic metabolism is supplemented by anaerobic metabolism is called the "anaerobic threshold" (AT).[25,27,28] *The determination of the anaerobic threshold is fundamental in the interpretation of a properly performed cardiopulmonary exercise study.*

## METHODOLOGY

Actual CPX testing is performed by exercising an individual against an increasing workload. The testing is accomplished by using either a treadmill or bicycle ergometer, and various protocols are available. The treadmill is more natural for most people because walking is a more familiar task than riding a bike.[29] However, it is our opinion that the treadmill is cumbersome and potentially dangerous because it may not be stopped quickly in emergency situations. The risk of serious injury is worrisome. Unlike the treadmill, the bicycle ergometer is safer and permits the workload to be easily quantified.[2] A system of hand signals should be prearranged between patient and physician, so the test subject can signal to stop the test when he or she is in distress. During CPX testing, the patients should be frequently asked if they are all right.

Testing using a bicycle ergometer involves two basic protocols. The first is an incremental test, where workload (measures in watts) is increased by a fixed amount at the beginning of every minute.[25] A simple example would be to increase the workload 25 W every minute, with the entire increase coming at the beginning of each minute (25 W at the start of the test, 50 W at 1 min, 75 W at 2 min, etc.). An alternative method is the Ramp Method, where the workload is continuously increased throughout the test.[24] The Ramp Method tends to give "smoother" data and does not have the sudden increase in workload that some individuals find difficult in the latter stages of a test.

Although both methods give valid information, we prefer the Ramp Method. Table 3 displays workload and $O_2$ consumption for a 70-kg, normal male using the Ramp Method on a bicycle ergometer.[30]

The time for completion of the test should be 8 to 10 min.[25] This is easily achieved by dividing the predicted maximum workload of the patient in watts by 10. The resulting value gives the workload increase in watts/minute. The formula for predicted maximum watts is:[31]

$$\text{Watts (males)} = 4.2 \text{ Ht[cm]} - 1.5(\text{Age}) - 359 \quad (5)$$

$$\text{Watts (females)} = 1.6 \text{ Ht[cm]} - 1.5(\text{Age}) + \text{Wt[kg]} - 126 \quad (6)$$

Tests shorter than 8 min produce crowded data points and are difficult to interpret. When the test is longer than 10 min, the test subject may fatigue before reaching maximum exercise capacity.

The *best end point* of CPX is where the patient is no longer able to continue because of fatigue or shortness of breath. Other reasons for stopping the CPX test include chest pain, mental status changes, multiple premature ventricular contractions or ischemic changes on the EKG, and a drop in the systolic pressure.[27,28,31]

Table 3   Oxygen Consumption for a Normal 70-kg Person
on a Bicycle Ergometer

| VO$_2$ml/min[a] | VO$_2$ ml/min/kg | W |
|---|---|---|
| 640 | 9.1 | 12 |
| 760 | 10.9 | 25 |
| 1050 | 15.0 | 50 |
| 1620 | 23.1 | 100 |
| 2160 | 30.9 | 150 |
| 2720 | 38.9 | 200 |
| 3280 | 46.9 | 250 |
| 3850 | 55.0 | 300 |

[a]VO$_2$ = oxygen consumption.
Adapted from American College Sports Medicine, *Guide for Graded Exercise Testing and Prescriptions,* Lea & Febiger, Philadelphia, 1975. With permission.

Oxygenation can be followed by measuring the PaO$_2$ from frequent sampling via an arterial line. It may also be monitored noninvasively with an ear or finger oximeter and by observation for evidence of hemoglobin desaturation.

## MEASUREMENTS

### Oxygen Consumption
The VO$_2$ when the CPX is terminated (due to fatigue or the other reasons listed above) is defined as maximum VO$_2$.[25] It is usually expressed in ml/min but can be adjusted for body weight by dividing by the weight (in kg) of the subject (see Table 3).

A less attractive and more cumbersome way to describe VO$_2$ per body weight is by a unit or measure called a "met." This unit was derived from the resting VO$_2$ of a 70-kg, 40-year-old normal male and was given a value of 3.5 ml/min/kg.[25] Many laboratories do not measure VO$_2$ directly but assume a relationship between work rate and VO$_2$. This estimated VO$_2$ is divided by body weight and then divided again by 3.5 to determine the number of "mets" performed. It is obvious that this method is much less accurate than directly measuring VO$_2$, and we do not recommend its use. Unfortunately, normal values for VO$_2$ may vary from laboratory to laboratory and reflect different corrections for body size. These corrections and differences in methodology are a cause for concern; a common methodology should be established in the near future.

### Respiratory Quotient
The relationship between VO$_2$ and VCO$_2$ is given by the ratio VCO$_2$/VO$_2$ which is the respiratory quotient (RQ).[25,27,28] At rest, the RQ is normally 0.8 but will vary depending on diet composition. If fat alone is the metabolic fuel, the RQ is 0.7, while glycogen (carbohydrate) alone gives an RQ of 1.0.

During exercise, since a higher proportion of carbohydrate is metabolized, the RQ tends to rise toward 1.0. When AT is reached, the buffering of lactate by HCO$_3^-$ produces increased CO$_2$, which causes even a further rise in the RQ and results in values above 1.0.[25,28]

It is possible to determine the VO$_2$ and VCO$_2$ on a breath-by-breath basis with rapid gas analyzers.[29] Several systems have computer software, which permits averaging of variable time intervals. This reduces the scatter seen on breath-by-breath plots and presents the data for an easier and more meaningful interpretation. We currently use an average of 6 to 10 s.

Figure 3 shows some of the graphs Wasserman and colleagues use in their CPX interpretation for anaerobic threshold.[25] The measurements used in the development of these graphs will be discussed later.

## Oxygen Delivery, Stroke Volume, and Heart Rate

As workload (in W) increases, the body must increase oxygen delivery($O_2D$) to maintain aerobic metabolism. The determinates of $O_2D$ are:[1]

$$O_2D = CO \times O_2 \text{ content}$$
$$= [SV \times HR] \times [SaO_2(Hb \times 1.38) + PaO_2(0.0031) \tag{7}$$

where $O_2D$ = oxygen delivery; $CO$ = cardiac output; $SV$ = stroke volume; $HR$ = heart rate; $SaO_2$ = hemoglobin oxygen saturation; $Hb$ = hemoglobin; $PaO_2$ = arterial partial pressure of oxygen; 1.38 = oxyhemoglobin-combining coefficient; 0.0031 = solubility coefficient for oxygen in blood.

If it is assumed that hemoglobin is fully saturated, then the only remaining way to increase $O_2D$ (while breathing room air) is to increase cardiac output. This is done by increasing SV and HR. Initially with exercise, SV increases, but the maximum increase is no greater than 50% above resting values. This peak in SV occurs at approximately 40% of an individual's maximum $VO_2$.[25] Any further increase in cardiac output must be accomplished by increasing the HR. The HR increases linearly with $VO_2$ until a maximum HR is reached.

Maximum heart rate is given by the following equations.

$$\text{Maximum Heart Rate} = 220 - 0.65 \text{ Age (in years)}^{28} \tag{8}$$

or

$$\text{Maximum Heart Rate} = 210 - \text{Age (in years)}^{33} \tag{9}$$

In normal individuals, the cardiac output is the limiting factor to maximum exercise.[33] The maximum $VO_2$ decreases with age because the maximum predicted HR decreases with age.[24]

## Minute Ventilation and Breathing Reserve

During exercise, minute ventilation (Ve) increases linearly with $VO_2$ until anaerobic threshold (AT) is reached.[27] At AT, the buffering of lactate increases $VCO_2$, which further stimulates Ve out of proportion to the increase in $VO_2$.[34]

In addition, metabolic acidosis increases Ve in an attempt to compensate by reducing blood $PCO_2$ levels. At the start of exercise, most of the gain in Ve results from increasing the tidal volume ($V_T$). Eventually, $V_T$ reaches a maximum value, and any additional increase is caused by an increased respiratory rate (RR). Maximum exercise $V_T$ is usually 60 to 70% of the vital capacity.[35]

The pulmonary system of normal individuals is not the limiting factor to maximum exercise.[25] When comparing maximum Ve during exercise to maximum minute ventilation (MVV), which is either directly measured or estimated by multiplying resting $FEV_1 \times 41$,[36] the maximum exercise Ve is usually no greater than 60 to 75% of the MVV, indicating at maximum exercise that significant ventilatory reserve exists.[37]

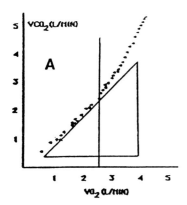

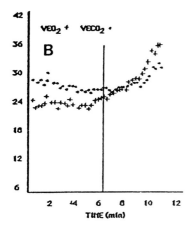

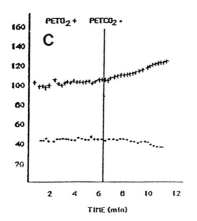

**Figure 3**  Graphs A,B,C show selected results of a cardiopulmonary exercise test in a healthy 27-year-old, 86-kg male. These plots represent the different noninvasive means to determine anaerobic threshold (AT). Graph A has $VO_2$ plotted against $VCO_2$ (V-slope method of Wasserman). As workload increases, the $VO_2$ and $VCO_2$ increase in proportion to each other, giving a line with a slope of 1 (hypotenuse of a right triangle). When AT occurs, the $VCO_2$ increases more rapidly than $VO_2$ (see text for details), and the slope of the line becomes >1. This point where the line increases slope is defined as AT. The right triangle has been drawn in for illustration. Graph B shows ventilatory equivalents for oxygen and carbon dioxide, and Graph C shows end-tidal readings for oxygen and carbon dioxide (see text for details). With AT, both $VEO_2$ and $PETO_2$ start to increase, while $VECO_2$ and $PETCO_2$ remain stable. These measurements may not be as precise as the V-slope method. The vertical line represents the AT as determined by V-slope method. ($VCO_2$ = carbon dioxide production, $VO_2$ = oxygen consumption, $VEO_2$ = ventilatory equivalent of oxygen, $VECO_2$ = ventilatory equivalent of carbon dioxide, $PETO_2$ = end-tidal pressure of oxygen, $PETCO_2$ = end-tidal pressure of carbon dioxide, AT = anaerobic threshold.)

This volume remaining is termed the "breathing reserve" and is defined as:

$$[MVV - Ve\ max]\ (normal > 15\ l/min)^{25} \tag{10}$$

Eschenbacher et al.[38] use the term "ventilatory reserve" and define it as:

$$VR = [1 - (VE\ max/predicted\ MVV)] \times 100\%$$
$$(normal < 80\%) \tag{11}$$

Wasserman's group,[25] in their interpretation of cardiopulmonary stress testing, use the breathing reserve as a major differentiating point between respiratory and circulatory disorders.

**Alveolar-Arterial Oxygen Tension**
The difference between predicted alveolar oxygen partial pressure ($PAO_2$) and the measured arterial oxygen partial pressure ($PaO_2$) is the alveolar-arterial oxygen tension gradient (A-a gradient). This relationship is predicted to be $2.5 + 0.21(Age)$[39] at rest. The SD for this regression equation is 7. If $2 \times SD$ (used to define our confidence limits) is added to the above regression equation, then any A-a gradient $>16.5 + 0.21(Age)$ would be considered abnormally high. An elevated A-a gradient at rest correlates with a DLCO $<25\%$ of predicted in most instances.[6] Above AT, increased ventilation causes the alveolar partial pressure of carbon dioxide ($PACO_2$) to fall, hence, the $PAO_2$ must rise, since total alveolar pressure is constant. As the $PAO_2$ increases, the A-a gradient rises because the $PaO_2$ usually remains constant or slightly increases with exercise. The maximum normal exercise A-a value is 30 mmHg.[25] To determine the A-a gradient at maximum exercise, the arterial blood gas (ABG) must be measured while exercise is still in progress. If not, the A-a gradient might not reflect exercise values.

**End-tidal $O_2$ and $CO_2$ Pressure**
Along with $VCO_2$, Ve, and $VO_2$, other measurements that are obtained noninvasively are the end-tidal oxygen tension ($PETO_2$) and end-tidal carbon dioxide tension ($PETCO_2$).

The $PETO_2$ at rest is normally 90 mmHg or greater and increases with heavy exercise. $PETCO_2$ at rest is 36 to 42 mmHg and increases 3 to 8 mmHg with mild to moderate exercise; then, above AT it decreases when Ve increases.[25]

**Anaerobic Threshold (Lactate, V-slope, End-tidal Plots, Ventilatory Equivalents)**
When anaerobic threshold occurs, serum lactate levels start to rise above resting values. Determining serum lactate levels requires either arterial or venous catheter placement prior to initiation of the study.

Anaerobic threshold can also be estimated noninvasively by the V-slope method.[25] The $VO_2$ is plotted on the X-axis and the $VCO_2$ is plotted on the Y-axis (Figure 3, plot A). As workload is increased, the $VO_2$ and $VCO_2$ increase in proportion to each other, but when anaerobic threshold is reached, $VCO_2$ increases more rapidly (as previously described, due to anaerobic metabolism and the resulting metabolic acidosis). As this occurs, the slope of the $VO_2$-$VCO_2$ plot develops an abrupt increase. The AT is defined by this abrupt increase in the slope of the line.

There are other noninvasive means to help confirm the V-slope method. These involve plots of $PETO_2$ and $PETCO_2$ against $VO_2$ (or time, W) (Figure 3, plot C).

With the increase in ventilation out of proportion to $VO_2$ at AT, the $ETO_2$ continues to rise, while the $ETCO_2$ is level.[25]

Likewise, ventilatory equivalents for $CO_2$ and $O_2$ can also be used to determine the AT noninvasively. At AT, the ventilatory equivalent of oxygen ($Ve/VO_2$) plotted against $VO_2$ (or time, W) increases, while the ventilatory equivalent for $CO_2$ ($Ve/VCO_2$) is stable, until it begins to increase due to respiratory compensation for metabolic acidosis (Figure 3, plot B).[25] In a normal individual at rest, approximately 25 l of external ventilation is required for 1 l of $O_2$ uptake ($Ve/VO_2 = 25$).[7]

The anaerobic threshold in normal individuals usually occurs at above 40% of the predicted maximum $VO_2$.[25]

$$AT > 0.4 \times \text{predicted } VO_2 \text{ max} \tag{12}$$

## Oxygen Pulse ($VO_2$/HR)

The oxygen pulse ($VO_2$/HR) is helpful in estimating the amount of oxygen that is delivered with each heartbeat. This ratio correlates with the adequacy of the stroke volume, because the arterial venous oxygen content difference remains stable with exercise.[27]

$$VO_2 = CO \times \text{(arterial-venous) } O_2 \text{ content}$$
$$VO_2 = SV \times HR \times \text{(arterial-venous) } O_2 \text{ content}$$
$$VO_2/HR = SV \times \text{(arterial-venous) } O_2 \text{ content} \tag{13}$$

The predicted maximum $O_2$ pulse is calculated from the predicted maximum $VO_2$ (ml/min) divided by the predicted HR maximum.

$$\text{Max } VO_2/HR = \text{predicted max } VO_2/\text{predicted max HR} \tag{14}$$

## Dead Space Ratio (VD/VT)

The dead space ratio (VD/VT) can be calculated during CPX by dividing the physiologic dead space (VD) by the tidal volume (TV). At rest the, VD/VT is normally 0.30.[40] In individuals with normal lungs, during exercise the recruitment of new vascular channels, especially in the upper lobes, reduces dead space often to <0.20.[41] In pulmonary vascular, parenchymal, and airway disease, the increased VD/VT partially explains why these individuals have an abnormally high minute ventilation (Ve) with exercise.[2] In the above conditions, increased VD/VT during exercise is due to inability to recruit additional vascular channels.

## DISCUSSION AND INTERPRETATION
## OF RESULTS

### ALGORITHMS

Wasserman et al. have provided an in-depth description and reasoned approach to the evaluation of CPX testing.[25] His group has derived algorithms where maximum $VO_2$ and AT are used as main differentiating points (Figure 4).[25] $VO_2$ max is considered low if it is <83% of predicted (95th confidence interval).[22] AT is considered low if it is <40% of the predicted $VO_2$ max.[42]

### PERFORMANCE

Evaluation of patient performance is important to show that a maximum effort was given, since the data derived in CPX testing is often used in impairment evaluation

180

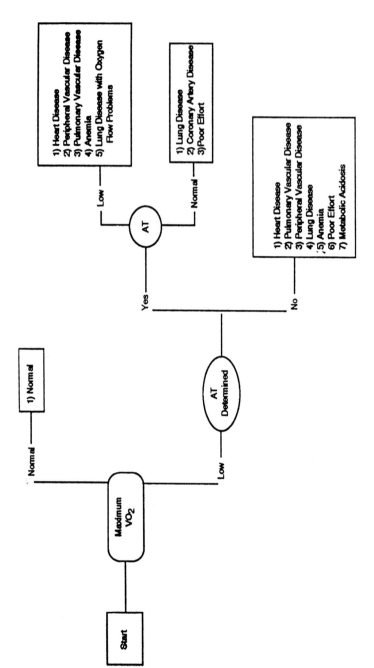

**Figure 4** Algorithm of primary flow chart used in interpretation of cardiopulmonary exercise results. (Adapted from Wasserman, K., Hansen, J. E., Sue, D. Y., et al., *Principles of Exercise Testing and Interpretation*, Lea & Febiger, Philadelphia, 1987. With permission.)

that may lead to awards for disability. There are several ways to determine whether the patient has provided a maximal effort. First, the pattern of ventilatory response should be examined. Except for the beginning of CPX testing, where ventilation may be somewhat uneven due to anxiety and anticipation, ventilation *rises* linearly with increasing workload.[34] When the ventilatory response is markedly uneven throughout the test, especially near termination, suspect a submaximal effort. Second, frequent coughing should be noted, since it gives an uneven ventilation pattern. Third, with a maximal exercise effort, the serum $HCO_3^-$ drops 6 Meq/l below the preexercise values when arterial blood is sampled 2 min after completion of the test. This drop may not be seen in peripheral vascular disease, where claudication forces termination of the test before there is a significant drop in $HCO_3^-$.[25] In individuals with lung disease, dypsnea may cause cessation of exercise before anaerobic threshold is reached. However, in subjects with severe lung disease, breathing reserve will be low. Finally, if the predicted maximum (also called target) heart rate is reached, a maximal effort was probably given. The calculation for the predicted maximum heart rate is given in Equations 8 and 9.

The concept of heart rate reserve (HRR), defined as the predicted maximum adult heart rate minus the observed maximum heart rate, is stressed by Wasserman et al.[25] in evaluating the cardiovascular system during exercise. However, although a high HRR >15 may be due to suboptimal effort, it may also be secondary to medications (such as beta-adrenergic blockers) or to coronary, peripheral vascular, lung, endocrine, or musculoskeletal diseases.

## ABNORMAL PATTERNS

Basic patterns exist for the different major categories of disorders that can be identified by CPX: Table 4.[2] A subject may have more than one type of disorder, and mixed patterns can result. Uncomplicated and easily identified patterns are as follows:

### Cardiac[25]

In heart disease, a decrease of max $VO_2$ is seen, often with plateau formation (non-increasing $VO_2$ despite increased workload). Because all types of severe cardiac disease (severe coronary artery disease, cardiomyopathy, valvular disease, tachyarythmias) decrease stroke volume, the maximum amount of oxygen that is delivered to exercising muscles is reduced. This leads to early anaerobic metabolism and lactic acid formation with early fatigue. Even though AT is in proper proportion to the $VO_2$ max, it is <40% of the predicted $VO_2$. The heart rate (HR) response is out of proportion to the $VO_2$ to compensate for a decreased stroke volume, and the HR often reaches the predicted maximum. This heart rate response may be blunted if the patient is on a beta-adrenergic blocking drug. Because of the decrease in stroke volume, the $O_2$ pulse is reduced.

The dead space ratio (VD/VT) falls normally in cardiac disease (except if concomitant pulmonary hypertension or pulmonary edema is present) since the pulmonary vasculature is able to dilate and recruit new vascular channels.

Electrocardiogram changes are commonly seen with coronary heart disease and have been described at length in many reviews. Dysrythmias are common when the myocardium is ischemic and are an indication to stop the test.

### Lung

#### *Obstructive and Restrictive Disease* [2,25]

If exercise is limited due to ventilatory dysfunction, the patient will have excessive ventilation for the amount of work performed, due to abnormal matching of ventilation and perfusion. This excessive ventilation causes a low breathing reserve at

Table 4  **Patterns of Abnormality in CPX**

I. Cardiac Disease

- Decreased maximum $VO_2$, often with plateau formation
- AT reduced as a percentage of predicted max $VO_2$
- Reduced $O_2$ pulse
- High heart rate
- VD/VT falls normally
- $PaO_2$ does not decrease
- Normal breathing reserve

II. Lung Dysfunction

A. Ventilatory Dysfunction

- Low breathing reserve
- High respiratory rate (restrictive disease)
- Decreased max $VO_2$
- AT either not detected or reduced as a percent of predicted max $VO_2$
- Normal $O_2$ pulse (except cor pulmonale, pulmonary vascular disease)
- VD/VT often elevated and does not fall normally with exercise
- $PaO_2$ may fall with exercise

B. Pulmonary Vascular Disease

- VD/VT increased at rest and does not fall with exercise
- Decreased max $VO_2$
- AT reduced as a percent of predicted max $VO_2$
- High heart rate
- Increased minute ventilation
- $PaO_2$ may fall with exercise

*Notes*: $VO_2$ = Oxygen consumption.
AT = Anaerobic threshold.
$PaO_2$ = Partial pressure of arterial oxygen.
VD/VT = Dead space ratio.
Adapted from Miller, A., *Pulmonary Function Tests in Clinical and Occupational Lung Disease,* Grune & Stratton, Orlando, FL, 1986, 353–373, 413–424, and from Wasserman, K., et al., *Principles of Exercise Testing and Interpretation,* Lea & Febiger, Philadelphia, 1987. With permission.

maximal exercise and is usually associated with a high respiratory rate. In this situation, the max $VO_2$ is decreased, since the individual will stop the test early because of dypsnea. When this occurs, the maximum predicted heart rate is not reached, and AT is either absent or occurs very close to the peak $VO_2$. The $O_2$ pulse is normal since stroke volume in pulmonary disease is usually maintained except in the presence of severe right-sided heart failure.

In individuals with ventilatory limitation, the VD/VT may be elevated at rest and does not fall normally with exercise. The $PaO_2$ is frequently below normal at rest in these patients and may fall even further with exercise. If an arterial blood gas is not measured, the fall in $PaO_2$ is reflected by a decreased $SaO_2$. The A-a gradient is usually elevated at rest and may increase to very high levels with exercise. This is due to overperfusion of poorly ventilated alveoli.

## Vascular Disease [2,25]

The results from exercise testing on individuals with primary pulmonary vascular disease have characteristics that overlap the patterns for cardiac and ventilatory dysfunction. Because of limited blood flow to ventilated alveoli, the VD/VT increases.

The max $VO_2$ is reduced because of a decreased cardiac output. In pulmonary vascular disease, the right ventricle must work against an increased resistance and cannot, during progressive exercise, deliver sufficient blood to the left-sided circulation. This results in a decreased cardiac output and a reduced maximum $VO_2$. With the decreased oxygen delivery to exercising muscles, anaerobic metabolism begins early, and the AT is reduced. The heart rate is high in compensation for the decreased stroke volume and therefore is high relative to oxygen consumption.

In individuals with primary pulmonary vascular disease, arterial hypoxemia may not be present at rest but will occur with progressive exercise. Several mechanisms may account for this observation. First, with destruction of capillary beds, the remaining vascular channels will have an increased blood flow, thus resulting in a shortened time for oxygen equilibrium. Second, increased right-side vascular pressures (high pulmonary artery pressure) may open a potentially patent foramen ovale, resulting in increased right-to-left shunting of blood. Repeating the exercise test while the individual is breathing 100% oxygen will distinguish right-to-left shunting of blood from other causes of hypoxemia.

## OXYGEN CONSUMPTION IN VARIOUS OCCUPATIONS

The American Thoracic Society (ATS)[4] includes exercise criteria in its evaluation of impairment when there are conflicting or questionable results from resting pulmonary function data. The ATS uses the $VO_2$/kg of body weight as the basic measure of function. Some values for normal individuals engaged in varying activities are as follows:[8]

- Office work: 5 to 7 ml/kg/min
- Walking 2 mi/h: 6 to 10 ml/kg/min
- Bricklaying: 14 to 16 ml/kg/min
- Cycling 13 mi/h: 25 to 30 ml/kg/min

See Table 5 for a more complete listing.

Based on ATS criteria,[3] a person with a max $VO_2$ of $\geq 25$ ml/kg/min will be able to do continuous heavy exertion for an 8-h shift for all but the most strenuous jobs. Short bursts of very demanding work may also be accomplished due to anaerobic metabolism, but these need to be followed by an adequate recovery period.

When $VO_2$ max is 15 to 25 ml/kg/min, an individual can perform a job only if 40% of his observed maximum $VO_2$ is greater than or equal to the average metabolic requirement of his or her occupation. For example, if a bricklayer needs 14 to 16 ml/kg/min to perform the activity but has only a $VO_2$ max of 25 ml/kg/min, he would not be able to sustain this job, since 40% of 25 ml/kg/min is only 10 ml/kg/min. Work more strenuous than 40% of a person's $VO_2$ max results in undue discomfort and fatigue.

A $VO_2$ max below 15 ml/kg/min is incompatible with the performance of most jobs, and the subject may even be uncomfortable with the energy demands of commuting to and from work.

184

Table 5  **Energy Requirements of Various Activities**

| Activity | Approximate $VO_2$[a] | |
|---|---|---|
| | ml/min[b] | ml/kg/min |
| Personal | | |
|     sleeping | 260 | 3.5 |
|     sitting | 400 | 5.3 |
|     dressing | 580 | 7.7 |
|     walking (3 mi/h) | 780 | 10.4 |
| Domestic | | |
|     scrubbing | 950 | 12.6 |
|     bedmaking | 1,020 | 13.6 |
| Light to moderate work (sitting) | | |
|     clerical | 420 | 5.6 |
|     operating heavy equipment | 660 | 8.8 |
|     truck driving | 950 | 12.6 |
| Moderate work (standing) | | |
|     janitorial work | 790 | 10.5 |
|     assembly line | 920 | 12.3 |
| Heavy work | | |
|     general heavy labor | 1,190 | 15.8 |
|     using heavy tools | 1,580 | 21.0 |
|     lift and carry 60–80 lb. | 1,970 | 26.2 |

[a]$VO_2$ = oxygen consumption.
[b]Requirements of a 75-kg male.
Adapted from Becklake, M. R., *Am. Rev. Respir. Dis.,* 129, S96, 1984. With permission.

The Canadian Thoracic Society is somewhat more strict in its interpretation of $VO_2$ max criteria.[43] Greater than 25 ml/kg/min is normal, 15 to 25 ml/kg/min is mild, 7 to <15 ml/kg/min is moderate, and <7 ml/kg/min is severe impairment. When dealing with disability claims, various agencies have specific criteria. This leads to a confusing array of "standards" for impairment determination based on cardiopulmonary exercise testing.

## SMOKING AND IMPAIRMENT

Finally, we suggest a note of caution when performing any type of impairment evaluation that may be related to air pollutants and other pulmonary toxins where social habits play a major role in the results of pulmonary testing. For example, the measured impairment (decreased spirometry and arterial $PaO_2$) of the majority of coal workers most probably reflects the cumulative effects of cigarette smoking and not occupational exposure to coal dust.[44] Although the separation of these and other factors is difficult, it is always important to urge anyone with a pulmonary impairment who smokes tobacco to quit.

## ACKNOWLEDGMENTS

We wish to thank Brad V. Spagnolo for his editorial assistance in review of the manuscript and Dr. Ann E. Medinger for her insightful comments and suggestions regarding our interpretation of physiological measurements during exercise testing and for her inspired teaching of pulmonary physiology.

# REFERENCES

1. **Spagnolo, S. V. and Medinger, A.,** *Handbook of Pulmonary Emergencies,* Plenum Press, New York, 1986.
2. **Miller, A.,** *Pulmonary Function Tests in Clinical and Occupational Lung Disease,* Grune & Stratton, Orlando, FL, 1986, 353–373, 413–424.
3. **American Thoracic Society,** Evaluation of impairment/disability secondary to respiratory disorders, *Am. Rev. Respir. Dis.,* 133, 1205, 1986.
4. **American Medical Association, Committee on Rating of Mental and Physical Impairment,** Guides to evaluation of permanent impairment. The respiratory system, *JAMA,* 194, 177, 1965.
5. **American Medical Association,** *Guides to the Evaluation of Permanent Impairment,* 2nd ed., AMA, Chicago, 1984, 85.
6. **Epler, G. R., Saber, F. A., and Gaensler, E. A.,** Determination of severe impairment (disability) in interstitial lung disease, *Am. Rev. Respir. Dis.,* 121, 647, 1980.
7. **Gaensler, E. A. and Wright, G. W.,** Evaluation of respiratory impairment, *Arch. Environ. Health,* 12, 146, 1966.
8. **Becklake, M. R.,** Organic or functional impairment: overall perspective, *Am. Rev. Respir. Dis.,* 129, S96, 1984.
9. **Battista, M.,** Disability: review of guides to the evaluation of permanent impairment, *JAMA,* 253, 3173, 1985.
10. **Miller, W. F. and Scacci, R.,** Pulmonary function assessment for determination of pulmonary impairment and disability evaluation, *Clin. Chest Med.,* 2, 327, 1981.
11. **Hankinson, J. L. and Bang, K. M.,** Acceptability and reproducibility criteria of the American Thoracic Society as observed in a sample of the general population, *Am. Rev. Respir. Dis.,* 143, 516, 1991.
12. **Pratter, M. P., Curley, F. J., Dubois, J., and Irwin, R. S.,** Cause and evaluation of chronic dyspnea in a pulmonary disease clinic, *Arch. Intern. Med.,* 149, 2277, 1989.
13. **American Thoracic Society,** ATS Statement—Snowbird Workshop on Standardization of Spirometry, *Am. Rev. Respir. Dis.,* 119, 831, 1979.
14. **Hepper, N. G., Black, L. F., and Fowler, W. S.,** Relationships of lung to height and arm span in normal subjects and in patients with spinal deformity, *Am. Rev. Respir. Dis.,* 91, 356, 1979.
15. **Abramowitz, S., Leiner, G. C., and Lewis, W. A.,** Vital capacity in the Negro, *Am. Rev. Respir. Dis.,* 93, 287, 1966.
16. **Social Security Administration,** Disability Evaluation Under Social Security. A Handbook for Physicians, HEW Publ. No. (SSA) 79-10089, August 1979.
17. **Harber, P.,** Alternative partial respiratory disability rating schemes, *Am. Rev. Respir. Dis.,* 134, 481, 1986.
18. **Sorbini, C. A.,** Arterial oxygen tension in relation to age in healthy subjects, *Respiration,* 25, 3, 1968.
19. *Clinical Pulmonary Function Testing,* 2nd ed., Intermountain Thoracic Society, Salt Lake City, 1984.
20. **Becklake, M. R., Rodarte, J. R., and Kalica, A. R.,** NHLBI workshop summary: scientific issues in the assessment of respiratory impairment, *Am. Rev. Respir. Dis.,* 137, 1505, 1988.
21. **Cruver, N. and Solliday, N.,** Fundamentals of exercise stress testing for lung disease, *Respir. Care,* 27, 1050, 1982.
22. **Oren, A., Sue, D. Y., Hansen, J. E., et al.,** The role of exercise testing in impairment evaluation, *Am. Rev. Respir. Dis.,* 135, 230, 1987.

23. **Wiedemann, H. P., Gee, J. B. L., Balmes, J. R., and Lake, J.,** Exercise testing in occupational lung diseases, *Clin. Chest Med.,* 5, 157, 1984.

24. **Neuberg, G. W., Friedman, S. H., Weiss, M. B., and Herman, M. V.,** Cardiopulmonary exercise testing; the clinical value of gas exchange data, *Arch. Intern. Med.,* 148, 2221, 1988.

25. **Wasserman, K., Hansen, J. E., Sue, D. Y., et al.,** *Principles of Exercise Testing and Interpretation,* Lea & Febiger, Philadelphia, 1987.

26. **Layzer, R. B.,** Editorial, How muscles use fuel, *N. Engl. J. Med.,* 324, 411, 1991.

27. **Astrand, P. O.,** Quantification of exercise capability and evaluation of physical capacity in man, *Prog. Cardiovasc. Dis.,* 19, 51, 1976.

28. **Jones, N. L. and Campbell, E. J. M.,** *Clinical Exercise Testing,* 2nd ed., W. B. Saunders, Philadelphia, 1982.

29. **Weber, K. T., Janiki, J. S., McElroy, P. A., and Reddy, H. K.,** Concepts and applications of cardiopulmonary exercise testing, *Chest,* 93, 843, 1988.

30. **American College of Sports Medicine,** *Guide for Graded Exercise Testing and Prescriptions,* Lea & Febiger, Philadelphia, 1975.

31. **Jones, N. L., Makrides, L., Hitchcock, C., et al.,** Normal standards for an incremental progressive cycle ergometer test, *Am. Rev. Respir. Dis.,* 131, 700, 1985.

32. **Astrand, P. O. and Radahl, K.,** *Textbook of Work Physiology,* 2nd ed., McGraw-Hill, New York, 1977.

33. **Bruce, R. A.,** Normal values for VO2 and the VO2-HR relationship, *Am. Rev. Respir. Dis.,* 129, S41, 1984.

34. **Wasserman, K.,** The anaerobic threshold measurement to evaluate exercise performance, *Am. Rev. Respir. Dis.,* 129, S35, 1984.

35. **Jones, N. L. and Rebuck, A. S.,** Tidal volume during exercise in patients with diffuse fibrosing alveolitis, *Bull. Eur. Physiopathol. Respir.,* 15, 321, 1979.

36. **Lindgren, I., Muller, B., and Gaensler, E. A.,** Pulmonary impairment and disability claims, *JAMA,* 194, 111, 1965.

37. **Sue, D. Y. and Hansen, J. E.,** Normal values in adults during exercise testing, *Clin. Chest Med.,* 5, 89, 1984.

38. **Eschenbacher, W. L. and Mannina, A.,** An algorithm for the interpretation of cardiopulmonary exercise tests, *Chest,* 97, 263, 1990.

39. **Mellemgaard, K.,** The alveolar-arterial oxygen difference: its size and components in normal man, *Acta Physiol. Scand.,* 67, 10, 1966.

40. **Bates, D. V.,** *Respiratory Function in Disease,* 3rd ed., W. B. Saunders, Philadelphia, 1989.

41. **Jones, N. L., McHardy, G. J. R., Naimark, A., et al.,** Physiologic deadspace and alveolar-arterial gas pressures during exercise, *Clin. Sci.,* 31, 19, 1966.

42. **Hansen, J. E., Sue, D. Y., and Wasserman, K.,** Predicted values for clinical exercise testing, *Am. Rev. Respir. Dis.,* S49, 1984.

43. **Ositguy, G. L.,** Summary of task force report on occupational respiratory disease (pneumoconiosis), *Can. Med. Assoc. J.,* 121, 414, 1979.

44. **Richman, S. I.,** Why change? A look at the current system of disability determination and worker's compensation for occupational lung disease, *Ann. Intern. Med.,* 97, 908, 1982.

# Chapter 8

# Immunological Mechanisms, Asthma, Hypersensitivity Pneumonitis, and Related Disorders

*Gail M. McNutt, M.D. and Jordan N. Fink, M.D.*

## CONTENTS

## INTRODUCTION

There is increasing evidence for immune pathogenesis as a basis for many pulmonary diseases. Furthermore, the underlying immunologic mechanisms have been defined in many instances. The Gell and Coombs[1] classification provides a useful framework for classifying the immune mechanisms of pulmonary disease.

Gell and Coombs Type I, or immediate hypersensitivity reactions, have long been known to be mediated by the various factors released when the Fab epitopes of mast cell-bound IgE are crosslinked by specific antigen. Anaphylaxis and allergic asthma are two well-defined examples of immediate hypersensitivity reactions involving the lung.

Antibody-dependent cytotoxic reactions, or Gell and Coombs Type II reactions, are involved in a variety of human diseases. Goodpasture's syndrome, in which antibodies are directed against basement membrane components common to lung and kidney, is the best example involving the lung.

Extensive experimental evidence supports the role of immune complex-mediated injury, Gell and Coombs Type III reactions, in the lung. Immune complex injury has been implicated in idiopathic interstitial pneumonitis,[2,3] pulmonary eosinophilic granuloma,[4] pulmonary systemic lupus erythematosus,[5,6] rheumatoid arthritis,[7,8] hypersensitivity pneumonitis,[9] and sarcoidosis.[10,11]

Table 1   **Mediators of Immediate Hypersensitivity**

| | |
|---|---|
| **Preformed** | |
| Histamine | Increased vascular permeability, small muscle contraction, prostaglandin stimulation |
| Eosinophil chemotactic factor of anaphylaxis (ECF-A) | Chemotactic for eosinophils, neutrophils |
| Neutrophil chemotactic factor | Chemotactic for eosinophils, neutrophils |
| Kininogenase | Kinin generation |
| **Newly generated** | |
| Prostaglandins (E, F, D series) | Predominant bronchoconstriction in lung |
| Leukotrienes (LTC$_4$, D$_4$, E$_4$) | Increased vascular permeability, small muscle contraction, neutrophil chemotaxis |
| Platelet activating factor (PAF) | Platelet aggravation, neutrophil chemotaxis, vasodilation |
| **Preformed/granule associated** | |
| Trypsin | C$_3$ cleavage |
| Heparin | Antithrombin action |

Finally, there is strong evidence to support cell-mediated immunologic injury, Gell and Coombs Type IV reactions, in chronic lung diseases such as interstitial pulmonary fibrosis, chronic mycobacterial infections, sarcoidosis, chronic berylliosis, and emphysema. The neutrophil and its products have been implicated in the development of emphysema,[12] while the T-lymphocyte and macrophage play a role in granuloma formation and/or fibrosis.[13]

## IMMUNOLOGIC MECHANISMS IN IMMEDIATE HYPERSENSITIVITY

The pulmonary manifestations of immediate hypersensitivity reactions may vary from very mild bronchoconstriction to severe obstruction. Pathologic alterations in bronchial smooth muscle tone, mucus secretion, and vascular permeability result from a variety of mediators released when antigen combines with mast cell-bound specific IgE. Mediator release is responsible not only for the immediate reaction but also for the late phase reaction occurring 2 to 8 h after mast cell degranulation. The time course of the reaction is determined by the type and amount of mediator, its duration of action, and the rate of granule dissolution.

Mast cells are abundant in the lung, accounting for 2% of alveolar tissue. Degranulation of mast cells is accomplished within 1 to 2 min after activation.[15] The granules, which initially are identified in the immediate cell vicinity, are found at a distance at 30 to 60 min after degranulation. Mast cell mediators may be classified as preformed, newly generated, or preformed/granule associated (Table 1). Preformed mediators are contained in the granule matrix and are rapidly released upon antigen-antibody combining. Newly generated mediators are those generated by the interaction of arachidonic acid with adjacent cell wall membrane enzymes, resulting in products of cyclooxygenase and lipooxygenase pathways. Granule-associated mediators, while preformed, are released slowly as dissolution of the granule occurs.

Recently, there has been substantial progress in identifying the antigen with which anti-GBM antibody results. Basement membrane structure is complex, consisting of a type IV collagenous backbone, which supports a variety of other determinants. Among these determinants are the glycoproteins, fibroneitin, laminin, and intactin. Wieslander has localized the relevant antigen in Goodpasture's syndrome to the non-collagenous region of the type IV collagen in glomerular basement membrane.[43] It has been postulated that tubular, glomerular, and alveolar basement membrane all contain the same distinct type IV collagen chain, in which the Goodpasture's antigen resides.[44] The antigen is not demonstrable in neonates,[45] and its expression appears to be polymorphic.[46]

## TYPE III—IMMUNE COMPLEX-MEDIATED DISEASE

Many forms of lung pathology have been associated with the presence of circulating immune complexes, particularly since methods for detecting immune complexes have become more sensitive. The demonstration of tissue deposition of immune complexes is much less frequent. Furthermore, the presence of circulating immune complexes does not imply causality, and the evidence for immune complex pathogenesis in lung disease is indirect.

### IMMUNE COMPLEX FORMATION

Immune complex formation in the lung may occur through several mechanisms. Bronchus-associated lymphoid tissue (BALT)[47] may be responsible for the production of the specific antibody to the inhaled antigen, resulting in immune complex formation at the site. Alternatively, systemically formed antigen-antibody complexes may be deposited in lung tissue. Cochrane[48] has demonstrated deposition of circulating immune complexes in pulmonary vasculature, as well as the predominant glomerular deposition. Pulmonary vascular deposition, however, did not result in disease. Finally, antigenic structures exposed in the lung, perhaps through epithelial denudation, may result in antibody production and immune complex localization.

### MEDIATORS IN TISSUE INJURY

Experimentally, in acute serum sickness, pulmonary vasculitis has been demonstrated, while in chronic serum sickness, pulmonary fibrosis has been seen. In a variety of experimental models, complement has been found to be essential to immune complex lung injury. Depletion of complement via cobra venom factor reduced injury to the lung by $>90\%$.[49,50] It is well documented that antibody-antigen complexes activate the classical complement cascade. The ratio of antigen to antibody, which is a determinant of complex size and, therefore, of complement-binding activity, is a significant factor in the induction of inflammation. Complexes formed at equivalence are most active in complement fixation.[51] Complexes formed in marked antigen excess, being smaller, have less complement-fixing ability and diminished inflammatory potential. Activation of the complement cascade results in inflammation through multiple mechanisms. The C3a and C5a anaphylatoxins produced cause increased vascular permeability. In addition, C5a is chemotactic for both eosinophils and neutrophils. Eosinophils and neutrophils are each capable of releasing a variety of tissue-damaging enzymes. The specific granules of the eosinophil contain lysosomal hydrolases, as well as the cationic protein, eosinophil-derived neurotoxin, and eosinophil peroxidase.[52] Neutrophils generate oxygen-derived metabolites such as superoxide ion ($O_2^-$), singlet oxygen ($.O_2$), hydroxyl radical ($.OH$), hydrogen peroxide ($H_2O_2$), and hypochlorous ion (HOCl). Oxygen-derived metabolites are believed to play a significant role in tissue injury.[53] In addition, the lysosomal granules of neutrophils contain enzymes capable of hydrolyzing collagen, elastin, and basement membrane. Complement- and neutrophile-dependent lung injuries have been demonstrated in the sensitized guinea pig[54] and the rabbit[55] following inhalation challenge.

The macrophage is among the phagocytic cells attracted by immune complexes. The functions of the macrophage are multiple and diverse. Macrophage receptors for the Fc fragment of IgG and for complement protein promote macrophage phagocytosis of antibody- or complement-coated molecules. In addition, the macrophage is a highly secretory cell, with a very broad range of secretory products. The macrophage is capable of secreting a variety of tissue-damaging enzymes, cytokines, tumor necrosis factor, fibronatin, multiple growth factors, platelet-activating factor, and arachadonate derivatives, among others. Clearly, the macrophage has a significant role in tissue injury.

## PULMONARY DISEASES ASSOCIATED WITH IMMUNE COMPLEXES

As previously discussed, while there has been no conclusive evidence that immune complexes play a pathogenic role in any human lung disease, immune complexes, either in circulation or tissue, have been associated with a variety of lung pathology (Table 2). Some of the more noteworthy examples are discussed below.

### Systemic Lupus Erythematosus (SLE)

Both parenchymal lung disease and pleural involvement are common in SLE. In fact, most patients with systemic lupus demonstrate some degree of parenchymal pulmonary disease.[56] The pneumonitis may range from a clinically silent to a fulminant hemorrhagic process and frequently parallels the overall severity of the systemic disease. In acute active pneumonitis, intrapulmonary DNA anti-DNA complexes in association with C3 have been demonstrated.[57] Much less frequently, the same histopathological features are seen in the chronic pneumonitis of SLE.[58] Electron-dense deposits under the alveolar basement membrane have also been identified in one case.[59] Of some significance, in the histologically normal lung tissue of patients with systemic disease, immune complexes have not been identified.[58,60] In pleural involvement in SLE, immune complexes are frequently demonstrated in pleural fluid, though at no increased concentration over that in paired serum.[61]

Despite the presence of circulating immune complexes in SLE, their pathogenesis in lung injury remains in question. As previously discussed, complement and neutrophils are felt to be two of the primary mediating factors in tissue inflammation, and while each is often present in association with immunoglobulin in the injured lung, neither is invariably present.[58] Even in the more fulminant forms of pneumonitis associated with SLE, neutrophils may be absent.[58]

### Rheumatoid Disease

As in SLE, pulmonary involvement is common in rheumatoid disease. Both rheumatoid factor[62] and cryoglobulin[63] levels have been found to correlate with the development of lung parenchymal disease. Immunoglobulin deposits, which contain rheumatoid factor, have been demonstrated in the lungs of affected patients.[64]

Pleural involvement is also common in rheumatoid disease, and immune complexes are almost always found in rheumatoid pleural fluid.[65] Furthermore, the concentration of immune complexes in rheumatoid pleural fluid often exceeds that of paired serum,[61] unlike the analogous situation in SLE. Complement levels in rheumatoid pleural fluid are lower than those in SLE, suggesting significant complement

Table 2   **Immune Complex-Associated Lung Disease**

Collagen vascular
   1) Systemic lupus erythematosus
   2) Rheumatoid disease
   3) Sjogren's syndrome
   4) Scleroderma
   5) Mixed connective tissue disease
   6) Polymyositis/dermatomyositis
Vasculitic
   1) Wegener's granulomatosis
   2) Polyarteritis nodosa
Other
   1) Hypersensitivity pneumonitis
   2) Idiopathic interstitial pneumonitis
   3) Sarcoidosis
   4) Eosinophilic granuloma

consumption. Taken together, these two factors are strong though indirect evidence for immune complex pathogenesis in rheumatoid pleural disease.

## Other Collagen Vascular Diseases

Circulating immune complexes have been demonstrated in a variety of collagen vascular diseases with associated pneumonitis, including Wegener's granulomatosis,[66] mixed connective tissue disease,[67] Sjogren's syndrome,[68] scleroderma,[69] polyarteritis nodasa,[70] and polymyositis/dermatomyositis.[71] The precise role of immune complexes in this group of diseases has yet to be defined.

## Hypersensitivity Pneumonitis

Hypersensitivity pneumonitis is characterized by fever, cough, myalgia, dyspnea, and leukocytosis, occurring 4 to 6 h after exposure to any of a variety of offending organic dust antigens. The histologic features of the acute disease include neutrophil, plasma cell, and mononuclear cell infiltration into the alveolar and interstitial wall. Noncaseating granuloma and fibrosis can be found in more chronic disease.[72,73] The majority of patients with hypersensitivity pneumonitis have high titer circulating precipitating antibody to the relevant antigen; however, so do a large number of asymptomatic people similarly exposed. Precipitating antibody has not been found to be predictive of disease activity and may not be relevant to pathogenesis.[74,75] Circulating complexes tend to be higher in chronic disease and show no correlation with time to last antigen exposure.[76] Immune complexes have rarely been demonstrated in the lung, suggesting that the lung process is different from that of the precipitating antibody.

Recent evidence also suggests a role for cell-mediated immunity and defective suppressor mechanisms in hypersensitivity pneumonitis.[77] Further research may reveal a combined immune complex and cell-mediated immunopathogenesis in this disease.

## Idiopathic Interstitial Pneumonitis

Idiopathic interstitial pneumonitis resembles that of the connective tissue disease, both clinically and histologically. Early in the process, the initial injury to the interstitium allows leakage of fluid and fibrin into the alveolar spaces, causing a patchy alveolitis. Various inflammatory cells are present, including macrophages, lymphocytes, and plasma cells. With ongoing inflammation, there is extensive destruction of the acinar structure and eventual fibrosis. Circulating immune complexes, as well as rheumatoid factor and antinuclear antibodies,[78] are frequently identified, even in the absence of clinical disease. Dreisen[79,80] reported a correlation between circulating immune complexes and disease activity. This issue remains controversial; however, Haslam[81] reported no association between complex levels and disease activity.

## TYPE IV—CELL-MEDIATED IMMUNE DISEASE

Gell and Coombs Type IV reactions are mediated through sensitized T-lymphocytes. Antigen, after processing by a phagocytic cell, is presented in association with Class II major histocompatibility determinants, to the T-lymphocyte. T-cell activation results from the recognition of the specific antigen in combination with major histocompatibility proteins. Activation of the T-cell leads in turn to clonal proliferation and cytokine secretion. Cytokines participate in cell recruitment, macrophage activation,[82] fibroblast proliferation, and inflammatory processes.[83,84] This sequence of events, designed to protect the host, can result in chronic inflammation, as demonstrated by granuloma formation or fibrosis, perhaps through continuous antigen presentation or dysregulation at some point in the system. The alveolar macrophage may play a key role in this disease through activation and secretion of pro-inflammatory cytokines and other mediators.

Cell-mediated immunity predominates in several forms of lung disease, all of which are characterized by granuloma formation and/or fibrosis. Chronic mycobacterial infections, sarcoidosis, and chronic berylliosis are the most notable examples.

### MYCOBACTERIAL DISEASE

Very early on in the study of tuberculosis, it was recognized that granuloma formation was necessary to contain the organism and provide resistance. Recently, granuloma formation was recognized as a manifestation of T-cell-mediated immunity.[85] The T-cell and macrophages play heavily interdependent roles in cell-mediated immunity. The activated macrophage has been clearly correlated with increased but nonspecific microbicidal activity,[86] which is not surprising given its role in antigen presentation and immunoregulation via numerous secretory products. Microbial resistance and presumably macrophage activation cannot, however, be induced in T-cell-deficient animals.[87] In turn, it is the macrophage presentation of antigen in conjunction with major histocompatibility proteins that serves as an immunogen for the T-lymphocyte. Macrophage secretion of interleukin 1 is also crucial to the T-lymphocyte response to antigen. T-lymphocyte proliferation may amplify the cellular immune response and play a role in immunologic memory.

### SARCOIDOSIS

Sarcoidosis is a multisystem granulomatous disorder of unknown etiology, with thoracic involvement demonstrated in 90% of cases at diagnosis. Histologically, noncaseating epithelioid granulomas are the hallmark of this disease. An initial mononuclear cell alveolitis precedes the development of granulomas.[88]

Several abnormalities in cell-mediated immunity are characteristic in sarcoidosis, including anergy to recall antigens on skin testing, depressed proliferative response of lymphocytes to mitogen, and lymphocytopenia.

Bronchoalveolar lavage, however, presents an entirely different picture. Increased numbers of helper lymphocytes have been demonstrated consistently in lavage fluid in patients with sarcoidosis.[89] Furthermore, these T-lymphocytes display decreased surface T-cell receptor molecules and express increased numbers of T-cell receptor B-chain mRNA transcripts, as well as secrete IL-2, all evidence of activation.[90]

Taken together, these findings indicate that sarcoidosis is not a result of generalized depression of cell-mediated immunity but rather the result of heightened, localized cell-mediated immunity. While the etiology of sarcoidosis remains unknown, current evidence certainly suggests an ongoing cellular immune process responding to an unknown antigen.

## CHRONIC BERYLLIUM DISEASE

Chronic beryllium disease produces a noncaseating granulomatous pneumonitis. Histologically, as well as noncaseating granulomas, there is a marked diffuse lymphocytic interstitial infiltrate. On light microscopy, the granulomas are indistinguishable from those of sarcoidosis. With more sophisticated techniques, beryllium has been identified within the granulomas.[91]

Antibodies to beryllium have not been identified, but mononuclear cell blast transformation on exposure to beryllium has been demonstrated.[92] Bronchoalveolar lavage studies in berylliosis are similar to those of sarcoidosis, demonstrating increased numbers of T-lymphocytes. As in sarcoidosis, these T-lymphocytes show evidence of activation. Furthermore, incubation of this population of lymphocytes with beryllium salts results in proliferation.[93] These features certainly point to a cellular immune response with beryllium as the antigenic stimulation. Given its extremely low molecular weight, it is most likely that beryllium is acting as a hapten, although the carrier protein(s) remain unknown.

## CLINICAL DISORDERS

### ASTHMA

Asthma is a complex disorder with immunologic, neurogenic, and inflammatory mechanisms as its basis. It is manifested as airway hyperresponsiveness, with reversibility of increased airflow resistance.

Asthma affects all races, ages, and both sexes to about 6% of the population.[94] In children and young adults, boys are more commonly affected than girls, and the prevalence is increased in blacks and Hispanics.[95] Recent evidence suggests that the mortality from asthma is increasing in much of the Western world.[96] This increase may be related to socioeconomic factors, to availability of health care, or to overuse of drugs for self-medication.

#### Histopathology

Asthma occurs as an inflammatory bronchitis comprising eosinophils, lymphocytes, neutrophils, and respiratory cells.[97] There may be denuding of the respiratory epithelium, with a thickened underlying basement membrane and fibrosis. Edema of the submucosa and infiltration with round cells may be prominent. Mucus-secreting glands are hypertrophied, and the muscular layer may be thickened. Recent evidence suggests that even in mild asthma, the inflammatory process may be demonstrated, pointing to the importance of the late phase allergic inflammatory response in the genesis of the disease.[98]

#### Pulmonary Function

Asthmatics may present with normal pulmonary function parameters.[99] On aerosol challenge with methacholine or histamine in increasing doses, however, there is a rapid reduction in airflow, with decreased 1 sec forced expiratory volume, which may be rapidly reversed with bronchodilators. This hyperreactivity of the airways is clinically manifest in the respiratory response to environmental stimuli, such as smoke, paint fumes, cold air, and pungent odors, and the degree of hyperreactivity often correlates with the severity of the asthma.[99,100]

During the acute episode, there is, aside from the obstruction, hyperinflation of the lungs, increase in work of breathing, and changes in distribution of ventilation and perfusion, resulting in reduced arterial oxygen tension due to the gas exchange abnormalities. During acute episodes, carbon dioxide tension is reduced due to hyperventilation and respiratory alkalosis results. With progression of the obstruction, there may be alveolar hypoventilation with resulting hypercapnia.[97] Following treatment, the abnormalities improve, but small airway flow as manifested by decreased forced expiratory flow at 25 to 75% may persist, even as the patient becomes asymptomatic.

## Clinical Presentation

Patients with asthma may present with overt episodes of cough and wheezing during contact with an allergen, whether it be airborne, such as pollens, molds, danders, or mites, or ingested as foods. Other patients may present with persistent wheezing after a viral respiratory infection or only with nocturnal cough or chest tightness. Physical examination may be completely normal, wheezes may be heard on expiration, or breath sounds may be generally decreased. Airflow obstruction may be confirmed by simple peak flow measurements with a hand-held peak flow meter or by spirometry before and after a bronchodilator. If pulmonary function is normal, hyperresponsiveness of the airways may be demonstrated by methacholine aerosol challenge, exercise on a treadmill, or hyperventilation of cold, dry air.[99]

In order to define possible immunologic aspects of the disease, skin tests with common inhalant allergens may be carried out. A wheal and flare reaction occurring in 20 min at the site of a prick test indicates the presence of IgE antibody. An alternate method of detecting such antibody is a radioallergosorbent test (RAST), using patient's serum and specific antigen coupled to a insoluble particle. However, the demonstration of IgE antibody must be correlated with the clinical picture to determine the role of the allergen-antibody reaction in the disease.

## Associated Disorders

In the evaluation of the asthmatic, other coexisting disorders need to be considered and, if found, treated. Sinusitis often accompanies asthma, especially if the patient has allergic rhinitis or aspirin sensitivity.[101] Gastroesophageal reflux may aggravate the asthmatic, as may cystic fibrosis, immune deficiency, or pulmonary infiltration with eosinophils due to fungal colonization of the airways.[102] Appropriate studies and therapy of such associated diseases may be necessary in order to bring the asthma under control.

## Treatment

The primary goal of therapy of asthma is to control the patient's disease with as little interference in lifestyle as possible. This may be achieved in most asthmatics by pharmacotherapy and in those with demonstrable allergic responses by immunologic intervention. Because of recent evidence implicating a role for inflammation, therapy is now directed toward control of that process at an early stage following diagnosis. Thus, the early use of cromolyn sodium or inhaled corticosteroids has increased in order to control inflammation which is believed to be the most important pathogenetic mechanism in the disease. Bronchodilators, such as beta-agonists are used for control of symptoms, and an anticholinergic agent, such as ipatroprium bromide, may also be added.[102] With increasing severity of the disease, increasing inhaled drug may be used or oral corticosteroid may be added. Sustained-release theophylline may be useful in control of nocturnal symptoms. If the patient has more severe disease, more sustained-release drug or longer courses of oral corticosteroids may be needed, accepting the side-effects of the drug over control of the disease.

Immunologic manipulations may include avoidance measures and immunotherapy. House dust mite sensitivity may be controlled by environmental measures to reduce mite concentration. These include the use of airtight covers on mattresses, box springs, and pillows, hot washing of bedding to kill mites, and reduction of humidity to below 40% to reduce mite growth.[103] Reduction of animal dander contact, however, is often difficult because of the attachment of man to his pets.

Immunotherapy is the systematic injection of the allergen to which the patient is sensitive over a period of time in order to change the immune response to the environment. Injections are given at increasing intervals in increasing amounts over time, reaching the highest dose possible, depending on the patient's sensitivity. Measurable immunologic changes include reduction in IgE levels, reduction of mediator release, blunting of the late allergic response, and reduction in symptoms during the season.[102]

## OCCUPATIONAL ASTHMA

Approximately 15% of asthma is estimated to be due to occupational exposures, although the prevalence usually depends on the occupation and the exposure involved.[104] For example, 5% of workers exposed to Western red cedar develop asthma,[105] and up to 50% of workers exposed to certain enzymes may become ill.[106]

Asthma may occur in occupational settings as a result of reflex bronchoconstriction when, in cold air, low concentrations of noxious fumes and irritants directly stimulate pulmonary irritant receptors.[107]

Bronchoconstriction may also follow exposure to high concentration of irritant chemicals, such as chlorine, sulfur dioxide, or smoke. In such cases, symptoms of asthma with cough, wheeze, and dyspnea may occur within hours of exposure, and airway hyperresponsiveness may result.[108] This syndrome has been referred to as "reactive airways dysfunction syndrome," and the symptoms may persist for years, often stimulated by previously tolerated environmental agents. The treatment includes avoidance of irritants and pharmacotherapy for the bronchoconstriction and inflammation underlying the syndrome.

Occupational agents may also induce bronchoconstriction in a pharmacologic fashion, perhaps through nonimmunologic mediator release from mast cells or by direct action on irritant receptors.[107] Agents such as cotton dust, grain dust, and organophosphate insecticides may act in such a manner. Vegetable dusts are complex mixtures that may contain histamine-like substances, endotoxin, or microorganisms as contaminants.[109,110] Inflammation and airway hyperresponsiveness are not usual in such patients.

Allergic mechanisms appear to be responsible for the majority of individuals developing occupational asthma, and over 200 agents have been identified.[104,105] Affected patients may present differently, depending upon whether the agent involved is of high or low molecular weight.

High molecular weight antigens include proteins, peptides, and polysaccharides and induce demonstrable IgE-mediated reactions. Such patients usually have an allergic background, and immediate wheal and flare skin reactions can be demonstrated to the occupational allergen. These patients usually present with immediate or dual-phase asthmatic reactions. Low molecular weight antigens often act as haptenes, combining with carrier proteins, often respiratory proteins, to act as antigens. Such reactive chemicals include anhydrides, isocyanates, and plicatic acid. About 5 to 10% of workers are affected, and the late asthmatic reactions predominate. Such workers are less likely to have an allergic background.

## Diagnosis

The diagnosis of occupational asthma rests on a detailed history relating the asthma to the occupational exposure. Often there are symptoms associated with the work place, with relief in weekends or vacation periods. Late responders may manifest symptoms in the evening or in the early hours of the morning. As exposure continues, the symptoms may progress in severity and in duration until there are few periods of relief; the patient then appears to have chronic asthma.

Objective documentation of the relationship between the patient's symptoms and workplace exposure can be accomplished by inhalation challenge tests, either by measuring lung function before and after a work shift, serial recording of the peak expiratory flow at home and at work, or purposeful challenge with the suspected offending agent in the pulmonary function laboratory. The tests in the laboratory should include careful monitoring of the patient following inhalation of a substance at a concentration approaching that of the workplace. Monitoring of the patient may then demonstrate an immediate late or dual respiratory response.[111] In some cases, the respiratory response is minimal, but there is a shift in the methacholine-induced hyperresponsiveness.[112] This suggests that the offending agent may induce airway inflammation and resulting hyperresponsiveness without demonstrable functional change.

## Therapy
Treatment of occupational asthma includes both immunologic and pharmacologic intervention. Immunotherapy is limited to avoidance measures, as injection of potentially toxic chemicals is not warranted. Pharmacotherapy of occupational asthma is similar to nonoccupational asthma.

## HYPERSENSITIVITY PNEUMONITIS
Hypersensitivity pneumonitis is an immunologically mediated pulmonary inflammation due to repeated exposure and sensitization to any of a wide variety of environmental organic dusts. The disease affects the interstitium and peripheral airways, and the patients develop both systemic and respiratory symptoms following inhalation of the offending dust.[113] The acute form of the disease is the most common form and presents as acute episodes of chills, fever, cough, malaise, and myalgia, occurring 4 to 6 h after exposure to an antigen to which an individual has become sensitized, and persisting for up to 24 h.[114] Physical examination characteristically reveals end inspiratory bibasilar rales, and the chest X-ray may demonstrate interstitial infiltrates. Laboratory studies reveal a leukocytosis with a shift to young forms. Pulmonary function demonstrates a restrictive impairment, with a decrease in diffusion capacity and hypoxemia. All features return to normal as the patient recovers, only to recur with subsequent exposures.

A smaller number of exposed individuals develop insidious disease, which may present as a subtle, productive cough, with or without associated anorexia and weight loss. Such patients do not usually have acute episodes but develop gradual pulmonary impairment, which may progress to irreversibility.[115] The chest X-ray may demonstrate evidence of fibrosis or hyperinflation, and laboratory studies are similar to the acute episodes. With continued exposure, irreversible pulmonary fibrosis or high grade obstruction may occur with resulting permanent pulmonary impairment.

## Organic Dusts Associated with Disease
A wide variety of organic dusts have been shown to be associated with hypersensitivity pneumonitis. In most cases, occupational exposures with inhalation of environmental antigen induce disease. Farming and sugar cane and mushroom growing may generate thermophilic actinomycetes, which contaminate vegetable compost. Avian proteins may be inhaled in pigeon- or other bird-breeding situations, and Penicillium species may contaminate cheese- or cork-processing plants. Low molecular weight chemicals, especially plasticizers, such as anhydrides or isocyanates, may also

induce hypersensitivity pneumonitis, as may bacteria and fungi contaminating forced air heating, cooling, and ventilation systems.[113]

## Diagnosis

As with occupational asthma, a careful and detailed history is essential if the diagnosis is to be entertained. Characteristically, the patient's history is that of repeated flu-like episodes shortly after exposure to the offending agent. Immune studies reveal serum-precipitating antibodies to that agent, and bronchoalveolar lavage reveals a marked lymphoptosis, with predominance of suppressor T-cells.[116] Chest X-ray and lung function studies may provide additional clues to the diagnosis. Lung biopsy will provide the diagnosis of the disease, with a lymphocytic alveolitis, interstitial infiltration with lymphocytes, plasma cells, and granulomas.[117] In late stages of the disease, fibrosis may predominate.

As in occupational asthma, purposeful inhalation challenge may relate a particular environment or agent to the disease. Patients may be challenged by measuring pulmonary function pre- and postexposure to the environment or by purposeful aerosol challenge in the laboratory. In either case, repeated evaluation following exposure may detect the characteristic systemic and respiratory response, with pulmonary function and laboratory abnormalities.

## Therapy

Avoidance continues to be the mainstay of therapy of hypersensitivity pneumonitis. If pulmonary parenchymal damage is not severe, there is complete recovery, with return of all parameters to normal. If pharmacotherapy is to be used to hasten recovery, corticosteroids remain the drug of choice. Other antiallergic drugs have no substantial clinical effect on hypersensitivity pneumonitis.

## REFERENCES

1. **Gell, P. G. H., Coombs, R. R. A., and Lachmann, P. J.,** *Clinical Aspects of Immunology,* 3rd ed., Blackwell, Oxford, 1975.
2. **Dreisin, R. B., Schwarz, M. I., Theofilopoulos, A. N., et al.,** Circulating immune complexes in idiopathic interstitial pneumonias, *N. Engl. J. Med.,* 298, 535, 1978.
3. **Haslam, P., Thompson, B., Mohammed, I., et al.,** Circulating immune complexes in patients with cryptogenic fibrosing alveolitis, *J. Immunol.,* 37, 381, 1979.
4. **King, T. E., Jr., Marvin, I., Schwarz, M. I., et al.,** Circulating immune complexes in pulmonary eosinophilic granuloma, *Ann. Intern. Med.,* 91, 3, 1979.
5. **Inoue, T., Kanayama, Y., Ohe, A., et al.,** Immunopathologic studies of pneumonia in systemic lupus erythematosus, *Ann. Intern. Med.,* 91, 30, 1979.
6. **Rodriguez-Iturbe, B., Garcia, R., Rubio, L., et al.,** Immunohistochemical findings in the lung in systemic lupus erythematosus, *Arch. Pathol. Lab. Med.,* 101, 342, 1977.
7. **Frank, S. T., Weg, J. G., Harkleroad, L. E., et al.,** Pulmonary dysfunction in rheumatoid disease, *Chest,* 63, 27, 1973.
8. **Wallser, W. C. and Wright, V.,** Pulmonary lesions and rheumatoid arthritis, *Medicine,* 47, 501, 1968.
9. **Roberts, R. L. and Moore, V. L.,** Immunopathogenesis of hypersensitivity pneumonitis, *Am. Rev. Respir. Dis.,* 116, 1075, 1977.
10. **Gupta, R. C., Kueppers, F., DeRemee, R. A., et al.,** Pulmonary and extrapulmonary sarcoidosis in relationship to circulating immune complexes, *Am. Rev. Respir. Dis.,* 116, 261, 1977.

11. **Hedfors, E. and Nordberg, R.,** Evidence for circulating immune complexes in sarcoidosis, *Clin. Exp. Immunol.,* 16, 493, 1974.

12. **Kueppers, F. and Block, L. E.,** Alpha 1 antitrypsin and its deficiency, *Am. Rev. Respir. Dis.,* 110, 176, 1974.

13. **Wahl, S. M., Wahl, L. M., and McCarthy, J. B.,** Lymphocyte mediated activation of fibroblast proliferation and collagen production, *J. Immunol.,* 121, 942, 1978.

14. **Wasserman, S. I.,** The lung mast cell: its physiology and potential relevance to defense of the lung, *Environ. Health Perspect.,* 35, 153, 1980.

15. **Metcalfe, D. D., Kaliner, M., and Donlin, M. A.,** The mast cell, *CRC Crit. Rev. Immunol.,* 3, 23, 1981.

16. **Kaliner, M., Orange, R. P., and Austen, K. F.,** Immunologic release of histamine and slow reacting substance of anaphylaxis from human lung. IV. Enhancement by cholinergic and alpha adrenergic stimulation, *J. Exp. Med.,* 136, 1972.

17. **Konig, W., Zarnetzki, B. M., and Lichtenstein, L. M.,** Eosinophil chemotactic factor (ECF). II. Release from human polymorphonuclear leukocytes during phagocytosis, *J. Immunol.,* 117, 235, 1976.

18. **Lewis, R. A., Soter, N. A., and Diamond, P. T.,** Prostaglandin $D_2$ generation after activation of rat and human mast cells with anti-IgE, *J. Immunol.,* 129, 1627, 1982.

19. **MacGlushan, D. W., Schleimer, R. P., Peters, S. P., et al.,** Generation of leukotrienes by purified human lung mast cells, *J. Clin. Invest.,* 70, 747, 1982.

20. **Borgeat, P., Hambeng, M., Samuelson, N., et al.,** Transformation of arachidonic and homo gamma linolenic acid by rabbit polymorphonuclear leukocytes, *J. Biol. Chem.,* 251, 7816, 1976.

21. **Jorg, A., Henderson, W. R., Murphy, R. C., et al.,** Leukotriene generation by eosinophils, *J. Exp. Med.,* 155, 390, 1982.

22. **Rouzer, C. A., Scott, W. A., Cohn, Z. A., et al.,** Mouse peritoneal macrophages release leukotriene C in response to a phagocytic stimulus, *Proc. Natl. Acad. Sci. U.S.A.,* 71, 4928, 1980.

23. **Drazen, J. M., Austen, K. F., Lewis, R. A., et al.,** Comparative airway and vascular activities of leukotrienes C-1 and D *in vivo* and *in vitro, Proc. Natl. Acad. Sci. U.S.A.,* 77, 4354, 1980.

24. **Manoni, Z., Shelhammer, J. H., and Bach, M. K.,** Slow reacting substances, leukotrienes C4 and D4 increase the release of mucus from human airways *in vitro, Am. Rev. Respir. Dis.,* 126, 449, 1982.

25. **Camussi, G., Aglieita, M., Cocla, R., et al.,** Release of platelet activating factor (PAF) and histamine, *Immunology,* 42, 191, 1981.

26. **Roubin, R., Mencia-Huereta, J. M., and Benveniste, J.,** Release of platelet activating factor (PAF-acether) and leukotrienes C and D from inflammatory macrophages, *Eur. J. Immunol.,* 12, 141, 1982.

27. **Lotner, G. Z., Lynch, J. M., and Betz, S. J.,** Human neutrophile derived platelet activating factor, *J. Immunol.,* 124, 676, 1980.

28. **McManus, L. M., Hanahan, D. J., and Pinckard, R. N.,** Human platelet stimulation by acetyl glycerol ether phosphorylcholine, *J. Clin. Invest.,* 67, 903, 1980.

29. **Halonen, M., Palmer, J. D., and Lohman, K.,** Differential effects of platelet depletion on the physiologic alterations of IgE anaphylaxis and acetyl glycerol ether phosphorylcholine infusion in the rabbit, *Am. Rev. Respir. Dis.,* 124, 416, 1981.

30. **Shaw, J. O., Pinckard, R. N., Ferrigni, K. S., et al.,** Activation of human neutrophils with 1-0-hexadecyl/octadecyl-2-acetyl-sn-glycerol-3-phosphorylcholine (Platelet Activation Factor), *J. Immunol.,* 127, 1250, 1981.

31. **Halonen, M., Palmer, J. D., Lohman, I. L., et al.,** Respiratory and circulatory alterations induced by acetyl glycerol ether phosphorylcholine, a mediator of IgE anaphylaxis in the rabbit, *Am. Rev. Respir. Dis.,* 122, 915, 1980.

32. **Schwartz, L. B., Kawakra, M. S., Hugli, T. E., et al.,** Generation of C3a anaphylatoxin from human C3 by human mast cell tryptase, *J. Immunol.,* 130, 1891, 1983.

33. **Weiler, J. M., Yurt, R. W., Fearon, D. T., et al.,** Modulation of the formation of the amplification convertase of complement c3b,BB by native and commercial heparin, *J. Exp. Med.,* 147, 409, 1978.

34. **Border, W. A., Bachler, R. W., Bhathena, D., et al.,** IgA anti-basement membrane nephritis with pulmonary hemorrhage, *Ann. Intern. Med.,* 91, 21, 1979.

35. **Loughlin, G. M., Taussig, L. M., Murphy, S. A., et al.,** Immune complex mediated glomerulonephritis and pulmonary hemorrhage simulating Goodpasture's syndrome, *J Pediatr.,* 93, 181, 1978.

36. **Beirne, G. J., Kopp, W. L., and Zimmerman, S. W.,** Goodpasture's syndrome, *Arch. Intern. Med.,* 132, 261, 1983.

37. **Pasternack, A., Tornroth, T., and Linder, E.,** Evidence of both anti-GBM and immune complex mediated pathogenesis in the initial phase of Goodpasture's syndrome, *Clin. Nephrol.,* 9, 77, 1978.

38. **Wilson, C. B. and Dixon, F. J.,** The renal response to immunologic injury, in *The Kidney,* Brenner, B. M. and Rector, F. C., Eds., W. B. Saunders, Philadelphia, 1981.

39. **Wilson, C. B., Holdsworth, S. R., and Neule, T. J.,** Anti-basement membrane antibodies in immunologic renal disease, *Aust. N.Z. J. Med.,* 11 (Suppl.), 94, 1981.

40. **Lockwood, C. M., Pearson, T. A., Rees, A. J., et al.,** Immunosuppression and plasma exchange in the treatment of Goodpasture's syndrome, *Lancet,* 1, 711, 1976.

41. **Johnson, J. P., Whitman, W., and Briggs, W. A.,** Plasmapheresis and immunosuppressive agents in anti-basement membrane antibody induced Goodpasture's syndrome, *Am. J. Med.,* 64, 354, 1978.

42. **Jennings, L., Rohot, O. A., Pressman, D., et al.,** Experimental anti-alveolar basement membrane antibody mediated pneumonitis, *J. Immunol.,* 127, 129, 1981.

43. **Wieslander, J., Barr, J. F., Butkowski, R. J., et al.,** Goodpasture's antigen of the glomerular basement membrane: localization to noncollagenous regions of type IV collagen, *Proc. Natl. Acad. Sci. U.S.A.,* 81, 3838, 1984.

44. **Pusey, C. D., Dash, A., Kershaw, M. J., et al.,** A single autoantigen in Goodpasture's syndrome identified by a monoclonal antibody to human glomerular basement membrane, *Lab. Invest.,* 56, 23, 1987.

45. **Anand, S. K., Landing, B. H., Heuser, E. T., et al.,** Changes in glomerular basement membrane antigen(s) with age, *J. Pediatr.,* 92, 952, 1978.

46. **McCoy, R. C., Johnson, H. K., Stone, W. J., et al.,** Variations in glomerular basement membrane antigens in hereditary nephritis, *Lab. Invest.,* 34 (Abstr.), 325, 1976.

47. **Bienenstock, J., Clancy, R. L., and Percy, D. Y. E.,** Bronchus-associated lymphoid tissue (BALT): its relationship to mucosal immunity, in *Immunologic and Infectious Reactions of the Lung,* Kirkpatrick, C. H. and Reynolds, H. Y., Eds., Marcel Dekker, New York, 1976, 29.

48. **Cochrane, C. G.,** Studies in the localization of circulating antigen-antibody complexes and other macro-molecules in vessels. II. Pathogenetic and pharmacodynamic studies, *J. Exp. Med.,* 118, 503, 1963.

49. **Johnson, K. G. and Ward, P. A.,** Acute immunologic pulmonary alveolitis, *J. Clin. Invest.,* 54, 349, 1974.

50. **Scheizer, H. and Ward, P. A.,** Lung and dermal vascular injury produced by preformed immune complexes, *Am. Rev. Respir. Dis.,* 117, 551, 1978.
51. **Scheizer, H. and Ward, P. A.,** Lung injury produced by immune complexes of varying composition, *J. Immunol.,* 121, 947, 1978.
52. **Peters, M. S., Rodriguez, M., and Gleich, G. J.,** Localization of human eosinophil granule major basic protein, eosinophil cationic protein and eosinophil-derived neurotoxin by immunoelectron microscopy, *Lab. Invest.,* 54, 656, 1986.
53. **Klebanoff, S. J.,** Oxygen metabolism and the toxic properties of phagocytes, *Ann. Intern. Med.,* 93, 480, 1980.
54. **Richerson, H. B.,** Acute experimental hypersensitivity pneumonitis in the guinea pig, *J. Lab. Clin. Med.,* 79, 745, 1972.
55. **Zavala, D. C., Rhodes, M. L., Richerson, H. B., and Oskvis, R.,** Light and immunofluorescent study of the Arthus reaction in rabbit lung, *J. Allergy Clin. Immunol.,* 56, 450, 1975.
56. **Silberstein, S. L., Barland, P., Grayzel, A., et al.,** Pulmonary dysfunction in systemic lupus erythematosus. Prevalence, classification and correlation with other organ involvement, *J. Rheumatol.,* 7, 187, 1980.
57. **Inoue, T., Kanayama, Y., Ohe, A., et al.,** Immunopathologic studies of pneumonitis in systemic lupus erythematosus, *Ann. Intern. Med.,* 91, 30, 1979.
58. **Eagen, J. W., Roberts, J. L., Schwartz, M. M., et al.,** The composition of pulmonary immune deposits in systemic lupus erythematosus, *Clin. Immunol. Immunopathol.,* 12, 204, 1979.
59. **Kuhn, C.,** Systemic lupus erythematosus in a patient with ultrastructural lesions of the pulmonary capillaries previously reported in the review as due to idiopathic pulmonary hemosiderosis, *Am. Rev. Respir. Dis.,* 106, 931, 1972.
60. **Eagen, J. W., Memoli, V. A., Roberts, J. L., et al.,** Pulmonary hemorrhage in systemic lupus erythematosus, *J. Med.,* 57, 545, 1978.
61. **Halla, J. T., Shrohenlober, R. E., and Volanakis, J. E.,** Immune complexes and other laboratory features of pleural effusions, *Ann. Intern. Med.,* 92, 748, 1980.
62. **Tomasi, T. B., Fredenberg, H. H., and Finby, H.,** Possible relationship of rheumatoid factors and pulmonary disease, *Am. J. Med.,* 33, 243, 1962.
63. **Weisman, M. and Zvaifler, N.,** Cryoimmunoglobulinemia in rheumatoid arthritis, *J. Clin. Invest.,* 56, 725, 1975.
64. **Deltoratius, R. J., Abruzzo, J. L., and Williams, R. L.,** Immunofluorescent and immunologic studies of rheumatoid lung, *Arch. Intern. Med.,* 129, 441, 1972.
65. **Hunder, G. G., McDuffie, F. L., Huston, K. A., et al.,** Pleural fluid complement, complement conversion and immune complexes in immunologic and non-immunologic diseases, *J. Lab. Clin. Med.,* 90, 971, 1977.
66. **Shasby, D. M., Schwarz, M. I., Forstat, J. Z., et al.,** Pulmonary immune complex deposition in Wegener's granulomatosis, *Chest,* 81, 338, 1982.
67. **Parker, M. D. and Marion, T.,** Circulating complement-fixing immune complexes in mixed connective tissue disease, *Clin. Res.,* 26, 374A, 1978.
68. **Lawley, T. J., Moutsopoulos, H. M., Ketz, S. I., et al.,** Demonstration of circulating immune complexes in Sjogren's syndrome, *J. Immunol.,* 123, 1382, 1979.
69. **Cunningham, P. H., Andrews, B. S., and Davis, J. S.,** Immune complexes in patients with progressive systemic sclerosis and mixed connective tissue disease, *Clin. Res.,* 26, 374A, 1978.
70. **Prince, A. M. and Trepo, L.,** Role of immune complexes involving SH antigen in pathogenesis of chronic hepatitis and polyarteritis nodosa, *Lancet,* 1, 1309, 1971.

71. **Schwarz, M. I., Matthay, R. A., Sahn, S. A., et al.,** Interstitial lung disease in polymyositis and dermatomyositis: analysis of six cases and review of the literature, *Medicine,* 55, 89, 1976.
72. **Wenzel, F. J., Emanuel, D. A., and Gray, R. L.,** Immunoflourescent studies in patients with farmer's lung, *J. Allergy Clin. Immunol.,* 48, 224, 1971.
73. **Ghose, T., Landrigan, P., Kileen, R., et al.,** Immunopathologic studies in patients with farmer's lung, *Clin. Allergy,* 4, 119, 1974.
74. **Fink, J. N., Barboriak, J. J., Sosman, A. J., et al.,** Antibodies against pigeon serum proteins in pigeon breeders, *J. Lab. Clin. Med.,* 71, 20, 1968.
75. **Moore, V. L., Schanfield, M. S., Fink, J. N, et al.,** Immunoglobulin allotypes in symptomatic and asymptomatic pigeon breeders, *Proc. Soc. Exp. Biol. Med.,* 149, 307, 1975.
76. **Dorral, G., Yang, W. H., Osterland, C. K., et al.,** Circulating immune complexes in hypersensitivity pneumonitis and bronchopulmonary aspergillosis. A longitudinal study, *J. Allergy Clin. Immunol.,* 63, 204, 1979.
77. **Roberts, R. C. and Moore, V. L.,** Immunopathogenesis of hypersensitivity pneumonitis, *Am. Rev. Respir. Dis.,* 116, 1075, 1977.
78. **Wallser, W. C. and Wright, V.,** Pulmonary lesions and rheumatoid arthritis, *Medicine,* 47, 501, 1968.
79. **Dreisin, R. B., Schwarz, M., Theofilopoulos, A. N., et al.,** Circulating immune complexes in the idiopathic interstitial pneumonias, *N. Engl. J. Med.,* 298, 535, 1978.
80. **Daniele, R. P., Henson, P. M., Fantone, J. C., III, et al.,** Immune complex injury of the lung, *Am. Rev. Respir. Dis.,* 124, 738, 1981.
81. **Haslam, P., Thompson, B., Mohammed, I., et al.,** Circulating immune complexes in patients with cryptogenic fibrosing alveolitis, *Clin. Exp. Immunol.,* 37, 331, 1979.
82. **Hadden, J. W., Sadler, J. R., and Warfel, A. H.,** Characterization of lymphokines acting in macrophages, in *The Lymphokines,* Hadden, J. W. and Steward, W. E., Eds., Humana, Clifton, NJ, 1981, 73.
83. **Schmidt, J. A., Green, I., and Secondary, M. L. R.,** Supernatants stimulate fibroblast proliferation, *Fed. Proc.* 39, 1158, 1980.
84. **Neilson, E. G., Jimenez, S. A., and Phillips, S. M.,** Cell mediated immunity in interstitial nephritis. III. T lymphocyte mediated fibroblast proliferation and collagen synthesis: an immune mechanism for renal fibrogenesis, *J. Immunol.,* 125, 1708, 1980.
85. **Boros, D. L. and Warren, K. S.,** Specific granulomatosis hypersensitivity elicited by bentonite particles coated with soluble antigens from schistosome eggs and tubercle bacilli, *Nature,* 229, 200, 1971.
86. **King, G. W., Bain, G., and LoBuglio, A. F.,** The effect of tuberculosis and neoplasia on human monocyte staphylocidal activity, *Cell Immunol.,* 16, 389, 1975.
87. **North, R. J.,** T Cell dependence of macrophage activation and mobilization during infection with mycobacterium tuberculosis, *Infect. Immun.,* 10, 66, 1974.
88. **Rosen, Y , Athanassiades, T. J., Moon, S., et al.,** Nongranulomatous interstitial pneumonitis in sarcoidosis: relationship to the development of epithelioid granulomas, *Chest,* 74, 122, 1978.
89. **Crystal, R. G., Roberts, W. C., Hunninghake, G. W., et al.,** Pulmonary sarcoidosis: a disease characterized and perpetuated by activated lung T-lymphocytes, *Ann. Intern. Med.,* 94, 73, 1981.
90. **duBois, R. M., Balbi, B., Kirby, M., et al.,** T-lymphocyte accumulation in pulmonary sarcoidosis: evidence for persistent stimulation of sarcoid lung t-lymphocyte via the T-cell antigen receptor, *Am. Rev. Respir. Dis.,* 139, A61, 1989.

91. **Williams, W. J.,** The diagnosis of chronic beryllium disease by laser ion mass analysis. Presented at 3rd Int. Conf. on Environmental Lung Disease, Montreal, Canada, October 17, 1986.

92. **Deodhar, S. D., Barna, B., and VanOrdstrand, H. S.,** A study of the immunologic aspects of chronic berylliosis, *Chest,* 63, 309, 1973.

93. **Epstein, P. E., Dauber, J. H., Rossman, M. D., et al.,** Bronchoalveolar lavage in a patient with chronic berylliosis: evidence for hypersensitivity pneumonitis, *Ann. Intern. Med.,* 97, 213, 1982.

94. **Dodge, R. R. and Burrows, B.,** The prevalence of asthma and asthma-like symptoms in a general population sample, *Am. Rev. Respir. Dis.,* 122, 567, 1980.

95. **Gergen, P. J., Mullaly, D. I., and Evans, R.,** National survey of prevalence of asthma among children in the United States 1987–1980, *Pediatrics,* 81, 1, 1988.

96. **Sheffer, A. L. and Brust, S.,** Eds., Asthma mortality workshop, *J. Allergy Clin. Immunol.,* 80, 361, 1987.

97. **Kleinerman, J. and Adelson, L.,** A study of asthma deaths in a coroner's population, *J. Allergy Clin. Immunol.,* 80, 406, 1987.

98. **Dujkanovic, R., Roche, W. R., and Wilson, J. W.,** Mucosal inflammation in asthma, *Am. Rev. Respir. Dis.,* 142, 434, 1990.

99. **McFadden, E. R.,** Asthma, airway dynamics, cardiac function, and clinical correlates, in *Allergy: Principles and Practice,* Middleton, E. R., Jr., Reed, C. E., Ellis, E. F., Adkinson, N. F., and Yunginger, J. W., Eds., C. V. Mosby, St. Louis, 1988, 1018.

100. **Jines, A., Pare, P., and Hogg, J.,** The mechanics of airway narrowing in asthma, *Am. Rev. Respir. Dis.* 139, 242, 1989.

101. **Rachelefsky, G. S., Katz, R. M., and Siegel, S. C.,** Chronic sinus disease with associated reactive airway disease in children, *Pediatrics,* 73, 526, 1984.

102. **Reed, C. E.,** Asthma: chronic desquamating eosinophilic-bronchitis, in *Immunologic Mediated Pulmonary Diseases,* Lynch, J. P., III and DeRemee, R. A., Eds., J. B. Lippincott, Philadelphia, 1991, 322.

103. **Platts-Mills, T. A. E. and deWeck, A. L.,** Dust mite allergens and asthma: a world wide problem, *J. Allergy Clin. Immunol.,* 83, 416, 1989.

104. **Karr, R. M., Davis, R. J., Butcher, B. T., et al.,** Occupational asthma, *J. Allergy Clin. Immunol.,* 61, 54, 1978.

105. **Chan-Yeung, M., Ashley, M. J., Corey, P., et al.,** A respiratory survey of cedar mill workers. I. Prevalence of symptoms and lung function abnormalities, *J. Occup. Med.,* 20, 328, 1978.

106. **Brooks, S. M.,** Bronchial asthma of occupational origin, *Scand. J. Work Environ. Health,* 3, 53, 1977.

107. **Gandevia, B.,** Occupational asthma, *Med. J. Aust.,* 2, 332, 1970.

108. **Brooks, S. M., Weiss, M. A., and Bernstein, I. L.,** Reactive airways dysfunction syndrome (RADS). Persistent airway hyperreactivity after high level irritant exposure, *Chest,* 88, 376, 1985.

109. **Pernis, B., Viglani, E. C., Cavanga, C., et al.,** The role of bacterial endotoxins in occupational diseases caused by inhaling vegetable dusts, *Br. J. Ind. Med.,* 36, 299, 1979.

110. **Chan-Yeung, M., Enarson, D., and Grzybowski, S.,** Grain dust and respiratory health, *Can. Med. Assoc. J.,* 133, 969, 1985.

111. **Pepys, J. and Hutchcroft, B. J.,** Bronchial provocation tests in etiologic diagnosis and analysis of asthma, *Am. Rev. Respir. Dis.,* 112, 829, 1975.

112. **McNutt, G. M., Schlueter, D. P., and Fink, J. N.,** Screening for occupational asthma—a word of caution, *J. Occup. Med.,* 33, 19, 1991.

113. **Fink, J. N.,** Hypersensitivity pneumonitis, *J. Allergy Clin. Immunol.,* 74, 1, 1984.
114. **Salvaggio, J. E.,** Hypersensitivity pneumonitis, *J. Allergy Clin. Immunol.,* 79, 558, 1987.
115. **Braun, S. R., doPico, G. A., Tsiatis, A., et al.,** Farmer's lung disease. Long term clinical and physiologic course, *Am. Rev. Respir. Dis.,* 117, 185, 1979.
116. **Leatherman, J. W., Michael, A. F., Schwartz, B. A., et al.,** Lung T-cells in hypersensitivity pneumonitis, *Ann. Intern. Med.,* 100, 390, 1984.

Chapter 9

# Pneumoconioses, Chronic Interstitial Pulmonary Fibrosis, and Bronchiolitis

*Talmadge E. King, Jr., M.D., David W. Kamp, M.D., and Ralph J. Panos, M.D.*

## CONTENTS

## INTRODUCTION

This chapter will focus on the occupational and environmental disorders most often associated with diffuse lung disease. In particular, those processes caused by inorganic dust exposure will be presented. Although these processes are frequently termed ''interstitial lung disease'' because they commonly involve the pulmonary interstitium, this nomenclature may be misleading because the airways, vessels, and alveolar architecture are extensively involved in most of the processes. Nonetheless, these heterogeneous parenchymal lung disorders are conveniently classified together because of common clinical, roentgenographic, physiologic, and pathologic features. The goal of this chapter is to review the current understanding of these diseases and to present a guide to the work-up and management of a patient presenting with one of the interstitial lung disorders. In addition, the spectrum of occupational and environmental processes causing bronchiolitis (with or without obliterans) will be reviewed.

Approximately 10% of the pulmonary problems an internist or family practitioner encounters involve the interstitial lung disorders. The more common causes of interstitial lung disease (ILD) are presented in Table 1. Because the individual conditions are uncommon and their etiologies are frequently obscure, these patients often present

Table 1  **Interstitial Lung Disease Associated
with Occupational and Environmental Exposures**

**I. Inhaled inorganic dusts**

  Silica ("silicosis")
  Silicates
    Asbestos ("asbestosis")
    Talc (hydrated Mg silicates; "talcosis")
    Kaolin or "china clay" (hydrated aluminum silicate)
    Diatomaceous earth (Fuller's earth, aluminum silicate with Fe and Mg)
    Nepehiline (hard rock containing mixed silicates)
    Aluminum silicates (sericite, sillimanite, zeolite)
    Portland cement
    Mica (principally K and Mg aluminum silicates)
  Metals
    Beryllium
    Aluminum
      Powdered aluminum
      Bauxite (aluminum oxide)
      Hard metal dusts
        Cadmium
        Titanium oxide
        Tungsten
        Hafnium
        Niobium
        Cobalt
        Vanadium carbides
  Carbon
    Coal dust ("coal workers' pneumoconiosis")
    Graphite ("carbon pneumoconiosis")
  Barium (powder of baryte or $BaSO_4$; "baritosis")
  Iron ("siderosis")
  Hematite (mixed dusts of iron oxide, silica and silicates; "siderosilicosis")
  Tin ("stannosis")
  Antimony (oxides and alloys)
  Mixed dusts
  Mixed dusts of silver and iron oxide ("argyrosiderosis")
  Rare earths (cerium, scandium, yttrium, lanthanum)
  $CuSO_4$ neutralized with hydrated lime (Bordeaux mixture; "vineyard sprayer's
    lung")

**II. Inhaled organic dusts (hypersensitivity pneumonitis)[a]**

  Thermophilic bacteria (i.e., *Macropolyspora faeni, Thermactinomyces
    vulgaris, T. sacchari*)
    Farmer's lung
    Grain handler's lung
    Humidifier or air conditioner lung
  Other bacteria (i.e., *Bacillus subtilis, B. sereus*)
    Humidifier lung
  True fungi (i.e., aspergillus, crytostroma corticale, *Aureobasidium pullulans,*
    penicillin species)
  Animal proteins (e.g., Bird fancier's disease)

Table 1 **(continued)**

**III. Inhaled agents other than inorganic or organic dusts**

Chemical sources
Synthetic-fiber lung (orlon, polyesters, nylon, acrylic)
Bakelite worker's lung
Vinyl chloride, polyvinyl chloride powder

Gases
Oxygen
Oxides of nitrogen
Sulfur dioxide
Chlorine gas
Methyl isocyanate

Fumes
Oxides of zinc, copper, manganese, cadmium, iron, magnesium, nickel,
brass, selenium, tin, and antimony
Diphenylmethane diisocyanate
Trimellitic anhydride toxicity

Vapors
Hydrocarbons
Thermosetting resins (rubber tire workers)
Toluene diisocyanate (TDI—asthmatic reactions prominent)
Mercury

Aerosols
Oils
Fats
Pyrethrum (a natural insecticide)

[a]Partial list; see Chapter 8 for details.
From Crystal, R. G., Interstitial lung disease, in *Cecil Textbook of Medicine,* 19th ed.,
Wyngaarden, J. B., Smith, L. H., Jr., and Bennett, J. C., Eds., W.B. Saunders, Philadelphia,
1992. With permission.

a diagnostic and therapeutic challenge. Approximately two thirds of these diseases
have no known cause and are, therefore, classified by their clinical or histopathologic
features. Sarcoidosis and idiopathic pulmonary fibrosis are the most common ILD of
unknown etiology, whereas occupational and environmental exposures, especially to
inorganic dust, are the most common causes of ILD of known etiology. Silicosis and
asbestosis are among the most common inhalation exposures causing diffuse lung
disease. These processes usually occur in individuals with a long exposure history,
usually >10 years. Physical demonstration of the specific dust in the lung is not
essential to the diagnosis as long as a clear and unambiguous history of exposure can
be documented.

## GENERAL APPROACH TO THE EVALUATION OF DIFFUSE LUNG DISEASE

The evaluation of a patient with suspected diffuse parenchymal lung disease begins
with a thorough history. Occupational, environmental, and medication history must
be carefully reviewed with the patient and often with the spouse. Review of the
environment (home and work), especially as it relates to pets, air conditioners, humidifiers, hot tubs, saunas, evaporative cooling systems (e.g., swamp coolers), etc.,

is valuable as well. The temporal relationship between symptoms and work (especially times of day, types of exposure, etc.) should be determined. Details of the occupational history are given in Chapter 8.

Progressive breathlessness with exertion and/or a persistent nonproductive cough are the most common presenting symptoms. Often the patient attributes these symptoms to aging, deconditioning, obesity, or a recent upper respiratory tract illness. The duration and severity of the breathlessness with exertion, cough, and sputum production, should be carefully elicited. Connective tissue disorders may be difficult to exclude because the pulmonary manifestations of these diseases occasionally precede the more typical systemic manifestations by months or years.

Physical examination commonly reveals tachypnea, reduced chest expansion, and bibasilar end-inspiratory dry crackles (''velcro rales''). Clubbing of the digits is common in some patients (idiopathic pulmonary fibrosis) and rare in others (sarcoidosis). Signs of pulmonary hypertension and cor pulmonale are secondary manifestations of advanced ILD.

The routine laboratory evaluation is often not helpful in diagnosis or as a guide to staging or follow-up of the disease course. An elevated erythrocyte sedimentation rate and hypergammaglobulinemia are commonly observed but are not diagnostic. Antinuclear antibodies, anti-immunoglobulin antibodies (rheumatoid factors), and circulating immune complexes are useful in identifying a connective tissue disorder but may be present in other ILD. Elevation of the LDH may be noted but is a nonspecific finding common to other pulmonary disorders (e.g., alveolar proteinosis). An increase in the serum angiotensin converting enzyme (ACE) level may be observed in sarcoidosis but also is nonspecific, as elevated ACE levels have been noted in several interstitial diseases including hypersensitivity pneumonitis. Antibodies to organic antigens may be helpful when hypersensitivity pneumonitis is suspected, although they are not diagnostic of the illness. The electrocardiogram is usually normal in the absence of pulmonary hypertension or concurrent cardiac disease.

Diffuse parenchymal lung disease is often first suspected because of an abnormal chest roentgenogram (Table 2). The most common radiographic abnormality is a reticular or reticulonodular pattern, although mixed patterns of alveolar filling and increased interstitial markings are not unusual in many of these processes. The correlation between the roentgenographic pattern and the stage of disease (clinical, physiologic, histopathologic) is generally poor. Only the radiographic finding of honeycombing (small cystic spaces) correlates with histopathologic findings and with a poor prognosis. The chest roentgenogram may be normal in as many as 10% of patients with diffuse lung disease that are subsequently documented by histopathological evaluation. Conversely, the chest X-ray may be quite abnormal in an asymptomatic patient. Thus, the physician should not ignore or incompletely evaluate or treat a symptomatic patient with a normal chest X-ray or an asymptomatic patient with radiographic evidence of ILD. In either situation, irreversible and progressive disease may develop before the patient seeks additional medical attention. Thus, the chest X-ray is an inadequate method to follow disease course or response to therapy. High resolution CT is useful in defining extent and severity of disease and may be particularly useful in detecting interstitial lung disease noninvasively.

Any patient with suspected diffuse lung disease should have pulmonary function tests. The finding of an obstructive or restrictive pattern is important in narrowing the number of possible diagnoses. Among the few disorders that cause interstitial infiltrates and an obstructive airflow limitation are sarcoidosis, lymphangioleiomyomatosis, hypersensitivity pneumonitis, tuberous sclerosis, and COPD with superimposed interstitial lung disease. Most of the interstitial disorders lead to a restrictive defect with reduced TLC, FRC, and RV. Flow rates are decreased ($FEV_1$ and FVC),

Table 2   **Helpful Radiographic Patterns in Occupational and Environmental Lung Diseases**

**Normal**
  A. Hypersensitivity pneumonitis
  B. Asbestosis
  C. Chronic beryllium disease
**Upper zone predominance**
  A. Chronic hypersensitivity pneumonitis
  B. Silicosis
  C. Chronic beryllium disease
  D. Coal miners' pneumoconiosis
  E. Hard metal disease
  F. Aluminosis
**Pleural involvement**
  A. Asbestosis (common)
  B. Chronic beryllium disease (uncommon)
**Hilar or mediastinal lymphadenopathy**
  A. Chronic beryllium disease
  B. Silicosis (eggshell calcification may develop)
  C. Aluminosis (mediastinal widening most common)
**Miliary pattern**
  A. Silicosis
  B. Bronchiolitis obliterans secondary to toxic fume inhalation
**Cavitation**
  A. Complicated silicosis (especially complicated by tuberculosis)
  B. Coal workers' pneumoconiosis (Caplan's syndrome)
**Pneumothorax**
  A. Aluminosis
**Emphysema**
  A. Aluminosis
  B. Cadmium
  C. Hypersensitivity pneumonitis

but this reduction is related to the decreased lung volumes. The FEV$_1$/FVC ratio is usually normal or increased. Smoking history must be considered when interpreting the functional studies. Reduction in lung compliance is also common. Importantly, in symptomatic patients with a normal chest radiograph and minimal or no restrictive disease, measurement of elastic recoil (pressure–volume curve) may be helpful by identifying lung stiffness.

The assessment of gas transfer is useful. The diffusing capacity for carbon monoxide (DLCO), may be decreased, due in part to loss of the alveolar capillary units, but what is more important is the extent of mismatching of ventilation and perfusion of the alveoli. Lung regions with reduced compliance due either to fibrosis or excessive cellularity may be poorly ventilated but still be well perfused. It is often difficult to correlate the severity of the DLCO reduction with disease stage.

The resting arterial blood gas may be normal or may reveal hypoxemia and respiratory alkalosis. The hypoxemia is usually secondary to a mismatching of pulmonary ventilation to perfusion. Carbon dioxide retention is rare and usually occurs in far-advanced end-stage disease. Importantly, a normal resting PaO$_2$ does not exclude significant hypoxemia during exercise or sleep. Because resting hypoxemia is not always evident and because severe exercise-induced hypoxemia may go undetected,

Table 3 **Bronchoalveolar Lavage: Common Cellular Patterns in Selected Diseases**[a]

|  | Lymph | Neutro | Eosino | Mast Cells | Other Features |
|---|---|---|---|---|---|
| IPF | + | +++ | + | N | |
| Sarcoidosis | ++ | N or + | N | N | |
| Asbestosis | N | ++ | + | N | Ferriginous bodies |
| HSP | +++ | N or + | N | + | |
| Chronic beryllium disease | +++ | + | N | N | + LTT |
| Silica-exposed | +/− | N | N | N | Dust particles by polarized light microscopy |
| Coal workers' pneumoconiosis | + | + | N | N | |
| Cobalt pneumoconiosis | + | + | N | ? | Multinucleated giant cells |
| Aluminum potroom workers | N | N | N | ? | + LTT |

[a]IPF and sarcoidosis shown for comparison.

*Notes:* + = increase. N = normal values.
*Abbreviations:* lymph = lymphocytes; neutro = neutrophils; eosino = eosinophils, IPF = idiopathic pulmonary fibrosis; LTT = lymphocyte transformation test; HSP = hypersensitivity pneumonitis.

it is important to perform exercise testing with arterial blood gases. Arterial oxygen desaturation, a failure to decrease dead space (i.e., a high $V_D/V_T$ ratio), and an excessive increase in respiratory rate with a lower than expected recruitment of tidal volume during exercise provide useful information regarding physiologic abnormalities and may suggest the extent of disease. In general, resting and/or exercise gas exchange is the most useful method to establish the degree of impairment and to monitor the disease course and response to therapy.

Bronchoalveolar lavage is a technique for sampling the cellular and fluid constituents of the lower respiratory tract by fiberoptic bronchoscopy. The diagnostic usefulness of bronchoalveolar lavage remains to be defined. However, the pattern of recovered inflammatory and immune effector cells in bronchoalveolar lavage may aid in the differential diagnosis of ILD (Table 3).

After the initial evaluation, it is important to confirm the diagnosis and establish the stage of disease. Lung biopsy is indicated because (1) it provides a more accurate diagnosis, (2) it will exclude neoplastic and infectious processes that occasionally mimic chronic, progressive interstitial disease, (3) it will occasionally identify a more treatable process than originally suspected (for example, extrinsic allergic alveolitis), and (4) it will provide a better assessment of disease activity.[1,2] Fiberoptic bronchoscopy with transbronchial lung biopsy may be useful when sarcoidosis, lymphangitic carcinomatosis, eosinophilic pneumonia, Goodpasture's syndrome, or infection is suspected. Open lung biopsy is indicated if a specific diagnosis is not established by transbronchial biopsy. Thoracoscopic or open thoracotomy lung biopsy is the most definitive method to diagnose and stage most ILD, especially idiopathic pulmonary fibrosis, so that appropriate prognostic and therapeutic decisions can be made. Fortunately, open lung biopsy is not frequently required when there is a clear and confirmed exposure to an occupational or environmental agent known to cause diffuse lung disease. Open lung biopsy is a relatively safe procedure with little morbidity and less than 1% mortality. Relative contraindications to open lung biopsy include serious cardiovascular disease, roentgenographic evidence of diffuse "honeycombing," severe pulmonary dysfunction, or other major operative risks (especially in the elderly population).

## IDIOPATHIC PULMONARY FIBROSIS

### INTRODUCTION

Idiopathic pulmonary fibrosis (IPF) or cryptogenic fibrosing alveolitis is an interstitial lung disease of unknown etiology. It has an estimated prevalence of 3 to 5 individuals per 100,000 population.[3] It is included in this chapter because it is important to distinguish IPF from other diffuse parenchymal lung diseases, especially those caused by occupational and environmental exposures. The onset of disease is usually in the fifth or sixth decade, and the median survival is approximately 3 to 5 years after diagnosis. The natural and treated course of IPF varies from patient to patient. Most patients experience a slow and inexorable decline in pulmonary function. Although the clinical, physiologic, and radiographic features of IPF have been studied extensively and insights into the pathophysiology of IPF have been gained, the inciting event(s) and the pathogenesis of this disorder remain unknown.

### HISTOPATHOLOGIC FEATURES

The clinical, radiographic, and physiologic characteristics of IPF are neither sufficiently sensitive nor specific to establish definitively the diagnosis of IPF. Today, the only procedure that can confirm the diagnosis of IPF is the histopathologic examination of adequate lung tissue that is generally obtained by thoracotomy or thoracoscopy. Open lung biopsy not only helps to establish the diagnosis of IPF, but it also assesses the level of disease activity and excludes other diseases that may mimic IPF.

The gross pathologic findings of the lung in IPF are variable but appear to correlate with the stage and extent of disease. In early IPF, the lung may be normal, whereas in the latter stages of this disorder, the lung is firm and the pleural surface may have a nodular appearance. The disorder is usually heterogeneous, and areas of normal or mildly involved parenchyma may be interspersed throughout areas of extensive fibrosis and honeycombing. The histopathologic findings in IPF can be divided into three major patterns (Figure 1).[4–6] The cellular pattern is characterized by an accumulation of alveolar macrophages and lymphocytes in the alveolar space. The alveolar walls are lined by hypertrophic hyperplastic alveolar type II cells. The mixed cellular-fibrotic pattern is characterized by thickened alveolar walls containing increased numbers of mesenchymal cells and infiltration by lymphocytes, monocytes, and occasionally plasma cells and eosinophils. A fibrinous exudate is often found in the intraalveolar space. The end-stage or honeycomb pattern is characterized by aberrant deposition of extracellular matrix. The normal alveolar architecture is replaced by thickened alveolar septa and irregular cysts. Histopathologic progression does not occur uniformly throughout the lung, and often these patterns are present in adjacent sections of lung tissue from the same patient. The extent and severity of inflammation present on open lung biopsies correspond to the stage of disease, prognosis, and response to therapy in IPF.[5,7,8] A more cellular pattern on open lung biopsy correlates with an improved survival, whereas a more fibrotic pattern correlates with a worse

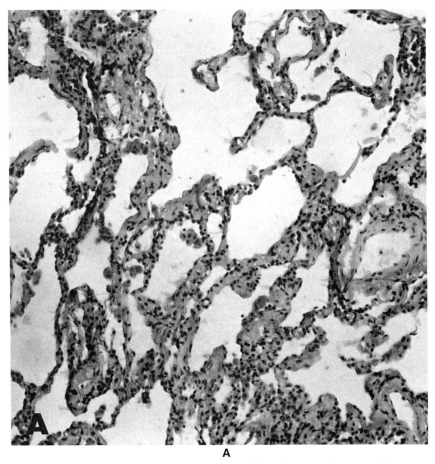

A

**Figure 1** Histopathologic findings in idiopathic pulmonary fibrosis. **A.** Low magnification view of the early stage of parenchymal involvement, with alterations of the alveolar walls and minimal mononuclear cell accumulation in the alveolar spaces. **B.** End-stage lung with intervening dense scarring of the parenchyma, smooth muscle proliferation, and absence of intact alveoli. Other areas of this patient's lung demonstrated less severe features of usual interstitial pneumonitis.

survival. The more cellular pattern also is associated with a greater response to corticosteroid therapy.

## SIGNS AND SYMPTOMS

Progressive dyspnea with exertion and nonproductive dry cough are the usual presenting symptoms in IPF. Weight loss, decreased energy, and lassitude occur commonly. Occasionally, patients appear to be asymptomatic but on further questioning may have adapted their lifestyle to compensate for impaired respiratory function by slowing or reducing their level of activity or no longer performing more vigorous tasks. Other patients may be truly asymptomatic and have an incidental chest roentgenogram that reveals an interstitial process or an abnormal physical examination.

Early in IPF, the physical examination may be normal. However, the most common physical examination findings are tachypnea and bilateral end inspiratory dry crackles (''velcro rales''). Digital clubbing may occur in the latter stages of IPF. Early in the disease, the cardiac examination may be normal, but as the disease

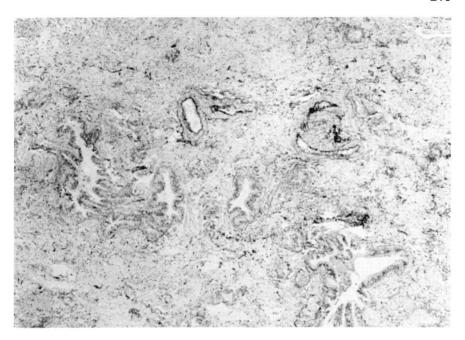

**Figure 1B**

progresses, signs of pulmonary hypertension and cor pulmonale, including increased P2, right-sided S3, and elevated jugular venous pressure may develop. Cyanosis may occur in the advanced stages of IPF.

## RADIOGRAPHIC FEATURES

The chest roentgenogram is a relatively insensitive indicator of disease activity in IPF.[9-11] Up to 10% of patients with a cellular histopathologic pattern on open lung biopsy have normal chest roentgenograms.[11] In the early stages of IPF, the chest roentgenogram may reveal only a mild reduction in lung volumes or a ground glass appearance or hazy homogeneous densities.[4,12] The most common radiographic findings are a reticular or a reticular nodular pattern. These processes usually involve both lungs equally and are generally more prominent in the lower lobes. As the disease progresses, a honeycomb pattern, that is, a coarse reticular pattern with translucencies or cysts measuring 0.5 to 1.0 cm in diameter, may develop. Progressive loss of lung volume, deviation of the trachea to the right, and tracheomegaly are also typical roentgenographic findings in the latter stages of IPF.[13] Pleural abnormalities, adenopathy, or localized parenchymal processes are uncommon in IPF, and their presence suggests another interstitial lung disease or an IPF-associated complication, such as an infection or neoplasm.

Alternative chest imaging studies, including high resolution thin section computed tomography (HRCT), have been used in the detection and staging of IPF.[14-18] In early (cellular) IPF, the HRCT findings of patchy, peripheral air space opacification appear to correlate with increased alveolar inflammation. Other patterns seen in more advanced disease include a lower zone predominant reticular pattern with thickened interlobular septa and intralobular lines, subpleural fibrosis and honeycombing (Figure 2), air space opacification, and ''picket fence'' subpleural septal thickening. HRCT can be used to assess the extent of these various patterns and, thus, determine the severity of disease. The use of HRCT appears promising in the noninvasive assessment of IPF.

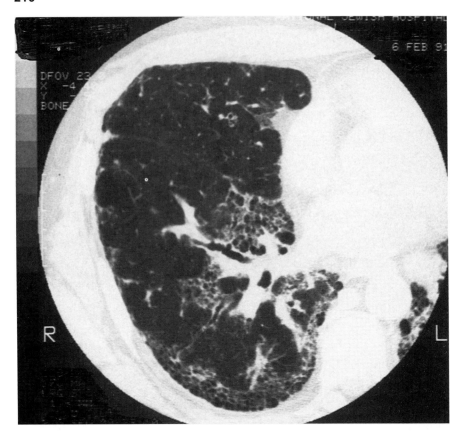

**Figure 2** High resolution, thin-section CT scan. The CT scan reveals extensive parenchymal involvement with irregular fibrotic changes containing small honeycomb-like air spaces in the lower lung zones.

## LABORATORY FINDINGS

Nonspecific blood and serologic abnormalities, including an elevated erythrocyte sedimentation rate, hypergammaglobulinemia, antinuclear antibodies, and rheumatoid factor, occur in many patients with IPF. Circulating immune complexes may also occur in IPF and may correlate with disease prognosis and stage.

Bronchoalveolar lavage (BAL) in IPF usually contains increased percentages of inflammatory cells, especially neutrophils, lymphocytes, and eosinophils.[7,8,19] Lymphocytosis is associated with an increased cellular histopathology, less honeycombing, and a higher rate of response to corticosteroids.[7,8,19] Neutrophilia and/or eosinophilia without lymphocytosis may suggest a decreased response to corticosteroids, but in general, increases in either neutrophils or eosinophils have limited diagnostic or prognostic value.[19,20] BAL has also demonstrated abnormalities in pulmonary surfactant phospholipid and protein components.[21–24]

## PULMONARY FUNCTION

Pulmonary function studies in patients with IPF may be normal in the early stages of the disease. As the disease process progresses, the lung volume components, including total lung capacity, functional residual capacity, and residual volume, are usually reduced. Spirometric measures of pulmonary function, forced expiratory volume in 1 second (FEV$_1$), and forced vital capacity (FVC) are usually reduced because of the reduction in lung volumes. The ratio of FEV$_1$ to FVC is usually normal or increased, due to the increased elastic recoil. Lung compliance is characteristically reduced, and the pressure volume curve is shifted down and to the right, due to the stiff, noncompliant pulmonary interstitium.

Although the diffusing capacity for carbon monoxide (DLCO) is reduced in IPF, this measurement of gas transfer does not correlate with the extent or severity of histopathological derangement, nor does it predict the clinical course or response to therapy.[25-28] Although the resting arterial oxygen tension (PaO$_2$) may be normal in early IPF, resting hypoxemia develops with disease progression. The PaO$_2$ usually decreases, and the alveolar-arterial oxygen (P(A-a)O$_2$) gradient widens with exercise. Significant hypoxemia may also occur during sleep in patients with IPF. The major cause of hypoxemia in IPF is ventilation perfusion mismatch.

Patients with IPF also demonstrate altered physiologic responses to exercise. Respiratory rate rather than tidal volume is increased to maintain minute ventilation during exertion. Also, the dead space to tidal volume (V$_D$/V$_T$) ratio may be increased at rest, due to the reduction in tidal volume. With exercise, the V$_D$/V$_T$ ratio may fail to decrease appropriately or may increase due to both an inability to increase the tidal volume and increased dead space secondary to regions of high ventilation and low perfusion.

## MANAGEMENT

Currently, there is no effective therapy to reverse the fibrotic changes. Therefore, early diagnosis of IPF is imperative so that therapy can be initiated prior to the development of irreversible pulmonary fibrosis. Delay in treatment until the patient becomes significantly symptomatic permits disease progression, development or worsening of pulmonary fibrosis, and may abrogate any potential response to therapy.

Corticosteroids and cytotoxic agents are the principal therapies for IPF. The goals of therapy are clinical improvement or stabilization. Although it is more difficult to quantitate, a decrease in the rate of clinical deterioration is also considered a positive response to treatment. Approximately 10 to 30% of patients with IPF improve objectively with corticosteroid treatment.[5,29,30] Cyclophosphamide is the cytotoxic agent that is used most frequently in the treatment of IPF.[31-33] Other cytotoxic medications that have been used in the treatment of IPF include azathioprine, chlorambucil, and vincristine.[32,34-36]

## COMPLICATIONS

Monitoring disease course and complications in the care of patients with IPF is difficult because this disorder has a variable natural and treated course. Some patients may have a stable course, whereas others may improve for variable periods of time after therapy,[5,8] and still others experience a rapid deterioration and are unresponsive to therapy. The most common cause of death in patients with IPF is respiratory failure due to pulmonary insufficiency. Cardiovascular complications, including heart failure, ischemic heart disease, and stroke, account for approximately 25% of deaths. Other causes of death include pulmonary embolism, infections, and bronchogenic carcinoma.[37]

## BRONCHIOLITIS AND OTHER LUNG DISEASES ASSOCIATED WITH CHEMICAL AGENTS

### BRONCHIOLITIS

The inhalation of many fumes, gases, vapors, and chemicals can lead to lung injury and, infrequently, bronchiolitis. This disorder is caused by damage to the bronchiolar

218

Table 4   Clinical Syndromes Associated with Histologic
Bronchiolitis, with or without Obliterans

**Inhalation injury (see Table 5)**
    Toxic fume inhalation
    Irritant gases
    Mineral dust exposure
**Postinfectious**
    Diffuse lesions
    Localized lesions
**Drug-induced reactions**
    Hexamethonium
    L-tryptophan
    Busulphan
    Free-base cocaine use
    Gold
    Cephalosporin
    Sulfasalazine
    Amiodarone
    Acebutolol
    Sulindac
**Idiopathic**
    No associated diseases
        Cryptogenic bronchiolitis
        Respiratory bronchiolitis-associated interstitial lung disease
        Cryptogenic organizing pneumonia
    Associated with other diseases
        Associated with organ transplantation (heart-lung, lung, bone marrow)
        Associated with connective tissue disease
            De novo process
            Drug reaction
        Hypersensitivity pneumonitis
        Chronic eosinophilic pneumonia
        Diffuse panbronchiolitis

epithelium of the small conducting airways. The alveoli adjacent to the small airways may be involved, but most of the interstitium is spared. The repair process leads to excessive proliferation of granulation tissue, either within the airway walls or lumens or both.

The terms bronchiolitis and bronchiolitis obliterans are often confusing because they describe both a clinical syndrome and a histopathologic pattern that may occur under many circumstances.[38] The clinical syndromes that are frequently associated with histologic evidence of bronchiolitis can be divided into broad groups as outlined in Table 4.[39] An acute illness or known exposure prior to disease onset is often useful in identifying inhalation injury, infection, or drug reaction. Idiopathic bronchiolitis usually has an insidious onset and can be confused with more common problems, such as chronic obstructive pulmonary disease.

Bronchiolitis is characterized by two broad histopathologic patterns (Figure 3): (a) proliferative bronchiolitis, fibrous tissue proliferation within the airways and (b) constrictive bronchiolitis, obliteration and permanent narrowing of small airways.[38,40] Proliferative bronchiolitis frequently occurs during the reparative phase of various types of lung injury.

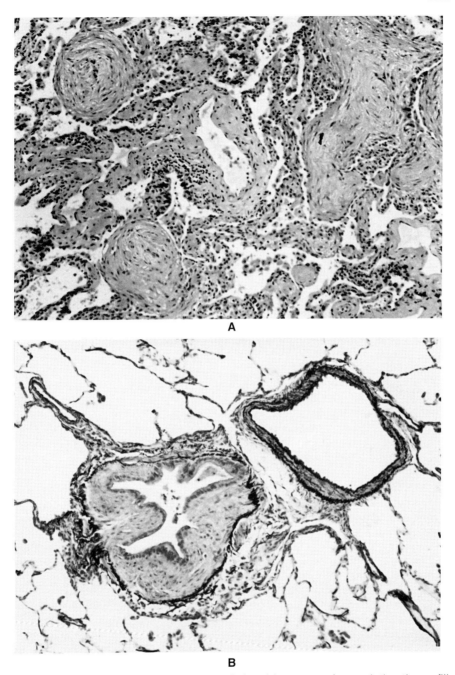

A

B

**Figure 3** **A.** Proliferative bronchiolitis. Polypoid masses of granulation tissue fill the lumens of respiratory bronchioles and alveolar ducts in a patient with crypto-genic organizing pneumonia (also called idiopathic bronchiolitis obliterans and or-ganizing pneumonia). Adjacent alveolar interstitium are broadened by a lymphoplasmacytic inflammatory infiltrate. **B.** Constrictive bronchiolitis. Patient with rapidly progressive dyspnea and obstructive airway disease with mild hyper-inflation on chest radiographs. Prominent concentric narrowing of the bronchiolar lumen is seen.

Proliferative bronchiolitis is characterized by an organizing intraluminal exudate.[38] This lesion is found in numerous pulmonary disorders but is most extensive and prominent in cryptogenic organizing pneumonitis (idiopathic bronchiolitis obliterans organizing pneumonia, or BOOP). Intraluminal fibrotic buds (Masson bodies) are seen in respiratory bronchioles, alveolar ducts, and alveoli. Proliferative bronchiolitis most frequently causes diffuse infiltrates on chest X-ray and a restrictive ventilatory defect on pulmonary function testing.

Constrictive bronchiolitis is a lesion characterized by concentric narrowing or complete obliteration of the airway lumen. Airway obliteration is not a constant feature of constrictive bronchiolitis and may be difficult to identify in many cases. The spectrum of alterations in the walls of membranous and respiratory bronchioles ranges from subtle cellular infiltrates around the small airways, to extensive cellular infiltrates with smooth muscle hyperplasia, bronchiolectasia with mucus stasis, and distortion and fibrosis, to total obliterative bronchiolar scarring. The alveolar ducts and alveolar walls are often spared. These lesions may be extremely subtle and only identified by careful step-sectioning of the lung biopsy and special staining (for example, elastic stains to identify remnants of airway walls). These lesions are typically seen in patients with progressive obstructive lung disease, often in the presence of a normal chest X-ray. Constrictive bronchiolitis is the lesion most commonly seen following inhalation injury. Its severity depends on the type, extent, and severity of the initial lung injury.

## INHALATION INJURY SECONDARY TO TOXIC FUMES, GASES, CHEMICALS, AEROSOLS, OR MISTS

Noxious and toxic fume and gas exposures are significant industrial and environmental hazards that occur in many occupations, including those of agricultural workers, firefighters, chemical workers, munitions and missile industry workers, gold and coal miners, and arc welders working in confined spaces (Table 5). These exposures, especially accidental inhalation of nitrogen dioxide ($NO_2$) or sulfur dioxide, may cause bronchiolitis, with or without obliterans.

### Clinical Manifestations

Silo-filler's disease is the classic but rare example of bronchiolitis obliterans that occurs after $NO_2$ exposure. $NO_2$ produces a yellow to brownish haze and an acrid, ammonia-like odor that is irritating and suffocating in heavily exposed individuals.[41] Unlike highly water-soluble gases, such as chlorine, ammonia, and sulfur dioxide, $NO_2$ is less irritating to the nasal and upper airway mucous membranes because it is relatively insoluble. After inhalation, $NO_2$ reaches the periphery of the lung, where it combines with water to form nitric and nitrous acids and nitric oxide, powerful oxidants capable of causing severe tissue injury.[42–45] The clinical manifestations of $NO_2$ exposure depend upon the concentration of $NO_2$ inhaled and the duration of exposure. Sudden death may occur in individuals exposed to high concentrations of $NO_2$. Death is due to bronchiolar spasm, laryngospasm, reflex respiratory arrest, or simple asphyxiation.[42] Three clinical patterns may follow less intense exposures.[46] Acutely, individuals with mild exposures may develop upper airway and visual disturbances, with cough, dyspnea, fatigue, cyanosis, vomiting, hemoptysis, arterial hypoxemia, vertigo, somnolence, headache, emotional difficulties, and loss of consciousness. These symptoms usually resolve in hours but may persist for several weeks before complete recovery occurs.[42,44,47] Pulmonary edema is a frequent complication at higher concentrations of $NO_2$ exposure. Frequently, these patients have deceptively mild symptoms initially but develop severe adult respiratory distress

Table 5 **Setting of Exposure to Toxic Gases, Fumes, and Mists That Have Been Associated with Bronchiolitis, with or without Obliterans**

Nitrogen dioxide ("nitrous fume")
    Spillage of nitric acid, component of jet and missile fuel
    Metal pickling
    Silo gas
    Chemical manufacturing: explosive, dyes, lacquers, celluloid
    Detonation of explosives
    Electric arc or acetylene gas welding
    Contamination of anesthetic gases (nitrous oxide gas cylinder)
    Nitrocellulose combustion
    Tobacco smoke
    Astronauts, firemen, or others exposed to burning materials
Sulfur dioxide
    Burning of sulfur-containing fossil fuels
    Bleaching of wool, straw, and wood pulp
    Sugar refining and fruit preserving
    Fungicide
    Refrigerant
    Ore smelting
    Acid production
Ammonia
    Fertilizer, refrigerator, explosive production
Chlorine
    Bleaching, disinfectant, plastic making, accidental exposure
Phosgene
    Chemical industry, dye, and insecticide manufacturing
Chloropircrin
Trichlorethylene
Ozone
    Arc welding, air, sewage, and water treatment
Cadmium oxide
    Ore: smelting, alloying, welding
Methyl sulfate
Hydrogen sulfide
    Natural gas making, paper pulp, sewage treatment, tannery work
Hydrogen fluoride
    Etching, petroleum industry, silk-working
Talcum powder (hydrous magnesium silicate)
Stearate of zinc powder
Oxygen toxicity
Asbestos (chrysotile and amphibole)
Iron oxide[a]
Aluminum oxide[a]
Silica[a]
Sheet silicates (talc, mica, etc.)[a]
Coal[a]
Activated charcoal
Talc
Cobalt

[a]These agents have been associated with the development of respiratory bronchiolitis.

From King, T. E., Jr., Bronchiolitis obliterans, in *Interstitial Lung Disease,* Schwarz, M. I. and King, T. E., Jr., Eds., B.C. Decker, Philadelphia, 1988. With permission.

syndrome within several hours (3 to 30 h).[47-51] Most patients recover with no significant sequelae, but death may occur at this stage due to progressive respiratory insufficiency. After recovery or in patients who had no acute illness after exposure, there may be a recurrence or onset of symptoms a few weeks later (2 to 6 weeks). This phase is characterized by progressive onset of cough and dyspnea. Rales are usually present on auscultation of the lungs, and cyanosis often occurs. Mild arterial hypoxemia may identify patients in the early asymptomatic stage.[44,51]

## Pulmonary Function

Arterial hypoxemia or a widened alveolar-arterial oxygen gradient is common and is due to ventilation and perfusion mismatching caused by altered airways dynamic and interstitial and alveolar edema. Methemoglobinemia, occurring when nitrate ions react with hemoglobin, contributes to the arterial hypoxemia.[42,52] Severe metabolic acidosis occurs because of the dissolution of $NO_2$ in body fluids, forming nitrous and nitric acid, and lactic acidosis secondary to tissue hypoxia.[42] Decreased diffusing capacity (Dco) is also common. In the acute phase, physiological studies reveal the simultaneous occurrence of restrictive (the static pressure–volume curve is shifted downward and to the right) and obstructive ventilatory defects.[42] These abnormalities gradually resolve in many patients. In those patients with respiratory sequelae after $NO_2$, the physiological disturbances include arterial hypoxemia at rest and/or with exercise and associated restrictive and/or obstructive pulmonary function.[49,51,53,54]

## Radiographic Features

During the acute stage, the chest roentgenogram often demonstrates pulmonary edema. In survivors, these changes often clear rapidly. The roentgenographic pattern in the late stage can be variable, either normal or hyperinflated. A miliary or discretely nodular pattern, thought to be characteristic of this type of bronchiolitis obliterans, is rarely seen.[42,43,54-59]

## Histopathologic Features

The histologic findings vary, depending on the time at which the tissue is examined following exposure.[38] Histopathologically, there may be severe pulmonary edema and evidence of diffuse alveolar damage during the stage of the adult respiratory distress syndrome. Autopsy studies reveal marked intra-alveolar edema and exudation, as well as thickening of the alveolar walls with lymphocytic cellular infiltrates.[42] An organizing diffuse alveolar damage, with a component of proliferative bronchiolitis obliterans, is seen in some patients who survive but have persistent lung injury.[38] Widespread constrictive bronchiolitis, with or without obliterans, is found in patients who develop progressive disease, especially in those with preceding pulmonary edema, but it may occur as the initial manifestation of previous exposure.

## Management and Control

Approximately one third of individuals acutely exposed to $NO_2$ die. The prognosis for survivors is generally good. Hospitalization for 24 to 48 h is required after acute exposure, and the patient should be followed weekly or biweekly for 6 to 8 weeks. If arterial blood gases or pulmonary function tests are abnormal, treatment with corticosteroids should be started immediately.[60] Corticosteroid therapy has been demonstrated to be useful in the management of both $NO_2$-induced pulmonary edema and bronchiolitis obliterans.[42-45,48,58,61] Treatment should be continued for a minimum of 8 weeks because relapses have been reported after the premature cessation of corticosteroid therapy.[42,50,62] Bronchodilators are occasionally helpful. Antibiotics should only be used when clinically indicated and directed at a specific pathogen.

Some authors have suggested that lasting pulmonary disability is uncommon in silo-filler's disease, whereas others have identified a wide variety of functional derangements.[42,49,60,63] Education has been a key in prevention of this disease. Simple measures to reduce the $NO_2$ levels and use of approved respiratory protection equipment have reduced the number of cases of silo-filler's disease. The pulmonary manifestations described above follow intense, acute, high-level exposures. It is not clear if chronic low-level exposure to $NO_2$ causes any functional abnormalities.

## Other Irritant Gases and Mineral Dusts

A number of other irritant gases have occasionally been associated with bronchiolitis, with or without obliterans. Lung biopsies have not been performed in most cases. Therefore, it is not always clear if the pulmonary injury associated with these inhalation exposures is bronchiolitis obliterans or another specific pulmonary injury pattern. However, sulfur dioxide, chlorine gas, "smoke inhalation" or inhalation burns, hydrogen chloride, ammonia, phosgene, and chloropicrin produce clinical, physiologic, and roentgenographic manifestations that are similar to $NO_2$-induced lung disease.[64,65]

Exposure to inorganic mineral dusts, including asbestos (chrysotile and amphibole), silica, iron oxide, aluminum oxide, several different sheet silicates (talc, mica, etc.), and coal, cause pathologic changes in the small airway (respiratory bronchiolitis).[66–71] The clinical relevance of these lesions is unclear. They occur most commonly in heavily exposed workers. Although the lesions are seen in workers who do not smoke, a role for cigarette smoke appears likely.[72–75] Airflow obstruction, rather than the classic restriction, can be found on lung function testing. Although the pathogenesis of these lesions is unclear, they appear to be caused by the inflammatory response that occurs after deposition of mineral particles or fibers in the walls of the small airways.[66] Histopathologically, these lesions are characterized by marked fibrosis in the walls of small airways, particularly the membranous and respiratory bronchioles (Figure 4).[66,76] Often, the lesions are accompanied by pigment deposition. Respiratory bronchiolitis following exposure to photochemical air pollutants and ozone have recently been reviewed.[77–80]

# THE PNEUMOCONIOSES

## DISEASES ASSOCIATED WITH THE INORGANIC DUSTS

### Silicosis

#### *Introduction*

Silicosis is a chronic lung disease characterized by the development of progressive parenchymal nodules and pulmonary fibrosis after the inhalation of crystalline free silica (see Chapters 2 and 4). Silica or silicon dioxide is the most abundant mineral in the earth's crust, and the most common form of crystalline silica is quartz. Other crystalline forms of silica include cristobalite, tridymite, stishovite, and coesite. Experimental evidence suggests that tridymite may be even more fibrogenic than quartz. However, quartz is the most important cause of silica-induced pulmonary disease because of its abundance. The silica content of different types of rock varies from almost 100% in sandstone and flint, to 20 to 70% in granite, to <10% in shale. Thus, the risk of silica-induced lung disease depends not only on the level of particle exposure but also on the content and type of silica within the inhaled dust.

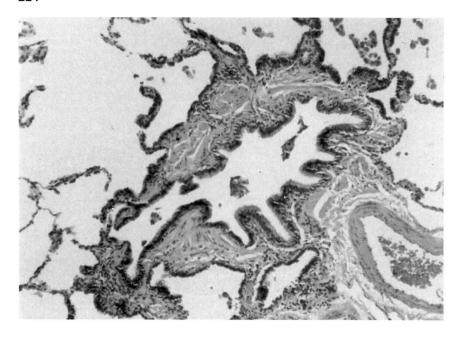

**Figure 4** Respiratory bronchiolitis. The wall of this small airway is thickened, and there is evidence of extension of the bronchiolar metaplastic epithelium into the immediately surrounding alveoli. Intraluminal pigmented macrophages are also present within the peribronchiolar alveolar spaces.

Lung disease due to silica exposure is one of the most common and best studied occupational diseases.[81,82] Agricola in his *Treatise on Mining* (1556) mentioned the effects of dust inhaled in the mining and production of metals, and van Dumerbroeck (1672) and Ramazzini (1713) described pulmonary disease in stone cutters.

### Work Environment

Silica dust exposure occurs in a wide spectrum of industries and occupations. Because of its ubiquitous abundance within the earth's crust, any occupation involving mining, tunneling, excavation, or quarrying poses a potential risk for the inhalation of silica dust. Work environments where secondary processing of earthen materials occurs, stone cutting, engraving, polishing, rock crushing, production of silica-containing abrasives and silica flour (finely milled silica used as an additive in paints, plastics, and surfacing materials) also may have high levels of silica dust. Sandblasting remains a potential cause of silicosis in the U.S. but has been banned in Great Britain and Europe. Workers in foundries where sand castings or molds are used, in ceramic- or glass-making factories, or in other industries that use free silica are at risk for silica dust exposure.

Most industrialized countries have established permissible exposure limits for silica dust in the workplace. These levels of exposure are based on the measurement of airborne dust as the number of respirable particles per vol air or the weight of dust per vol air. Dust is collected in either a stationary collector within the work environment or in a personal dust sampler and analyzed for its free silica content. Within the U.S., the Occupational Safety and Health Administration (OSHA) has established a control level that approaches 0.1 mg/m$^3$ of respirable quartz. The levels for cristobalite and tridymite are half those for quartz.[83]

### Histopathology

The Silicosis and Silicate Disease Committee of the National Institute for Occupational Safety and Health has reviewed the histopathologic findings of silica-induced lung disease.[83] The dust reticulation, a loose collection of dust-laden macrophages and reticular fibers, is the earliest abnormality in silicosis. These lesions are located within the pulmonary parenchyma in proximity to respiratory bronchioles and blood vessels and expand concentrically to form silicotic nodules. Three zones may be found in silicotic nodules: (1) a central area of whirled, dense, fibrous tissue, (2) a midzone of concentrically layered collagen fibers, and (3) a peripheral layer of dust-laden macrophages and lymphocytes intermixed with collagen fibers. At this stage of disease, silicotic nodules may be interspersed throughout an otherwise normal-appearing pulmonary parenchyma. As complex silicosis develops, the silicotic nodules coalesce, and the surrounding interstitium becomes fibrotic and begins to contract. Adjacent blood vessels and airways are deranged and eventually destroyed. Air spaces next to the silicotic nodules may become enlarged, leading to emphysematous lesions. Progressive massive fibrosis develops as the nodules contract and aggregate. The upper lung zones are preferentially involved at this stage of silicosis.

Silicotic nodules may also occur in the hilar and mediastinal lymph nodes, as well as in aortic and perihepatic nodes and within the liver, spleen, and bone marrow. Silicon nephropathy, a spectrum of renal disorders ranging from mild renal function abnormalities to rapidly progressive renal failure, has been described after chronic silica exposure.[84,85] The histopathologic findings in these cases range from crescentic mesangial nephropathy to necrotizing vasculitis.

In acute silicosis, the intraalveolar spaces are diffusely filled with a proteinaceous lipid-containing material that is enriched with surfactant phospholipids and apoproteins.[83,86-88] The adjacent interstitium may be slightly fibrotic with small focal nodules, and the epithelium is extensively damaged and replaced with hypertrophic and hyperplastic alveolar type II cells.

### Signs and Symptoms

Silica-induced lung disease may be chronic, accelerated, or acute, depending upon the intensity and duration of silica dust exposure. Chronic silicosis is the most common form and usually occurs decades after the onset of exposure to relatively low levels of dust. Accelerated silicosis occurs after prolonged high-level exposure and has clinical, physiologic, histopathologic, and radiographic findings similar to chronic silicosis. Acute silicosis develops after massive exposure to silica dust and is characterized by histopathologic and radiographic findings that resemble alveolar proteinosis.

Most patients with chronic silicosis have no symptoms, and the diagnosis is established by chest radiography and an occupational history of silica dust exposure. In the more advanced stages of disease, breathlessness with exertion develops and may progress to shortness of breath at rest. Cough and sputum production may also occur but are frequently due to chronic bronchitis in smokers or concurrent infections. Fever and weight loss suggest the development of infection, especially mycobacterial lung disease. Signs of cor pulmonale may develop in individuals with severe, long-standing silica-induced lung disease and interstitial fibrosis. Crackles on auscultation and finger clubbing are uncommon in silicosis and suggest another diagnosis or the development of bronchogenic carcinoma or infection. The signs and symptoms of accelerated silicosis are similar to those of chronic silicosis, except that they develop and progress more rapidly.

Acute silicosis generally develops within months to several years after intense, massive exposure to silica dust. In most cases, there is an occupational history of sandblasting, manufacture of silica-containing abrasives, hard rock drilling, or tunneling.[87] Progressive breathlessness occasionally accompanied by cough and sputum production are the usual symptoms.[86,87]

### Radiographic Features

The chest radiographic findings of silica-induced lung disease are usually classified as simple or complicated silicosis. Simple silicosis is characterized by diffuse nodular densities that are distributed symmetrically with a predilection for the upper lung zones. These parenchymal nodules are usually <1 cm in diameter and may calcify. Hilar and paratracheal adenopathy also occurs and may precede the development of parenchymal lesions. Eggshell calcification of these nodes is suggestive but not pathognomonic for silicosis. Pleural involvement is infrequent in silica-induced lung disease, but pleural thickening or calcification may occur. Complicated silicosis or progressive massive fibrosis occurs in the latter stages of silica-induced lung disease, as larger masses are formed by the coalescence of smaller nodules. As the disease progresses, these mass lesions contract, distorting the pulmonary vasculature and airways (Figure 5). The upper lung zones are most commonly affected. Cavitation occurs infrequently and suggests the development of mycobacterial disease, especially tuberculosis. In acute silicosis, the chest roentgenogram demonstrates an alveolar filling process that resembles the radiographic presentation of alveolar proteinosis. Hilar adenopathy may accompany the bilateral, diffuse alveolar infiltrates.

Chest computed tomography (CT) is more sensitive than conventional roentgenograms in detecting the small nodular opacities of simple silicosis and identifying coalescence of nodules (Figure 6).[89,90] Studies of the chest CT findings in patients with silicosis confirm the upper lobe predominance of nodules and suggest a posterior predilection.[91] Parenchymal derangement, especially interstitial fibrosis and distortion of blood vessels and airways, as well as emphysematous changes are more readily detected by CT than by conventional roentgenogram.[91] Thus, chest CT may be useful in evaluating patients with early, subtle radiographic changes and in determining the extent of silica-induced lung disease.

### Pulmonary Function

Pulmonary function testing varies with the stage of silicosis and with other superimposed lung disorders, especially those caused by cigarette smoking. In the early stages of disease, pulmonary function testing may be normal. As the disease progresses, restriction, obstruction, or a mixed process may develop and, in progressive massive fibrosis, restriction is the predominant process. With advanced fibrosis, the diffusing capacity is usually decreased and elastic recoil increased.[92] The $PaO_2$ is frequently normal at rest but may decrease with exercise.

Studies of the long-term effect of silica dust exposure on pulmonary function have been complicated by confounding factors, especially cigarette smoking and the inhalation of nonsilica dust.[93-95] In a study of granite shed workers, Musk and coworkers[93] found an excessive loss in FVC of 70 to 80 ml/yr and in $FEV_1$ 50 to 70 ml/yr. In contrast, Graham and colleagues,[94] in a subsequent study of the same population, found improvements in both FVC and $FEV_1$ and suggested that the earlier finding of accelerated pulmonary function loss may have been due to technical factors.

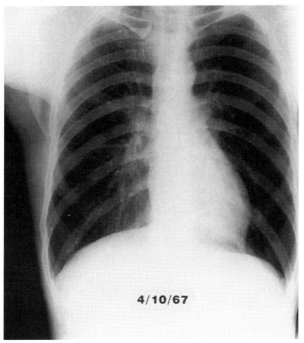

**A**

**Figure 5** Serial chest radiographs in a hard rock miner showing disease progression from simple to conglomerate silicosis. **A.** 4/10/67: Age 44, baseline chest radiograph, which appears normal. **B.** 8/13/70: Age 47, small and rounded opacities are present throughout both lungs, more evident in the upper and mid zones than in the bases. **C.** 7/7/77: Seven years later, a roentgenogram reveals marked extension of the disease with contraction of the upper lung zones. In the lower lung zones, there is a loss of markings suggesting the development of emphysema. The central pulmonary arteries appear enlarged. **D.** 10/21/87: Ten years later, similar findings are present with a suggestion of migration of the lesions medially. **E.** 3/26/92: Now at age 69, there is progressive enlargement of the confluent masses in the upper lung zones. Additional reticular nodular opacities are present in the perihilar regions. **F.** 4/30/92: Computed tomography scan in the upper lung zones reveals that the conglomerate masses contain large amounts of calcium. The underlying lung showed fine nodularity interspersed with areas of emphysema.

### *Laboratory Findings*

The laboratory findings in silica-induced lung disease are nonspecific and may include the presence of rheumatoid factor, antinuclear antibodies, serum immune complexes, and a polyclonal increase in immunoglobulins.[81,96] In a study of 53 patients with silicosis, Doll and colleagues [96] found antinuclear antibodies in a homogeneous or speckled pattern in 26% and elevated rheumatoid factor titers in 28%. Circulating immune complexes were detected in 13 of 43 patients by Raji-cell assay, and immunoglobulins, principally IgG, were elevated in over half the patients. Jones[81] showed that 17 of 39 (44%) silicotic sandblasters had serum antinuclear antibodies. The peripheral blood T-lymphocyte helper to suppressor ratio is increased in individuals with silicosis due to decreased numbers of suppressor T-cells.[97] Studies of

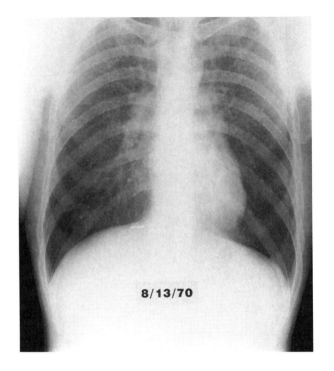

Figure 5B

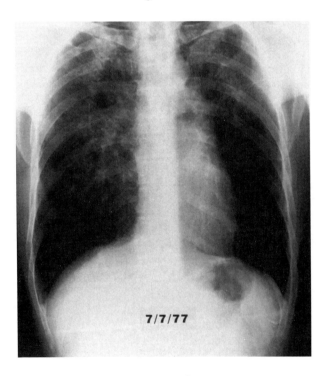

Figure 5C

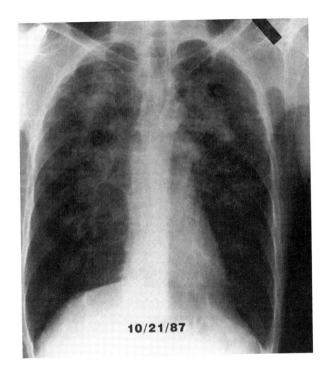

Figure 5D

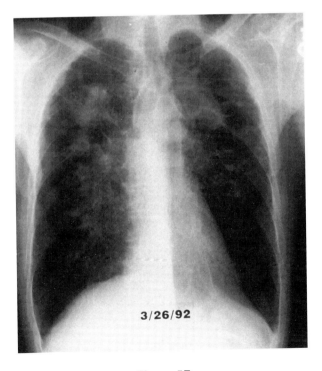

Figure 5E

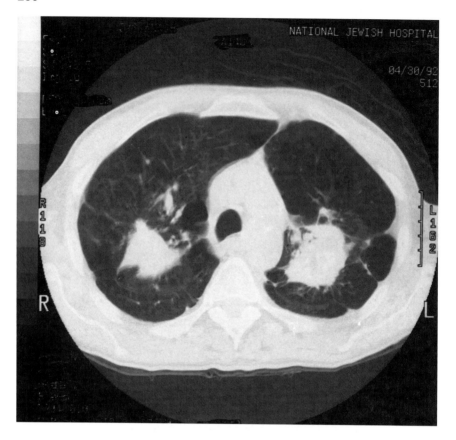

**Figure 5F**

bronchoalveolar lavage from granite workers demonstrated lymphocytosis and increases in IgG, IgA, and IgM.[98,99] These studies do not demonstrate a definitive correlation between serological autoantibodies and severity or progression of silica-induced lung disease. It remains unclear if immunological activation is involved in the pathogenesis of silicosis or if it is an epiphenomenon.

Because the human leukocyte antigen (HLA) genes of the major histocompatibility complex are associated with immunologically mediated diseases, several studies have evaluated the distribution of HLA antigens in individuals with silicosis.[100–102] The prevalence of B44 and A29 is increased in silicotic hard rock miners from Leadville, CO,[102] AW19 is increased in Finnish silicosis patients with mixed occupational exposures,[100] and BW54, DR4, and DRw53 are increased in Japanese patients with silicosis.[101] Thus, based on these studies, there is no definitive association between HLA phenotypes and the development of silicosis.

## Complications

Tuberculosis is a major infectious complication in patients with silicosis. Mycobacterial infection may occur at any time during the course of silicosis but generally is more likely in advanced disease.[103] Tuberculosis is the most common mycobacterial infection but *M. kansaii* and *M. avium*-intracellulare may also be pathogenic. The diagnosis of mycobacterial infection may be difficult because the clinical symptoms,

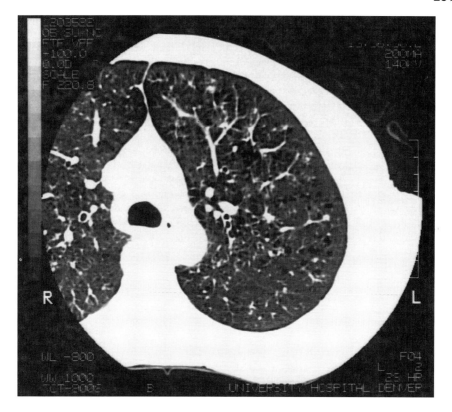

**Figure 6** High resolution CT scan. High resolution CT scan in a home ceramics worker with extensive silica exposure shows fine pulmonary nodules. (Courtesy of Dr. David Lynch, University of Colorado Health Sciences Center, Denver. With permission.)

fever, weight loss, malaise, or increased breathlessness, are not specific and may be due to bacterial pneumonia, bronchitis, or progressive silicosis. Other infectious complications include nocardiosis, sporotrichosis, cryptococcosis, and aspergillosis. Therapy for silicotuberculosis is complicated by diminished host defense mechanisms, especially impaired macrophage function, and decreased drug penetration into silicotic nodules.[103,104] Treatment with streptomycin, isoniazid, rifampin, and pyrazinamide (administered three times weekly) for at least 8 months is required for adequate treatment of silicotuberculosis.[104]

Pneumothorax is an infrequent complication of accelerated silicosis or progressive massive fibrosis and is usually due to the rupture of a subpleural bleb. Enlarged silicotic lymph nodes may cause segmental or lobar collapse due to bronchial compression. In addition, left vocal cord paralysis may be due to left recurrent laryngeal nerve involvement. Dysphagia may result from esophageal compression.

Several studies have reported an association between silicosis, silica dust exposure, and the development of scleroderma.[105–110] Zschunke and colleagues[109] estimated that men with silicosis who are over 40 years old have a 190-fold increased risk for scleroderma compared with similarly aged men without silicosis. Men exposed to silica dust have a 50-fold increased risk for the development of scleroderma compared with unexposed individuals. In a study of 12 patients with silicosis and scleroderma, 11 had Raynaud's phenomenon, 7 had positive rheumatoid factor, and 10 had antinuclear antibodies.[110] Silicosis may also be associated with the development of systemic lupus erythematosus,[111] and individuals with rheumatoid arthritis and silica dust exposure have more rapidly progressive silicosis and tend to have an increased risk for the development of progressive massive fibrosis.[112]

A two- to fourfold increased risk of lung cancer has been described in several cohort studies of men with silicosis who died from lung cancer.[113–116] In a study of granite industry workers, Costello and Graham[117] suggested that the risk of lung cancer was increased in workers exposed to high levels of silica and who also smoked cigarettes. In contrast, the International Agency for Research on Cancer working group reviewed epidemiological studies of silicosis and lung cancer and did not find a definitive causal relationship between silica exposure and lung cancer in man.[118] They did, however, find sufficient evidence to conclude that crystalline silica was carcinogenic in experimental animals. Thus, experimental evidence suggests that silica-induced lung disease may increase the risk of lung cancer. However, the association between silica and cancer remains controversial.[119]

### Treatment

There is no known effective therapy to prevent or ameliorate the course of silicosis after the inhalation of silica dust. Therefore, efforts have been directed at preventing silica-induced lung disease by decreasing the inhaled dust burden through protective ventilation, isolation, wet processes, and positive pressure protective respirators. In an uncontrolled 6-month study of 34 male silicotic workers, Sharma and colleagues[120] showed that lung volumes, $FEV_1$, diffusing capacity, and arterial oxygen tension improved after treatment with oral corticosteroids. The long-term effects of corticosteroids in silica-induced lung disease have not been examined in a prospective, controlled study.

Because there is no effective treatment for silica-induced lung disease, most therapeutic interventions are directed against the complications of silicosis. Supplemental oxygen is indicated for hypoxemia at rest or with exertion and may improve right heart failure due to hypoxic vasoconstriction and pulmonary hypertension. Airflow obstruction should be treated with bronchodilators, including both beta adrenergic agents and ipratropium bromide.

## Asbestosis and Other Asbestos Related-Lung Diseases

### Introduction

Asbestosis is a pneumoconiosis caused by the inhalation of asbestos fibers that is characterized by slowly progressive, diffuse pulmonary fibrosis. Asbestos is a general commercial term for a group of naturally occurring fibers composed of hydrated magnesium silicates. The tensile strength and physical and chemical properties of asbestos are ideally suited for a variety of construction and insulating purposes. More than 30 million tons of asbestos have been mined, processed, and applied in the U.S. since the early 1900s.[121,122] There are two types of asbestos fibers: serpentine and amphibole. Serpentine fibers, of which chrysotile is the only significant commercial variety, are long, curly strands, whereas amphibole fibers (crocidolite, amosite, tremolite, and others) are long, straight, rod-like structures. Chrysotile accounts for over 90% of the asbestos in commercial use in the U.S. and is generally considered less toxic than the amphibole fibers.[121,122] The epidemiological association between asbestos exposure and asbestosis, pleural disease (focal and diffuse pleural plaques), and malignancies (bronchogenic carcinoma and diffuse malignant mesothelioma) is

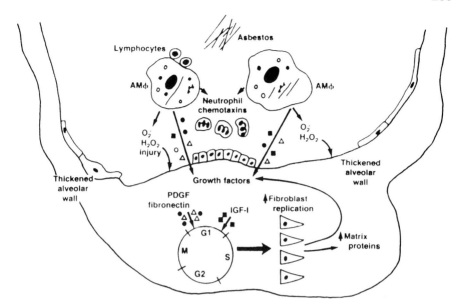

**Figure 7** Pathogenesis of asbestosis. Schema of hypothesized cellular mechanisms of asbestosis in the alveolus. Asbestos fibers deposit in respiratory bronchioles and alveolar ducts, leading to phagocytosis by and activation of alveolar macrophages (AMΦ). AMΦ exhibit morphological changes and release mediator that augment the cellular inflammation by recruitment of additional AMΦ and other cells (lymphocytes, neutrophils). Release of oxidants by AMΦ may injure adjacent cells, including type I alveolar epithelial cells, leading to proliferation of type II alveolar epithelial cells and areas of epithelial denudation and providing access to the interstitium for alveolar-macrophage-derived growth factors. Growth factors (PDGF, IGF-I, others) signal interstitial fibroblasts to replicate and modulate their production of connective tissue proteins. Fibroblasts can also produce growth factors, providing an amplification loop to the model. The accumulation of inflammatory cells, fibroblasts, and connective tissue matrix leads to thickening of alveolar and bronchiole walls and, hence, fibrosis. (From Rom, W. N., Travis, W. D., and Brody, A. R., *Am. Rev. Respir. Dis.,* 143, 408, 1991. With permission.)

development of bronchogenic carcinoma.[145] Further investigations are necessary to clarify the contribution of immune-mediated mechanisms to the pathogenesis of asbestosis.

The histopathologic diagnosis of asbestosis requires the presence of uncoated or coated asbestos fibers (asbestos body) in association with interstitial pulmonary fibrosis (Figure 8).[123,129] Asbestos bodies assessed by light microscopy are composed of transparent fibers surrounded by a coat of iron or protein. These structures are also referred to as ferruginous bodies, because this coating may surround a number of other particles, such as glass, talc, iron, or carbon. Definitive identification of asbestos at the core of these structures may require scanning electron microscopy, energy dispersive X-ray analysis of lung specimens. Amphiboles form asbestos bodies more readily than chrysotile fibers, in part due to the less efficient pulmonary clearance of amphiboles.[121,122]

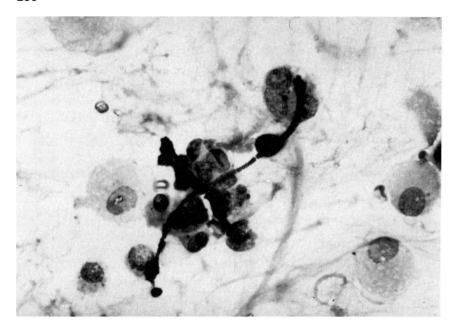

**Figure 8** Asbestos or ferruginous body in lung of patient with pulmonary fibrosis and a history of extensive asbestos exposure. Importantly, asbestosis bodies are only a marker of exposure and not disease.

The quantity of asbestos bodies and uncoated fibers in the lungs correlates with the severity of fibrosis and is generally 10- to 20-fold higher in patients with asbestosis compared with unexposed individuals.[146] However, asbestos bodies underestimate the total lung fiber burden. The number of asbestos bodies in digested lung is generally 10- to 10,000-fold less than the total number of uncoated fibers determined by electron microscopy.[146] The clinical utility of assessing fiber burden in individual patients is limited by marked interpatient variation and the significant number of asbestos fibers present in unexposed individuals.[146] De Vuyst and co-workers[147] reported increased numbers of asbestos bodies in the BAL fluid of asbestos workers with asbestosis, compared with asbestos workers with a normal chest roentgenogram (110 vs. 4 asbestos bodies/ml BAL, respectively). They also observed wide interpatient variation.[147] Sebastien and co-workers[148] showed that >1 asbestos bodies/ml of BAL fluid correlated with >1000 asbestos bodies/g of lung tissue in 69 patients with suspected asbestos-related disease. These levels are clearly associated with significant exposure (Figure 9).[148] Thus, the number of asbestos bodies in BAL fluid is proportional to the fiber burden in the lung. However, the clinical utility of this evaluation should remain experimental until the techniques used to quantitate asbestos bodies in BAL fluid and tissues are standardized.[149]

### Signs and Symptoms

Most patients are asymptomatic for at least 20 to 30 years after the initial exposure to asbestos.[121,122] The earliest symptom of asbestosis is usually insidiously progressive exertional dyspnea. A careful clinical history is often necessary to elicit symptoms of exertional breathlessness because many patients gradually modify their lifestyle to compensate for the slowly developing shortness of breath. Dyspnea commonly progresses inexorably, even in the absence of further asbestos exposure.

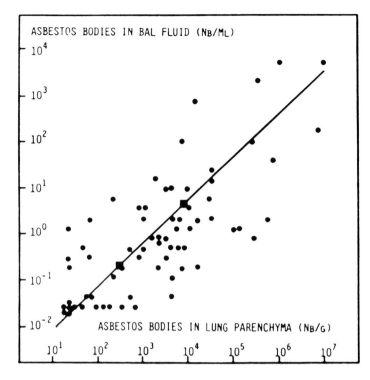

**Figure 9** Scattergram of the 69 pairs of asbestosis body concentration measured in lung parenchyma and in BAL fluid, both plotted on a log scale. The correlation between the log concentrations was highly significant ($r = 0.74$, $p < 0.0001$). The two square dots represent the geometric means for the 39 subjects who had "probable" exposure to asbestos, and the remaining 30, $n = 24$ "possible" exposure and $n = 6$ "unlikely" exposure. The estimated structural relationship between underlying concentrations in BAL fluid and in lung parenchyma is the straight line including these two square dots. (From Sebastien, P., et al., *Am. Rev. Respir. Dis.*, 137, 75, 1988. With permission.)

Cough, sputum production, and wheezing are generally due to cigarette smoking rather than asbestos-induced lung disease.[150] As asbestosis progresses, patients may develop bibasilar, fine, end-inspiratory crackles (32 to 64%) and clubbing (32 to 42%).[150] In advanced disease, signs of cor pulmonale may be evident, including peripheral edema, jugular venous distension, hepatojugular reflux, and a right ventricular heave or gallop.

### Radiographic Features
The chest roentgenogram in patients with asbestosis usually reveals small parenchymal opacities with a nodular and/or reticular pattern. In one study of 56 asbestos workers, the chest roentgenographic findings included pleural abnormalities without fibrosis (48%), combined pleural and interstitial fibrotic abnormalities (41%), and interstitial changes alone (11%).[151] Typically, the interstitial process begins in the lower lung zones (Figure 10) and is associated with bilateral mid-lung zone parietal pleural plaques. Pleural involvement is a hallmark of asbestos exposure that differentiates asbestos-induced pulmonary disease from other interstitial lung disorders (Figure 11). Pleural adhesions can cause atelectasis of a part of the peripheral lung.

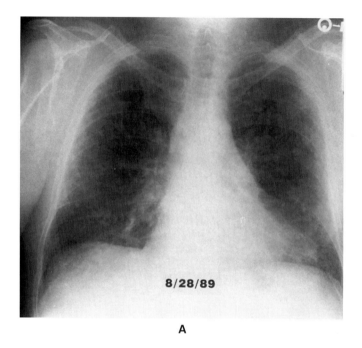

A

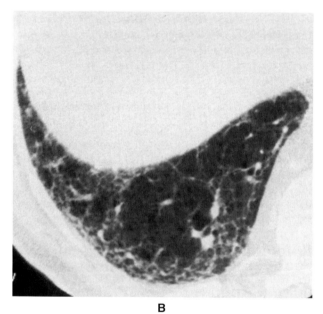

B

**Figure 10** 62-year-old man with extensive asbestos exposure. **A.** Serial chest roentgenograms over the preceding 7 years revealed a reduction in lung volumes associated with diffuse lower lung zone opacities. This film shows diffuse reticular opacities most pronounced at the lung periphery and lung bases. **B.** CT scanning revealed extensive subpleural honeycombing.

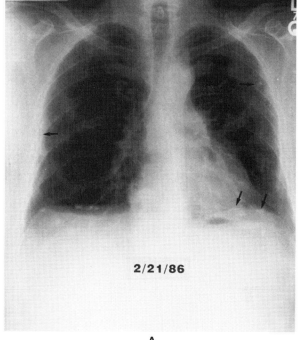

**A**

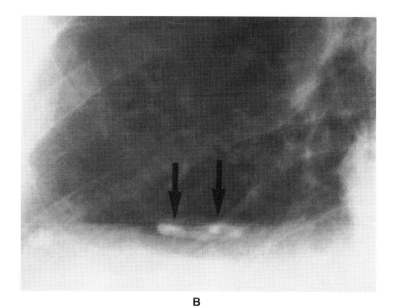

**B**

**Figure 11**  **A.** Chest radiograph showing hyaline plaques on pleural surface.
**B.** Calcified plaque on diaphragm.

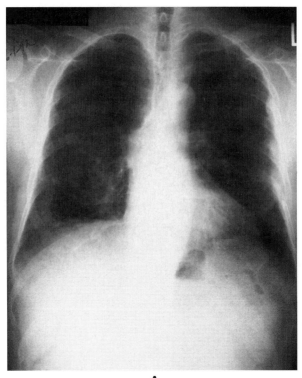

**A**

**Figure 12** Asbestosis exposure with rounded atelectasis. **A.** Chest radiograph shows bilateral noncalcified pleural plaques with a poorly defined opacity at the right base. **B.** CT scan shows a calcified pleural plaque anteriorly in the right chest. **C.** CT scan shows a soft tissue density mass adjacent to the area of the pleural thickening. Bronchi and vessels are seen curving into the medial and lateral aspects of the mass. These appearances are typical for rounded atelectasis. (Courtesy of Dr. David Lynch, University of Colorado Health Sciences Center, Denver. With permission.)

On the plain chest roentgenogram, this often takes on a rounded appearance, i.e., rounded atelectasis. Rounded atelectasis can occur as a result of any type of pleural inflammation; however, it is most often associated with asbestos exposure (Figure 12).[152,153] In the early stages of asbestosis, combined interstitial and pleural involvement may cause a hazy, ''ground glass'' appearance to the chest roentgenograms that may blur the heart border (''shaggy heart'' sign) and diaphragm.[151] Honeycombing and upper lobe involvement generally do not develop until the advanced stages of asbestosis. Hilar and mediastinal lymphadenopathy are not typically seen with asbestosis and suggest other pulmonary disorders.

The chest radiographic changes associated with asbestosis may be subtle. Kipen and associates[154] demonstrated that 18% (25/138) of asbestos insulation workers who died of bronchogenic carcinoma had histopathological evidence of pulmonary fibrosis yet had no interstitial abnormalities on their chest roentgenograms. Parenchymal opacities that are radiographically occult on the chest roentgenogram may be detected by high resolution computed tomography (HRCT).[155–157] Although not diagnostic, the characteristic HRCT findings of asbestosis (Figure 13) include (1) subpleural linear densities of varying length parallel to the pleura, (2) parenchymal bands

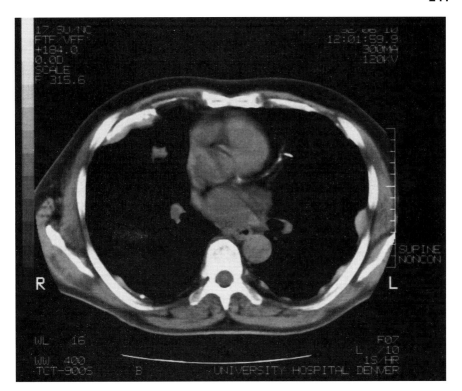

**Figure 12B**

2 to 5 cm in length, often contiguous with the pleura, (3) thickened interlobular septal lines, and (4) honeycombing.[156–160] Staples and co-workers[156] examined HRCT scans in 169 asbestos-exposed workers and demonstrated abnormalities consistent with asbestosis in 57 (33.7%), normal scans in 76 (45%), and indeterminate scans in 36 (21.3%). The group with HRCT findings of asbestosis had a lower vital capacity and diffusing capacity for carbon monoxide (DLCO) and increased dyspnea score compared with individuals with normal HRCT scans.[156] Thus, HRCT may be useful in identifying parenchymal and/or pleural opacities in patients with a history of asbestos exposure and normal or borderline abnormal chest roentgenograms or pulmonary function studies. Prospective longitudinal studies of asbestos-exposed individuals should determine if HRCT is more sensitive than physiologic studies in the detection of asbestos-induced lung disease.

Delclos and co-workers[143] demonstrated that 47% (15/32) of asymptomatic subjects exposed to asbestos had increased gallium uptake in the lung parenchyma, suggesting an active pulmonary inflammatory or immune process. However, the degree of gallium uptake did not correlate with findings on the chest roentgenogram, pulmonary function tests, or BAL cell profile.[143] Therefore, the use of gallium scan in the evaluation of patients with asbestosis remains experimental.

### Pulmonary Function Tests

The characteristic pulmonary function abnormalities in patients with asbestosis include (1) reduced lung volumes, especially the vital capacity and total lung capacity, (2) diminished single breath DLCO, (3) decreased pulmonary compliance, and (4) normal spirometry (ratio of $FEV_1$ to FVC; $FEV_1$%).[123,150] Arterial blood gases

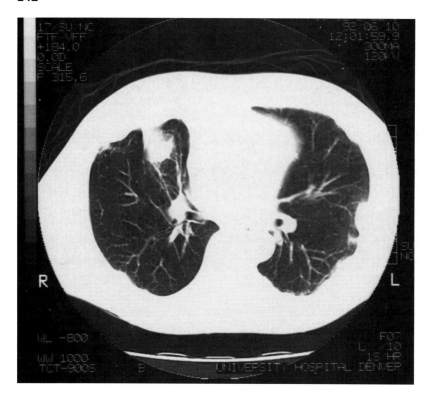

**Figure 12C**

often demonstrate hypoxemia with exercise early in the course of asbestosis and at rest with advanced disease.[123,150] The earliest physiologic abnormalities in asbestos-exposed patients are reductions in the DLCO, pulmonary compliance, or exercise-induced oxygen desaturation.[123,150] Although abnormal spirometry (FEV$_1$% <75) generally reflects concomitant exposure to cigarette smoke, Fournier-Massey and Becklake[161] showed that airways obstruction detected by spirometry in over a thousand Quebec chrysotile workers was not solely attributable to smoking. Airflow limitation in some of these patients may be due to large airways inflammation caused by asbestos deposition along the respiratory bronchioles and alveolar duct bifurcations.

### Laboratory Findings

Laboratory studies in patients with asbestosis are generally nonspecific and not useful. Antinuclear antibody, rheumatoid factor, and increases in the erythrocyte sedimentation rate may be present but do not correlate with disease severity or activity.[162]

### Management

Establishing the appropriate diagnosis is the most important aspect in the management of patients with asbestosis. Although direct tissue examination is the most definitive manner to diagnose asbestosis, it is not necessary for compensation purposes.[123] An Ad Hoc Committee of the Scientific Assembly on Environmental and Occupational Health concluded that, in the absence of lung tissue, a clinical diagnosis of asbestosis is established by (1) a reliable history of exposure, (2) an

compared with unexposed individuals included cancers of the larynx, oropharynx, kidney, esophagus, and gall bladder/bile duct.[165]

## Talc Pneumoconiosis

### Introduction
Talc is widely used in the ceramic, rubber, chemical, cosmetic, and pharmaceutical industries and as a filler in paint, paper, soaps, and roofing materials because of its properties as a dry lubricant, an absorbent, and a bulk filler.[174] Consequently, workers are exposed to talc in the mining and milling of the mineral as well as in secondary industrial processes. Talc pneumoconiosis, talcosis, develops in heavily exposed workers. Pure talc has been associated with disease, but most talc is contaminated with other fibers, especially amphibole and serpentine asbestos, or other minerals (including carbonates and quartz). Thus, the lung diseases seen in these workers may reflect the combined effects of these agents.

Talc is a term applied both to a heterogeneous group of hydrated magnesium silicates and a commercial term for a mixture of minerals in which the mineral talc may be a minor constituent.[174] Relatively pure talc is mined from several areas (for example, Vermont, the French Pyrenees, and Italy) and is used in the cosmetic industry. Talc from other areas of the U.S. (New York, Texas, California) is less pure and contains up to 60% talc. This low-grade talc is used for industrial purposes.

Talc deposits are usually mined in open pits because talc is found near the earth's surface. The rock is crushed into a fine powder, bagged, and shipped for commercial use. Modern dust control at the mines and in secondary industries appears to have reduced talc dust levels markedly. Consequently, talcosis is a rare disease today. When talcosis does occur, it is usually due to unusually intense and prolonged exposure. Cosmetic talc is usually very pure, >90%, and contains virtually no asbestos. Despite its wide use by consumers, it is rarely associated with disease. Nonetheless, respiratory distress and even death have been reported in babies.[175] Occasionally, adults who use massive amounts of baby powder develop pulmonary talcosis. Finally, a rare form of talc-induced disease occurs in drug abusers who crush tablets, dissolve them, and inject the fluid intravenously or who inhale powder from crushed tablets.

### Histopathologic Features and Pathogenesis
Talcosis is characterized by various degrees of pulmonary fibrosis.[175] The extent and severity of the fibrosis are somewhat related to the duration of exposure.[176] Early changes are characterized by collections of macrophages and fibroblasts around respiratory bronchioles and small blood vessels. Macrophages contain numerous talc particles. Ill-formed fibrotic nodules composed of collagen, connective tissue matrix, and occasional cellular infiltrates appear in a stellate pattern. Talc dust is rarely found within the center of these lesions but is present in macrophages at the edges of the lesion. Type II cell hyperplasia may be prominent.

In later stages of the process, focal and diffuse interstitial fibrosis is quite prominent and is located primarily near small airways and vessels.[176] Multinucleated giant cells are occasionally associated primarily with dust-containing scars. An endarteritis with intimal hyperplasia is present in small pulmonary arteries.[175] Because the pattern of diffuse fibrosis resembles asbestosis, it is thought to result from exposure to low-grade industrial talc contaminated with substantial amounts of asbestos. Interestingly, the patterns of ill-defined nodularity and diffuse parenchymal fibrosis may coexist in the same lung.

Foreign body granulomas are also found in talc-induced lung disease. In talcosis, these lesions are usually associated with nodular fibrosis. In intravenous drug abusers, the talc granulomas may be located predominantly within the pulmonary arteries and cause pulmonary hypertension, cor pulmonale, and right heart failure or may be within the interstitium and associated with interstitial fibrosis.[177,178] Intravenous drug abusers who inject talc powder develop pulmonary vascular obstruction and peri-vascular interstitial granulomatous reactions.[177–180] Recently, it has been reported that intravenous abusers develop profound emphysema that clinically, radiographically, and pathologically looks like alpha-1-antitrypsin deficiency, despite normal alpha-1-antitrypsin levels. In addition, these lungs have areas of granulomatous inflammation and fibrosis surrounding the talc particles in the pulmonary interstitium.[179,180] Grossly, multiple rounded nodules are seen, mostly in the mid-lung zones. Occasionally these appear as large masses and may cavitate. Pleural reaction can occur.

### Signs and Symptoms

In the nodular form of talcosis, there may be substantial radiographic changes without symptoms. When present, symptoms may be mild and develop only after many years of exposure. Nonproductive cough and dyspnea are the common symptoms.

With diffuse interstitial fibrosis, the clinical manifestations and the course of the disease are quite similar to those of asbestosis. Dyspnea and cough occur and may precede the radiographic changes. Crackles are often found on physical examination. Clubbing is a late finding.

A form of "industrial bronchitis," i.e., cough, mucus hypersecretion, and wheezing, has been reported in rubber workers exposed to ambient concentrations of talc that did not lead to pneumoconiosis.

### Radiographic Features

Small rounded opacities within the mid-lung zone are a common pattern of talcosis. As the disease advances, the lesions coalesce and distort the lung, causing the development of large mass lesions associated with hyperinflated areas in the adjacent lung zones. The chest X-ray in talcosis associated with diffuse pulmonary fibrosis shows a pattern of small irregular opacities distributed diffusely in the lower and mid-lung zones. Calcified, pleural plaques may be seen.

### Pulmonary Function

Pulmonary function may be normal, despite extensive nodular opacities in talcosis. With time, restrictive changes and hypoxemia develop. In the patients with talcosis complicated by asbestos exposure, the lung function is typical of that found in asbestosis.

### Treatment and Outcome

Prevention is the key to treatment. No clear role for corticosteroid or other therapy has been reported. Treatment is felt not to be indicated in workers with radiographic evidence of disease without symptoms or physiological abnormalities. Symptomatic patients with talcosis might benefit from a trial of corticosteroids if disease progression can be documented.

## Coal Worker's Pneumoconiosis

### Introduction and Work Environment

The chronic inhalation of coal dust (primarily composed of carbon) may cause an uncommon type of parenchymal lung disease—coal worker's pneumoconiosis (CWP). A similar process complicates heavy exposure to respirable graphite (a pure crystalline carbon) or carbon black. Coal is found both on the earth's surface and in underground seams. Men and women working in both areas, on the surface or underground, are considered miners and may be exposed to coal dust. The prevalence rate of CWP is influenced by the dust level, worksite, and length of exposure. The prevalence of CWP varies between 4% in Britain and 10% in the U.S. Less than 0.5% of workers with CWP develop complicated CWP. Importantly, all exposed workers are at risk for the development of coal dust-related illnesses, industrial bronchitis, silicosis, bronchogenic cancer, and tuberculosis. The diagnosis of CWP is based on a history of exposure and appropriate radiographic abnormalities.

### Histopathologic Features and Pathogenesis
Coal dust is not as fibrogenic as silica. However, particles <7 μm in size are retained in the lung, and accumulation of large amounts of radiopaque coal dust over a long period of exposure (usually over 20 years) causes the radiographic abnormalities in CWP. The coal dust is largely within alveolar macrophages and does not appear to be toxic to these cells.

Several studies have evaluated the pathologic manifestations of CWP.[181,182] The distinctive lesion of simple CWP is the coal macule, a discrete, small black nodule that is distributed predominantly in the mid- and upper-lung zones. Histopathologically, these lesions consist of dust-laden macrophages, and dust particles in the extracellular space and the pulmonary lymphatics. Loose collections of reticulin fibers may be present, but a marked fibrotic response is not seen. These lesions usually occur near respiratory bronchioles and frequently have a stellate pattern. Dilation of the respiratory bronchiole and associated focal centralobular emphysema may be seen but are not associated with significant respiratory impairment.

Complicated CWP is an uncommon response, characterized by progressive massive fibrosis (PMF). Davis and colleagues[182] demonstrated that the lesions of progressive massive fibrosis appear in two main forms: solitary masses that are not associated with fibrotic nodules and the fusion of closely associated groups of clear-centered fibrotic nodules. The development of CWP depends upon the size of the particles and the type of coal that are inhaled. Hard coal (anthracite) fragments more easily into smaller particles (<5 μm), compared with soft coal (bituminous). Consequently, anthracite is associated with a greater incidence of progressive massive fibrosis. The progression to progressive massive fibrosis does not require continued exposure to the coal dust. However, continuous exposure to anthracite increases the chance of the development of PMF in subjects with simple pneumoconiosis. Also, older subjects are more likely to develop PMF. Finally, the presence of typical or atypical mycobacterial infection and poorly defined immunologic host factors may also be important in the development of PMF.

Although the data are somewhat conflicting, there appears to be no relationship between the presence of autoantibodies, rheumatoid factor, or antinuclear antibody, and the development of CWP.[183] Antilung antibodies have been identified in the lungs and sera of some miners with CWP. Their role and significance in the pathogenesis and progression of disease are unclear.[183,184] No clear relationship has been demonstrated between the histocompatibility antigens (HLA) and CWP.[183,185]

Several recent studies have focused on potential markers of injury, inflammation, and fibrosis to identify individuals at risk for the development of lung damage after exposure to coal mine dust or silica. Superoxide anion release by alveolar macrophages,[186] release of tumor necrosis factor-alpha from blood monocytes,[187] and serum type III procollagen N-terminal peptide[188] are useful markers that have the potential for distinguishing coal workers with fibrotic lung disease from those without fibrosis. The value of these ''markers'' of potential injury/repair in clinical practice requires much more extensive study.

## Signs and Symptoms

Cough and phlegm are common symptoms in coal workers (so-called industrial bronchitis). Mild abnormalities in lung function may be present, such as reductions in forced expiratory flows; however, the presence of industrial bronchitis does not correlate with the presence of abnormal lung function or with the presence of parenchymal lung disease. More important, patients with radiographic abnormalities of simple coal workers' pneumoconiosis are frequently asymptomatic.

Patients with CWP may develop breathlessness with exertion. Progression to respiratory insufficiency occurs and is associated with either obstructive and restrictive patterns of pulmonary function abnormalities,[189] cor pulmonale, pulmonary hypertension, and/or right ventricular failure. Digital clubbing and melanoptosis (jet black respiratory secretions) are rarely seen in these subjects.

Caplan's syndrome, the development of rheumatoid arthritis in coal workers, is characterized by large, peripheral, round lung opacities. The nodular opacities may grow rapidly (over a few weeks) and cavitate. Subcutaneous nodules and active joint disease may occur. The nodules may precede the development of the joint disease by up to 10 years. Histopathologically, the nodules resemble the nodules found in rheumatoid lung disease. However, unlike in progressive massive fibrosis, they do not contain many dust particles. Similar findings have been described in rheumatoid patients with silicosis and asbestosis. The pathogenesis of Caplan's syndrome is unknown.[67] Gallium lung scans have revealed increased uptake in workers with CWP. The significance remains to be defined.

## Radiographic Features

Coal workers' pneumoconiosis is subdivided radiographically into simple and complicated pneumoconiosis. Simple CWP is based on the presence of small opacities (<1 cm in diameter) in the upper lung zones. In complicated CWP the chest radiograph reveals one or more large masses (>1 cm in diameter) surrounded by lucent regions presumably secondary to compensatory emphysema. These lesions usually appear in a posterior segment of an upper lobe or the superior segment of a lower lobe (Figure 14).

## Pulmonary Function

In simple CWP, routine pulmonary function tests may be normal or show slight reductions in FVC, $FEV_1$, and DLCO. This mild degree of airway obstruction may occur in smokers and nonsmokers.[190] In progressive massive fibrosis, both airway obstruction and lung restriction may be identified. The diffusing capacity for carbon monoxide may be reduced in these patients. Also, the $PaO_2$ may be lower and the alveolar-arterial $O_2$ difference higher in coal miners compared with controls.[190]

## Treatment and Outcome

The development of isolated simple pneumoconiosis does not require cessation of mining and leaving the work environment. However, the more severe the profusion of the opacities, the more likely is the development of PMF. Consequently, workers with simple pneumoconiosis should be monitored regularly for disease progression. Documented worsening of simple CWP should exclude a worker from further coal dust exposure. After cessation of exposure, simple CWP may remain stable or may even regress. Thus, the identification of a change in the lesions seen on chest radiograph should suggest another diagnosis, especially mycobacterial infection. Unfortunately, complicated CWP may progress despite stopping coal dust exposure.

No specific treatment for CWP exists. Prevention is possible and depends upon adequate ventilation and improved work practices at each step in the extraction and processing of the coal. Importantly, cigarette smoking is responsible for more of the

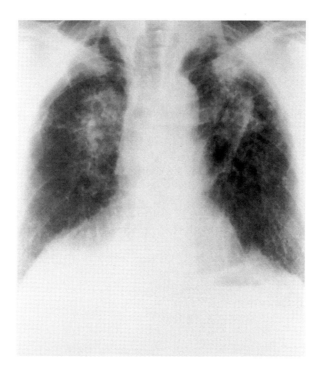

**Figure 14** Progressive massive fibrosis. Chest radiograph in an ex-coal miner with extensive silica exposure shows bilateral, symmetric upper lobe masses with hilar retraction, typical of complicated silicosis or progressive massive fibrosis. Moderately profuse nodules are seen throughout the remainder of the lung.

ventilatory impairment in these workers than is coal dust. Also, life expectancy in simple pneumoconiosis is similar in miners with and without radiographic changes of CWP.[183] However, progressive massive fibrosis can cause premature death.

## Uncommon Causes of Inorganic Dusts Pneumoconiosis

### *Mica Pneumoconiosis*
Mica refers to a family of complex silicates of aluminum and potassium with either alkaline metals or with iron, magnesium, lithium, and sodium. The geological names include muscovite, biotite, phlogopite, etc.[191] Mica is distinguished by the ability to split into thin, translucent sheets; it is used as window glass in stoves and furnaces because of its heat resistance and in powdered form as filler. Synthetic materials have replaced mica for many of its previous uses.

Because mica was shown to be inert in animal experiments, it was not considered a likely cause of human disease. This was particularly the case since human disease could have resulted from contaminating minerals, for example, quartz. However, several recent reports have demonstrated lung disease in millers of mica that contained little or no free silica and in workers after grinding and packing powdered muscovite.[192] Clinically, the patients developed dyspnea, crackles, finger clubbing, hypoxemia at rest and during exercise, reduced diffusing capacity for carbon monoxide, and a restrictive ventilatory defect. Radiographically, both diffuse, fine nodular and linear shadows occurred. Progression of disease occurred despite cessation of exposure. Pathological examination revealed widespread fine fibrosis and nodules up to 1.5 cm in diameter. Mineralogy of the lung tissue revealed only muscovite. Additional studies are required to fully understand the significance of mica exposure to lung disease.

### Kaolin Pneumoconiosis

Nonfibrous silicates other than talc (discussed in a preceding section) are associated with varying degrees of pulmonary abnormalities. Kaolin, or china clay, is a nonfibrous silicate largely composed of the aluminum silicate kaolinite. This clay is found in the U.S. primarily in central Georgia and west central South Carolina.[193] Some kaolin deposits may be contaminated by quartz and mica and therefore require considerable extraction procedures. Thus, the exposure may vary in different areas and countries. The major hazard, with the presence of respirable-sized particles ($<5$ μm in diameter), occurs in the processing of the clay in an aqueous slurry form where spray drying to remove water results in exposure to dry, respirable dust. Kaolin's industrial uses include paper products, refractory materials, ceramics, and as a filler in plastics, rubber, and paints.[193]

Dust-induced lung disease associated with kaolin processing has been known since the 1930s. The prevalence of pneumoconiosis among kaolin workers varies in published reports from 0.7 to 19%.[193–195] It appears that high dust exposure of $>5$ years is required for the development of respiratory disease. Dyspnea with exertion is the most characteristic symptom. The radiographic abnormalities are quite similar to those associated with silicosis. Rounded opacities or combined rounded and irregular opacities are observed. Complicated pneumoconiosis with conglomerate upper lobe lesions is seen in workers with radiographic evidence of pneumoconiosis, especially in those workers with 5 years or more heavy kaolin exposure. There may be no direct correlation between the presence of respiratory symptoms, i.e., cough, sputum production, wheezing, and dyspnea, and chest roentgenographic findings. Similarly, lung function changes do not consistently follow the severity of the radiographic changes.[195] This lack of parallel between symptoms, lung function, and chest radiographic changes are similar to that described for coal miners with uncomplicated pneumoconiosis and workers with simple silicosis. The histopathologic findings resemble those found in coal workers' pneumoconiosis, i.e., dust-laden macrophages, alveolar wall fibrosis, nodular fibrosis, and massive lesions not showing features of either silicosis or rheumatoid pneumoconiosis.[196,197]

The pathogenesis of the pneumoconiosis found in kaolin workers is unknown. Exposure to free silica appears to be an important yet variable component of the dust exposure. The pathogenic potential of kaolin is debated. Animal models of lung injury using kaolin have failed to produce a fibrotic response in the absence of another form of injury.[193] It is known that kaolin can cause injury to cells, erythrocytes, and macrophages *in vitro,* but what significance this has to human disease is unclear. Thus, kaolin has been considered a "nuisance particulate," with little fibrogenic potential. Nevertheless, since cumulative work exposure (and age) is a risk factor for kaolin pneumoconiosis, it is important that the industry maintain adequate industrial hygiene measures to reduce potential exposure of these workers to this dust. Additional studies are required to determine the exact significance and risk of kaolin exposure.

### Fuller's Earth, Bentonites, and Montmorillanite

This is an ill-defined group of poorly crystallized and amorphous silicates of sodium, potassium, aluminum, and magnesium. They are used in many building materials.[191] They have been loosely associated with the development of pneumoconiosis, but adequate data are not available.

Table 8 **Some Nuisance Particulates**[a,b,c]

| | |
|---|---|
| alpha-Alumina (Al$_2$O$_3$) | Mineral wool fiber |
| Calcium carbonate | Pentaerythritol |
| Calcium silicate | Plaster of Paris |
| Cellulose (paper fiber) | Portland cement |
| Emery | Rouge |
| Glycerin mist | Silicon |
| Gypsum | Silicon carbide[c] |
| Kaolin[c] | Starch |
| Limestone | Stearates |
| Magnesite | Sucrose |
| Marble | Titanium dioxide |
| Vegetable oil mists (except castor oil, | Zinc stearate |
| cashew nut, or similar irritant oils) | Zinc oxide dust |

[a]Defined as particulates that have a long history of little adverse effect on lungs and do not produce significant organic disease or toxic effect when exposures are maintained below a reasonable limit. Importantly, all dusts may result in some cellular response in the lung when inhaled in sufficient amount; however, ''nuisance dusts'' do not alter air space architecture, scar tissue is not formed to a significant extent, and the tissue reaction is potentially reversible.[247]
[b]A threshold limit (TLV-TWA) of 10 mg/m$^3$ of total dust when toxic impurities (such as quartz <1%) are not present.
[c]As noted in the text, several of these particulates have recently been suggested to cause pneumoconiosis.
Adapted from Zenz, C., *Occupational Medicine*, 2nd ed., Year Book Medical Publ., Chicago, 1988, 1163. With permission.

### Gypsum, Plaster of Paris
Gypsum and plaster of Paris are generally considered nuisance particulates (Table 8). However, this hydrated calcium sulphate is found in many places. Some deposits are contaminated with quartz as well as other crystals that could be damaging to the lung.[191] Recent data suggest that simple pneumoconiosis with minor loss of lung function in heavily exposed workers can be found.[198]

### Rare Earth Pneumoconiosis
Rare earth-containing compounds are utilized in many industries: in the manufacture of optical glass and in ceramic glass as an abrasive, in the lithographic industry as an arc-stabilizer in cored carbon arc lamps, in the metallurgic industry for the preparation of special alloys and deoxidizer agents, in the gas-mantle industry, and in the electronic industries.[199] However, little is known about the potential role of these agents in human disease. Vocaturo and colleagues[199] reported the case of a photo-engraver who worked for 46 years being exposed to smoke emitted from carbon arc lamps. He developed pulmonary fibrosis complicated by cor pulmonale and death secondary to respiratory failure. Examination of his lung tissue revealed high concentrations of rare earth, with high levels of thorium. Clinically and physiologically, the patient had the manifestations typical of interstitial lung disease. The lung pathology in this patient revealed multiple changes, including focal hyperplasia of bronchiolar epithelium, chronic inflammatory peribronchiolar infiltrates, and foci of sclerotic thickening of the septal connective tissue.

## Manmade Mineral Fibers

Manmade mineral fibers or fiberglass may cause minor upper respiratory tract and skin irritation. However, there is not clear data that exposure to these fibers will result in excess lung cancer or nonmalignant respiratory disease.[200,201] Lung tissue examination from autopsy material in 20 chronically exposed, long-term workers revealed no effects attributable to fiberglass.[202]

## Cement

Cement is a mixture of mineral-like oxide compounds and is similar to the naturally occurring silicates. Calcium silicate and calcium oxide are the major components of cement. Cement workers are exposed to an extremely dusty environment. Asbestos cement is a mixture of asbestos and Portland cement. The asbestos cement worker is at risk for the development of asbestos-related health effects noted above (see asbestosis). Cement dust has been associated with mild impairment in lung function, primarily small airways dysfunction.[203] Cigarette smoking plays an additive role in the obstructive dysfunction. Radiographic findings in these cement workers include reticular and poorly defined micronodular opacities. However, most workers have normal chest X-rays, despite many years of exposure.[203]

## Silicon Carbide (Carborundum) Pneumoconiosis

Silicon carbide (trade name carborundum) is a synthetic abrasive used for sandpaper, for abrasive wheels, and as a refractory material in the making of boilers and foundry furnaces.[204] It is manufactured through a process of fusion of high-grade silica sand and finely ground carbon in an electric furnace at temperatures well over 2000° C.[204,205] Until recently, silicon carbide was considered a nuisance particulate because it was not known to cause a pneumoconiosis. Agreement regarding the fibrogenic potential of silicon carbide was questioned because the manufacturing process is associated with exposure to respirable dust containing crystalline silica, hydrocarbons, sulfur dioxide, other inorganic materials, and organic compounds that are more likely to result in lung disease. Epidemiological studies in the Quebec carborundum industry have now documented abnormalities consistent with pneumoconiosis among workers in the industry.[204,206–208] In addition, pulmonary function abnormalities were identified to be excessive in this population. Age, duration of exposure, years of smoking, and cumulative exposures were important factors.[206] The radiographic abnormalities included pleural thickening, round opacities (related to respirable particulates), and linear opacities (related to age and smoking). Pulmonary function abnormalities were also identified, but these were largely confined to smokers.[206] These smokers demonstrated significant reductions in $FEV_1$ and FVC that related to duration of employment and cumulative exposure to respirable dust. Also, the losses in $FEV_1$ were associated with cumulative sulfur dioxide exposure.[206] In a more recent follow-up study in this population, it was further demonstrated that a restrictive pattern of lung function loss was associated with duration of work and cumulative exposure to respirable dust and was independent of smoking habit.[209] Several other studies have associated respirable dust exposure to silicon carbide and radiographic and physiologic evidence of pneumoconiosis.[210–212] Phlegm, wheezing, and mild dyspnea are the most common respiratory symptoms.[213] Interestingly, symptoms were highly correlated with both cumulative and average exposures to sulfur dioxide.

The histopathology of cases of carborundum pneumoconiosis appear distinct.[204] Small silicotic-like nodules with myelinized center and anthracotic pigment deposition at the periphery and in surrounding tissue were seen. The alveolar space contained macrophages and ferruginous bodies with black central cores were present.

These ferruginous bodies contained abundant free silicon and silicon at the core of the bodies. Also, very few magnesium- and calcium-containing materials were identified, thereby ruling-out asbestos exposure.[204] This is to be expected, since asbestos is not widely used in the industry, and no evidence of asbestos exposure was evident.[204] The pathogenesis remains unknown. However, Begin and colleagues[214] demonstrated that raw and ashed silicon carbide particles were completely inert. However, silicon carbide fibers were found to be fibrogenic in a sheep model of lung injury. Therefore, they suggested that fibrous silicon carbide had significant biologic activity and can initiate a fibrosing lung disease.

## DISEASES ASSOCIATED WITH METALS

### Beryllium Lung Disease

#### *Introduction*
Beryllium is a rare, light metal of the alkaline earth group. Its toxic properties were recognized in the 1930s and 1940s, when it was used in phosphors for fluorescent lights. Beryllium is found in the ore beryl (a beryllium aluminum silicate) that is mined in the U.S., Brazil, and the People's Republic of China and refined in the U.S., Japan, and the Soviet Republics. Beryllium is used for heat shields and rotor blades because of its unusually high melting point, exceptionally low density, elasticity, low coefficient of thermal expansion, and high stiffness-to-weight ratio.[215] In addition, beryllium is permeable to radiation and has a low neutron absorption cross-section that makes it useful in X-ray tubes, in microwave equipment, and as a moderator in nuclear reactors.[215] Although beryllium occurs naturally in soil and coal, exposure usually occurs in industry (Table 9).

It has been estimated that up to 800,000 workers have current or past beryllium exposure,[216] and the prevalence of chronic disease is estimated to be between 1 and 3%. Importantly, applications for this unique metal continue to increase in aerospace, nuclear, computer, and electronics industries.[215,216] Although controversial at one time, it is now clear that inhaled beryllium fumes or dust induces two types of lung injury: an acute inflammatory pneumonitis caused by short exposures to high concentrations and chronic beryllium disease caused by lower level exposure over months to years.[217] The latter is a progressive granulomatous disorder that is principally limited to the pulmonary interstitium. It is believed to involve a beryllium-specific immune response.[218,219] Acute beryllium pneumonitis is now rare. This is due to the use of environmental controls in industry. Nonetheless, cases of chronic beryllium disease continue to be recognized.[218–222]

#### *Histopathologic Features and Pathogenesis*
Different forms of beryllium oxide have been shown to have varying degrees of toxicity. Smaller particles and more soluble grades are the most toxic.[215] Also, decreased pulmonary clearance of beryllium particles probably contributes to its toxicity.[218] Kriebel and colleagues[215] recently summarized the primary histopathologic changes in beryllium disease (Table 10).

Acute beryllium pneumonitis is likely caused by direct injury of the lung by the inhaled cytotoxic beryllium particles that induce a severe, dense cellular exudation.[215] Beryllium is also able to initiate an antigen-specific immune response that is responsible for the pathogenesis of chronic beryllium disease.[223] A diffuse accumulation of chronic immune cells, lymphocytes, and mononuclear phagocytes is present within the alveolar structures. This alveolitis leads to the formation of noncaseating granulomas.

**Table 9  Uses and Properties of Some Important Beryllium (Be) Compounds**

| Compound | Total Be Consumption (%) | Formula | Density (g/cm³) | Melting Point (Centigrade) | Uses |
|---|---|---|---|---|---|
| Beryllium (metallic) | 33 | Be | 1.84 | 1290 | Nuclear reactors and weapons, inertial guidance systems, brakes, X-ray tube windows, turbine rotor blades |
| Beryl | — | $3BeO \cdot Al_2O_3 \cdot 6SiO_2$ | 2.70 | 1410 | Principal beryllium ore |
| Bertrandite | — | $4BeO \cdot SiO_2 \cdot H_2O$ | 2.60 | — | Another beryllium ore |
| Beryllium oxide | 5 | BeO | 3.01 | 2530 | Spark plugs, laser tubes, electrical components, rocket engine liners, ceramic applications, intermediate in Be refining |
| Beryllium fluoride | [a] | $BeF_2$ | 1.99 | [b] | |
| Beryllium copper alloy (typically 2% beryllium in copper) | 50 | | | | Springs, bellows, gears, aircraft engines, bearings, welding electrodes, electrical contacts |

[a]Produced only as an intermediate.
[b]Indefinite; softens at approximately 800° C.

Adapted from Kriebel, D., Brain, J. D., Sprince, N. L., and Kazemi, H., *Am. Rev. Respir. Dis.*, 137, 464, 1988. With permission.

Table 10  **A Summary of Histopathologic Changes in Experimental Beryllium Lung Disease**

| | |
|---|---|
| Phase I. | Nonspecific inflammatory responses. If exposure is severe, dense cellular exudation, emphysema, and obliteration of normal architecture may progress to fibrosis. If exposure is mild, may have no permanent effects. |
| Phase II. | ''Foamy'' macrophages containing phagosomes with beryllium particles. Peribronchial lymphoid tissue proliferation, lymphocytes visible in intracellular septae and around dust particles. |
| Phase III. | Granuloma formation: focal accumulations of macrophages, epithelioid cells, and multinucleated giant cells. |
| Phase IV. | Interstitial fibrosis, hyperplasia of alveolar septae. |

Adapted from Kriebel, D., Brain, J. D., Sprince, N. L., and Kazemi, H., *Am. Rev. Respir. Dis.*, 137, 464, 1988. With permission.

Multiple exposures, over time, are required for the pulmonary inflammatory process to develop. Several laboratories have demonstrated that T-lymphocytes from the lung and blood of patients with chronic beryllium disease proliferate in response to beryllium salts *in vitro*. Also, intradermal injection of beryllium salts may cause a delayed hypersensitivity reaction.[224] These data suggest that beryllium serves as an antigen alone or as a hapten that stimulates the T-cell-mediated lung disorder.[219,223] Approximately 3 to 16% of workers exposed to beryllium are sensitized, as determined by peripheral blood lymphocyte transformation testing (LTT) (Figure 15).[219] (See Table 11 for other potential uses of the LTT in the diagnosis of occupational lung disease.)

As the disease progresses, a granulomatous pneumonitis develops that is frequently associated with interstitial infiltrates. At this stage, chronic beryllium disease is histopathologically (and clinically) similar to sarcoidosis. Fibrosis often associated with bullous changes may occur. Although these interstitial noncaseating granulomas are usually confined to the lung, they can be found occasionally in skin, liver, spleen, lymph nodes, myocardium, skeletal muscles, kidney, bone, and salivary glands.[215] Granulomas are not identified in the lungs of all patients.

Importantly, contact of beryllium salts with the skin may cause a primary dermatitis and may sensitize the skin to subsequent beryllium contact. The dermatitis usually resolves, but skin ulceration can occur if beryllium particles are retained in the skin.[215]

## *Signs and Symptoms*

Acute beryllium disease rarely occurs. However, direct acute exposure can cause conjunctivitis, periorbital edema, nasopharyngitis, tracheobronchitis, and pneumonitis. The acute pneumonitis is characterized by dyspnea, cough and sputum, chest pain, tachycardia, crackles, and cyanosis.[215]

Chronic beryllium disease is a progressive disabling and occasionally fatal process that is caused by chronic, low-level beryllium dust exposure lasting for months to years. Recognition of chronic beryllium disease usually occurs after a latency period that ranges from months to usually >10 years. Dyspnea is the most common presenting symptom, but cough, chest pain, weight loss, arthralgia, and fatigue may be present. Physical examination may be normal but occasionally reveals bibasilar

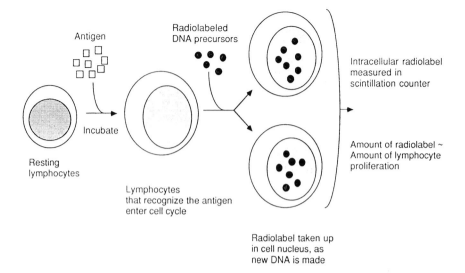

Antigen

Radiolabeled
DNA precursors

Intracellular radiolabel
measured in
scintillation counter

Incubate

Resting
lymphocytes

Amount of radiolabel ~
Amount of lymphocyte
proliferation

Lymphocytes
that recognize the antigen
enter cell cycle

Radiolabel taken up
in cell nucleus, as
new DNA is made

**Figure 15** Lymphocyte transformation testing. Measurement of lymphocyte proliferation in response to an antigen—called lymphocyte transformation testing or lymphocyte blastogenesis—provides an *in vitro* indication of a patient's state of cellular hypersensitivity to an agent of interest. Lung or peripheral blood mononuclear cells are placed in culture medium with the suspected antigen in a range of antigen concentrations and then incubated for 5 to 7 d. During this time, macrophages in the wells interact with the antigen and "present" it to the lymphocytes. Those lymphocytes that have previously "seen" the antigen will enter cell cycle and proliferate. The cells undergo a physical change, becoming blast-like in appearance. The preferred method of detecting lymphocyte transformation is the use of radiolabeled DNA precursors to detect proliferating cells. The amount of radiolabel that has been incorporated into proliferating cells is measured, and the results are expressed as a stimulation index. (From Newman, L., Storey, E., and Kreiss, K., *Occup. Med. State Art Rev.,* 2, 345, 1987. With permission.)

Table 11 **Antigen-Specific Lymphocyte Transformation Tests (LTT) Possibly Useful in Diagnosis of Occupational Lung Disease**

| | | Applications of Positive LTT and Sources of Lymphocytes | |
|---|---|---|---|
| **Metal Salt** | **Lung Disease** | **Blood** | **Bronchoalveolar Lavage** |
| Beryllium | Chronic beryllium disease | 70–100% positive | 100% positive |
| Cobalt | Hard metal disease | Few cases | Unpublished |
| Aluminum | Granulomatosis | 1 case | Unknown |
| Titanium | Granulomatosis | 1 case | Unknown |
| Gold | Drug-induced hypersensitivity | Lung, dermatitis, jaundice cases | Unknown |
| Nickel | Asthma | Dermatitis cases | Unknown |
| Chromium | Asthma | Dermatitis cases | Unknown |

From Newman, L., Storey, E., and Kreiss, K., *Occup. Med. State Art Rev.,* 2, 345, 1987. With permission.

crackles on chest examination, skin lesion, lymphadenopathy, and hepatosplenome-galy. Clubbing of the digits may occur. The final stages of this disease may be characterized by respiratory failure, cor pulmonale, and right heart failure.[215]

Importantly, it should be emphasized that subclinical lung disease may exist among individuals who demonstrate beryllium sensitization.[219] As a result of this finding, a redefinition of the diagnostic criteria for beryllium lung disease that reflects this wider range of beryllium effects is suggested (Table 12). By these criteria, any person suspected of having beryllium disease should undergo an evaluation that includes fiberoptic bronchoscopy with BAL to obtain cells for lymphocyte transformation testing and lymph node or lung biopsy to confirm histopathologic alterations. More traditional evaluations, including chest radiograph and pulmonary function testing, are useful for staging and determining impairment. These patients are classified into three groups on the basis of the beryllium-lymphocyte transformation test, histopathologic abnormalities, and clinical manifestations, i.e., "beryllium-sensitized," "subclinical beryllium disease," and "clinical beryllium disease," respectively. Individuals with beryllium exposure who have a positive blood and/or lung lymphocyte transformation test but who do not have compatible histopathologic abnormalities are beryllium-sensitized. Patients who are sensitized and have compatible histopathologic alterations on biopsy but who are asymptomatic and lack objective radiographic or physiologic evidence of impairment are classified as having subclinical beryllium disease. Individuals who meet all the diagnostic criteria and have clinical symptoms or measurable impairment would be classified as having clinical beryllium disease.

### Radiographic Features

In acute beryllium-induced pneumonitis, diffuse or localized opacities are present on the chest radiograph. In chronic beryllium disease, the chest radiographs usually reveal diffuse opacities and bilateral hilar adenopathy (Figure 16). However, in

**Table 12  Diagnostic Criteria for Clinical Beryllium Lung Disease,[a] Subclinical Beryllium Lung Disease,[b] and Beryllium Sensitization[c]**

1. History of exposure to beryllium

2. Beryllium-specific immune response
   - positive peripheral blood LTT and /or
   - positive bronchoalveolar lavage LTT

3. Histopathology on lung biopsy compatible with beryllium disease
   - noncaseating granulomas or
   - mononuclear cell infiltrates

4. Constellation of clinical findings that may include any of the following:
   - respiratory symptoms or signs
   - reticulonodular infiltrates on chest radiograph or other imaging technique
   - altered pulmonary physiology with restrictive and/or obstructive physiology, decreased diffusing capacity for carbon monoxide, ventilatory impairment, or altered gas exchange on exercise testing

[a]Diagnosis requires all four criteria.
[b]Diagnosis requires criteria 1 through 3.
[c]Diagnosis requires criteria 1 and 2. The meaning of a positive peripheral blood lymphocyte transformation testing (LTT) in the absence of a positive lavage requires further study.

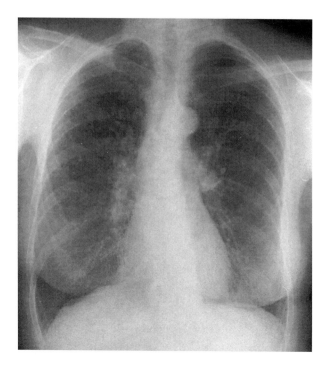

**A**

**Figure 16** Chronic beryllium disease. **A.** Posteroanterior chest radiograph of a 61-year-old woman with chronic beryllium disease demonstrates diffuse nodular opacities throughout the lung, predominantly in the upper lung fields (ILO perfusion of 2/1 of q/r small, rounded opacities). **B.** Bilateral hilar adenopathy is suspected on the PA film and is confirmed on the lateral film.

subclinical disease, the chest radiograph may be normal, yet the lung biopsy reveals granulomatous infiltration. Also, respiratory symptoms may precede the development of radiographic abnormalities by months to years. Although reticulonodular opacities may occur throughout all lung zones, they are typically distributed predominantly in the upper-lung zones.[225] Large, calcified masses may form as these opacities coalesce. A pleural reaction may be seen. Only one third of patients will have hilar adenopathy. High-resolution computed tomography (HRCT) may be more useful than plain chest X-ray in identifying early changes. Importantly, the radiographic abnormalities do not predict the extent of functional pulmonary disability.[225]

### Pulmonary Functions

In acute beryllium-induced pneumonitis, the lung volumes are reduced, and hypoxemia is present, even at rest. The physiologic changes in chronic disease are quite variable. In early disease, the lung function may be normal. Obstructive changes, independent of cigarette smoking, are found in approximately one third of patients. A restrictive or mixed pattern is found in most patients, especially those with advanced disease. Gas exchange abnormalities are common. The DLCO is usually decreased but is neither a sensitive indicator of disease or a useful marker of progression. Arterial hypoxemia may be present at rest but often is detected only with exercise. Arterial blood gases with exercise appear to be most sensitive in the detection of gas exchange abnormalities.

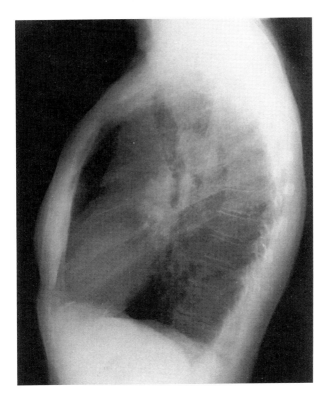

**Figure 16B**

## Laboratory Findings

A number of nonspecific abnormalities in laboratory tests have been found in patients with chronic beryllium disease: hyperuricemia, hypercalcemia, hypercalciuria, elevations in hepatocellular injury tests, elevated serum angiotensin converting enzyme activity, and elevated serum immunoglobulins. None of these are useful in diagnosis or monitoring the disease course.

## Treatment and Outcome

Management of the acute pneumonitis is confined mainly to removal from further exposure and treatment of the hypoxemia. Sometimes in severe disease, mechanical ventilation is required. Chronic disease will occur in approximately one fifth of patients with acute pneumonitis. Fortunately, acute beryllium pneumonitis is very rarely seen in the U.S. because of improved industrial hygiene.

The clinical course of chronic beryllium disease is variable. Most patients have a slow, progressive decline in lung function. If untreated, respiratory failure and right heart failure occurs in up to 35% of patients.[226] No controlled study of corticosteroid therapy has been reported. However, many patients improve clinically and radiographically after corticosteroid treatment. Often the disease recurs when treatment is discontinued, necessitating prolonged therapy.

Disease prevention is the primary mode of treatment. Newman and Kreiss have recommended screening beryllium-exposed individuals for beryllium hypersensitivity (by the blood beryllium lymphocyte transformation test) and surveillance chest radiographs.[219,227,228] The blood beryllium lymphocyte transformation test has been shown to be more sensitive than other clinical manifestations, chest radiograph, and lung function testing in identifying individuals with beryllium-associated lung disease.

Patients with chronic beryllium disease should be removed from exposure. Although cessation of exposure may not prevent the subsequent development of progressive disease, it seems prudent given our current understanding of the prognosis in these patients. A recent cohort mortality study on 689 patients listed by the Beryllium Case Registry demonstrated significant excess mortality from lung cancer and from nonmalignant beryllium disease.

## Cobalt Pneumoconiosis

### Introduction and Work Environment

Occupational exposure to cobalt-containing dust has been associated with the development of several forms of pulmonary disease: (1) an obstructive airways disease that resembles occupational asthma with an immediate and late asthmatic reaction, (2) a relatively acute interstitial lung disease characterized by features resembling hypersensitivity pneumonitis, and (3) a chronic diffuse interstitial lung disease that frequently progresses to end-stage fibrosis. The term "hard metal disease" has been previously applied to these processes because of their recognition in workers exposed to hard metal.

Hard metal is produced by powder metallurgy. The chief components are tungsten and titanium carbides (70 to 95%) and cobalt (5 to 25%). Various amounts of other metals, such as tantalum carbide, chromium carbide, vanadium, niobium, and nickel, may be added, depending on the desired properties of the final products. Tungsten carbide is inert; therefore, it is the cobalt that has been associated with the lung damage.[229,230] Consequently, the clinical processes associated with exposure to hard metal are best termed cobalt pneumoconiosis.

Exposure to cobalt occurs in many job environments. Cobalt is a widely used hardener and binder in metal alloys of tungsten, aluminum, chrome, molybdenum, beryllium, as well as others. Tungsten carbide, a unique hard metal, is almost as hard as a diamond. It is used in situations where strength, rigidity, resistance to heat, and longevity are required. Tungsten carbide is bound or glued to surfaces of plates or wheels to produce durable grinding tools for shaping and polishing of other material. Thus, any worker who makes tools or metal parts or who polishes or grinds a variety of materials with hard-metal abrasives should be suspect for cobalt exposure. Examples of its use include metal cutting and mining tools, dental drills, bearings, dies, seal rings, and many more applications.[231,232] Up to 25% cobalt is added to the mixture with tungsten carbide. The particles of this original mixture of powdered metals are extremely small; most are 1 to 2 μm in size. Consequently, it is easy for these inhaled particles to reach the air spaces in the lung. Therefore, the exposure may occur:[233]

1. as a result of work with the original powdered mixture of cobalt and other metals;
2. from inhalation of this dust during the manufacture of alloy ingots or the fabrication of alloy into secondary products;
3. in tool makers, who may experience significant exposures as they grind, bore, and shape hard-metal blanks into tools and parts;
4. as a result of exposure to coolants and cutting oils that are used to cool and lubricate the cutting points of tools during the fabrication process.[229] There is a tendency for the coolants to dissolve and accumulate cobalt (which is soluble in certain coolants) of variable amounts, depending on the composition of the coolant. The coolant forms a fine spray of droplets that may contain very high

concentrations of cobalt that can be inhaled into the worker's lung. It has been suggested that the cobalt dissolved in the coolants is in an ionized form that has a greater biological effect than the nonionized form;[234]

5. from inhalation of tungsten-carbide chips or cobalt binders that are chipped from the abrasive surfaces of grinding wheels or plates, especially as these surfaces decay and liberate an aerosol on their rapidly turning surfaces;

6. in dental technicians who work close to the grinding process for crowns and prostheses and who may be exposed to cobalt.

Hard metal disease was first described in the 1940s. The prevalence is not clear and is thought to be decreasing because of environmental controls in industry. There are approximately 30,000 hard metal workers in the U.S.[235] The prevalence of hard metal disease, specifically asthma, was only 5.6% in a Japanese plant with relatively high dust levels. The prevalence of interstitial lung disease has been estimated at <1% in hard metal workers in the U.S.[235] Unfortunately, up-to-date data regarding environmental conditions is limited.[234] In particular, there are limited data regarding the environmental conditions associated with the grinding of finished hard metal tools. This is of concern because there appears to be a clear risk of potential exposure to cobalt dust in any facility where hard metal tools are machined. Consequently, it has been recommended that strict dust control measures be employed in these work areas.

### Histopathologic Features

The histopathologic features of cobalt asthma have not been reported. Presumably, bronchitis that is reversible is present since most of these patients have a disappearance of their symptoms once they avoid exposure. It is not clear that the hypersensitivity pneumonitis associated with cobalt exposure is a distinct entity or part of the spectrum of the more chronic diffuse interstitial process. The classic histologic features of hypersensitivity pneumonitis have not been documented in patients with the clinical manifestations.[235] Specifically, although multinucleated giant cells are present, granulomatous inflammation has been described in only a few cases.[236]

Giant cell interstitial pneumoconiosis (GIP) is the most distinctive pathologic pattern in cobalt pneumoconiosis. GIP was first described by Liebow in 1975.[237] It was later demonstrated that these patients' tissue contained hard metal dust.[235,238–240] Austenfeld and Colby[235] described the major pathologic features of cobalt pneumoconiosis as:

1. prominent multinucleated giant cells within alveoli;
2. numerous mononuclear macrophages in alveolar spaces;
3. phagocytosis of mononuclear macrophages by giant cells ("cannibalism");
4. peribronchiolar chronic inflammation and fibrosis;
5. syncytial clusters of type II epithelial cells along alveolar walls (in some cases).

Dark staining dust particles may be found in the cytoplasm of the alveolar macrophages and giant cells. Interestingly, there are two types of giant cells: one derived from alveolar macrophage and another from type II epithelial cells.[235] There is some overlap between GIP and desquamative interstitial pneumonitis (DIP), with some patients at both ends of this "spectrum." In DIP the major features are alveolar filling by macrophages with occasional but not prominent formation of multinucleated giant cells, hyperplasia of type II epithelial lining cells, and chronic interstitial inflammation and mild fibrosis and uniformity of involvement throughout the biopsy. Also, a significant number of biopsies in these patients reveal the more common pattern of

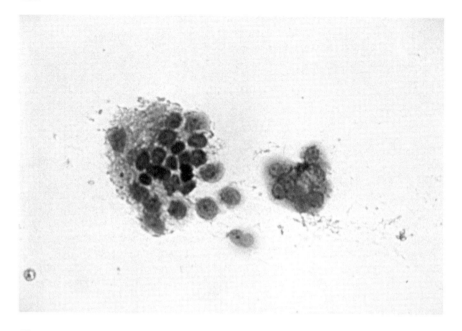

**Figure 17** Bizarre multinucleated giant cell in a patient with cobalt pneumoconiosis.

"usual interstitial pneumonitis" that is characteristic of pulmonary fibrosis of unknown etiology. With this lesion, there is patchy involvement of the lung and more areas of chronic fibrosis than are present in GIP or DIP patterns. Bronchiolitis obliterans may be seen occasionally.

Chemical analysis may be required to establish the relationship of these changes with hard metals because there are idiopathic cases (i.e., patients without known or clear-cut exposure) who have similar lesions, GIP, DIP, or UIP. The methods of chemical analysis include (1) atomic absorption spectrometry using digested tissue, (2) energy dispersive X-ray spectroscopy, a type of analytical electron microscopy, that can be applied to paraffin-fixed tissue, and (3) neutron activation analysis. Polarized light microscopy is usually not helpful in the detection of these metals.

Bronchoalveolar lavage (BAL) has been reported in patients with cobalt pneumoconiosis. The findings are in general, nonspecific, with a variety of cellular patterns reported. Importantly, bizarre multinucleated giant cells may be seen (Figure 17). The presence of giant cells helps support or even suggests the diagnosis of cobalt pneumoconiosis in a subject with ill-defined diffuse parenchymal lung disease. BAL may also be used to obtain material for chemical analysis of hard metal dust in the lungs.[241]

### Signs and Symptoms

Many subjects exposed to cobalt can have rhinitis, conjunctival irritation, and contact dermatitis. Occupational asthma is a recognized complication of cobalt exposure, presumably due to an immunologic reaction in the lungs of some individuals exposed. Most patients have a history of chest tightness, dyspnea, and wheezing. There is a strong relationship of asthmatic symptoms and obstructive airflow with cobalt exposure. These patients do not have evidence of diffuse parenchymal lung disease.[242,243] The patients with diffuse lung disease have the same clinical manifestations as are common with interstitial lung disease, i.e., breathlessness with exertion and nonproductive cough. Crackles are frequently present on chest examination, and digital clubbing may occur at late stages of the disease. Weight loss is seen in some subjects.

### Radiographic Features

There are no prominent radiographic features found in cobalt-associated occupational asthma. However, in a cross-sectional study of hard metal workers, slight radiographic abnormalities (0/1, 1/1 according to the ILO classification) were more frequent in exposed workers compared to controls.[244] Interestingly, it was those workers in the powder, press, and furnace workshops (exposed to ''soft carbide'') who had the most abnormalities compared to those of the sintering, finishing, or maintenance worker.[244] Also, these workers with radiographic abnormalities were more likely to have functional changes. In hard metal-associated interstitial lung disease, the chest X-ray usually shows small, irregular opacities predominantly in the lower-lung zones. With progression of the disease, more widespread and coarser patterns of opacities can be seen, and end-stage honeycombing may develop. In acute disease, a ground-glass or hazy pattern of shadowing is sometimes present. Finally, a normal chest X-ray does not preclude the presence of significant clinical and physiologic impairment. Therefore, in symptomatic patients, the diagnosis of hard metal disease should not be delayed until radiographic abnormalities are present, since these may lag behind other evidence of disease.

### Pulmonary Functions

Physiologic abnormalities can be demonstrated in workers with occupational asthma (chronic bronchitis and chest tightness), both cross-shift decrements in FVC, $FEV_1$, and maximal midexpiratory flow rates ($FEF_{25-75}$) and following exposure to the work environment. In the latter cases, symptoms and airflow obstruction may occur promptly upon exposure and may persist for many hours or several days after the challenge.[233] It appears that exposure to hard metal dust at a mean level of 1.6 mg/$m^3$ or cobalt in air at a mean concentration of $>126$ $\mu/m^3$ causes chronic bronchial obstruction, whereas lower exposures do not cause lung function change.[243] Acute exposure in previously unexposed healthy volunteers resulted in an irritant effect on the large airways characterized by cough, sore throat, and a slight reduction in FVC.[243]

In patients with diffuse interstitial lung disease, lung function abnormalities include a reduction in lung volumes, decreased diffusing capacity for carbon monoxide, and hypoxemia at rest that is worsened by exercise. In some patients, an obstructive defect may also be present but is not the predominant defeat. Occasionally, the physiologic changes may precede abnormalities in the chest roentgenogram.[231,236]

### Laboratory Findings

There are no specific abnormalities detected in the routine laboratory tests among exposed or diseased workers. On the other hand, several methods for monitoring exposure to cobalt utilize blood or urine analyses. Urinary levels of cobalt have been shown to reflect exposure. Inhaled cobalt is associated with a two phase elimination in urine:[234] a rapid phase lasting approximately 2 d, and a second phase of prolonged elimination at lower rate. The two phases of elimination often overlap. Blood cobalt concentration appears to have a linear relationship, with exposure to levels $>0.1$ mg/$m^3$. The usefulness of monitoring blood levels in not known. Serum IgE antibodies to cobalt have been demonstrated. In addition, the transformation of blood-lymphocytes by cobalt has been shown in patients with cobalt asthma[245] and cobalt-induced dermatitis.

## Management and Control

Prevention is the key to management. This can be accomplished by available industrial hygiene measures designed to reduce exposure to dust and aerosols. The recommended threshold limit value is 50 mg/m$^3$ as a time-weighted average for an 8-h d.[235] Periodic screening by respiratory questionnaires, chest roentgenograms, and pulmonary function testing can be employed to screen exposed workers to identify early, asymptomatic disease.

Both those patients with established occupational asthma and interstitial lung disease should be removed from further exposure. Appropriate asthma management should be provided patients with clinical disease. Patients with cobalt pneumoconiosis frequently require corticosteroid therapy to speed recovery and hopefully prevent disease progression. Recurrences have occurred in patients who returned to the work environment after being successfully treated by removal from exposure and corticosteroid therapy. A few subjects have required cytotoxic treatment, cyclophosphamide or azathioprine, when they did not improve with corticosteroid therapy alone. Several deaths have been reported from progressive disease, usually from chronic respiratory insufficiency and cor pulmonale that develop many years after the onset of clinically evident disease. A rapidly progressive fatal form of the disease has been reported.[246]

## Aluminum Pneumoconiosis

### Introduction and Work Environment

Aluminum is one of the most abundant elements in the earth's crust. The main ore is bauxite, hydrated aluminum oxide, and iron oxides.[247] Bauxite ($Al_2O_3$) must be melted with cryolite ($Na_3AlF_6$) and electrolytically reduced to produce free metal. Aluminum has vast uses in construction, in the manufacture of beverage and food containers, in automobiles, and in cooking utensils.[247] It is used in explosives, solid rocket propellants, and as a fine powder. It is also used as a desiccant, for antacids, in antiperspirants, and as a catalyst for organic synthesis.[247]

Occupational exposure to aluminum is primarily via inhalation associated with the manufacturing process. Aluminum is derived from the reduction of alumina in smelter. The process involves grinding bauxite and mixing it with iron and coke. This mixture is then shoveled into large metal pots, the electrolytic cell in which the reduction takes place. The building is called the potroom. Carbon electrodes are lowered to the surface, and fusion occurs at a temperature of 2000° C, generated by an arc between the electrodes.[203] During this process, workers in the potrooms are exposed to a number of air contaminants, such as fluoride (of gaseous, hydrogen fluoride, and particulate form), alumina, sulfur dioxide, carbon monoxide, carbon, and particulate polycyclic organic matter (coal tar).[203,248] Several reports have appeared that document acute and chronic pulmonary effects in workers in several industries. Skin and mucous membrane irritation may occur in response to exposure to aluminum chloride. The rate of absorption of aluminum via the lung is not known.

The role of aluminum smelting in the development of lung disease has been recently reviewed.[249] Several diseases have reportedly been associated with work in the aluminum smelting industry: potroom asthma, chronic obstructive lung disease, lung cancer, and pulmonary fibrosis. The fusing of alumina, coke, and iron in electric pots to form corundum, an extremely hard abrasive, causes dense white fumes that contain aluminum oxide, silica, and several substances.[203,233] This fume contains approximately 29 to 44% silica and 41 to 62% alumina. A potentially severe and fatal mixed-dust pneumoconiosis (''Shaver's disease'') can result from exposure to these fumes, particularly in furnace feeders or crane operators.[203]

Aluminum oxide workers exposed to finely divided alumina in the production and use of aluminum oxide abrasive grinding wheels and tipped tools[250] and workers exposed to finely divided aluminum powder used in the production of explosives ("pyro" powder)[251] may develop pulmonary fibrosis.[250] Aluminum welders and smelters exposed to aluminum fumes in poorly ventilated areas can develop metal fume fever. This process does not appear to lead to chronic fibrotic lung disease. Aluminum arc welding has also been associated with a granulomatous lung reaction,[252,253] pulmonary alveolar proteinosis,[254] and desquamative interstitial pneumonia.[255] Silicate compounds of aluminum may also produce pulmonary fibrosis.[256]

### *Histopathologic Features*

The pathology of Shaver's disease reveals, on gross examination of the lungs, a "gunmetal gray" color.[254] Subpleural bullae may be particularly prominent. Pleural thickening is common. Diffuse interstitial fibrosis and scar emphysema are seen in all cases. Obliterative endarteritis is also present, especially in the severely fibrotic areas. Tissue analysis reveals a high quantity of both silica (21 to 31%) and alumina (25 to 41%).[254]

The histopathologic features of aluminum oxide workers, including potroom workers, aluminum welders, and polishers, with interstitial lung disease are probably quite similar but have not been extensively studied. Most show features of the spectrum of changes characteristic of interstitial lung disease:

1. chronic accumulation of inflammatory and immune effector cells within the alveolar spaces (so-called desquamative interstitial pneumonia);[250,255]
2. areas of focal and diffuse interstitial infiltrates with fibrosis (i.e., usual interstitial pneumonia);[255,257]
3. multiple nodular lesions composed of dense, partially hylanized, collagenous tissue, with irregular areas of acellular, coagulative necrosis. Often dense accumulations of macrophages containing a granular, gray-brown, metallic material is present in both the nodular masses and the interstitium of the fibrotic tissue.[255,257] Phagocytosis of particles by type I alveolar epithelium is also observed;[255]
4. dense interstitial fibrosis and honeycomb changes.[250,251]

The presence of excessive quantities of aluminum (often 1000 times higher than controls) has been identified in the tissue or cells by many different analytic techniques:[249] neutron activation analysis, electron optical analysis, X-ray diffractions, scanning electron microscopy, X-ray spectrometry, electron microprobe analysis, and energy dispersive spectrometry. It has been suggested that bare aluminum metal or gamma aluminum oxide is the culprit in production of the pulmonary fibrosis.[251] However, this has been disputed. Gilks and Churg[251] have suggested that it might be the presence of aluminum fibers that is the key to the development of aluminum-induced interstitial injury and fibrosis. Bronchoalveolar lavage in exposed potroom workers demonstrated no abnormalities in the cellular profile, but the concentrations of fibronectin and albumin was increased and that of angiotensin converting enzyme was decreased. The significance of these findings to the pathogenesis is unknown.[258] A granulomatous reaction associated with interstitial fibrosis has been reported following crystalline aluminum silicate exposure[256] and exposure to an aluminum metal in a welder.[252]

Pulmonary alveolar proteinosis presumably induced by inhalation of aluminum particles in a 44-year-old aluminum rail grinder has been described by Miller and colleagues.[254] Crystalline silica is a known exposure associated with the development of pulmonary alveolar proteinosis.

Table 13  **Metal Fume Fever**[a]

| Metals | Common Name |
|---|---|
| Cadmium, hard-metals products | Braziers' disease |
| Aluminum, cadmium | Smelters' shakes |
| Cadmium | Brass chills, brass founders' ague |
| Zinc oxide | Zinc chills, zinc oxide chills |
| Galvanized steel, oxides of zinc, copper, magnesium | Galvanizers' poisoning, welders' ague |
| Copper | Copper fever |
| Teflon (plastic) | Teflon fume fever |

[a]Also called Monday Fever, Foundry Fever, The Smothers.

### Signs and Symptoms

Aluminum smelter workers (i.e., potroom workers) have more cough and wheeze than do unexposed subjects.[248] Most of these workers spent more than 50% of their work time in the potrooms. Despite these findings, it is unclear if potroom asthma occurs as a response to potroom exposure or merely precipitates asthma-like symptoms in a predisposed individual.[248,249] Importantly, the prognosis of potroom asthma is variable. However, some exposures may result in long-lasting bronchial hyperactivity, even after cessation of exposure.[249]

Like potroom asthma, the relationship of aluminum smelting exposure and the development of COPD is not clear. Cross-sectional studies have demonstrated more productive cough, dyspnea, and wheezing in these workers.[249] However, longitudinal studies are lacking to determine if these workers have a greater chance of developing chronic bronchitis or emphysema when other confounding factors, such as cigarette smoking, are controlled. The lung cancer risk, despite the exposure to coal tar and pitch volatiles, is very low if present at all.[249]

Interstitial lung disease, i.e., pulmonary fibrosis, has been reported in workers manufacturing alumina abrasives or explosives from stamped aluminum powders. There have been a few cases of pulmonary fibrosis in potroom workers.[251,253,255,257] Nonetheless, the fibrogenic potential of alumina dust is much less than that of other inorganic dusts such as asbestos, silica, and coal. In fact, $Al_2O_3$ dust has been inhaled in high doses as a preventative "treatment" against the development of pneumoconiosis in coal miners without apparent adverse effects.[250,255,259] In a recent study using an animal model of silicosis, it has been demonstrated that it is aluminum salts, not $Al_2O_3$, that prevents the development of silicosis and promotes silica clearance from the lung.[260]

In aluminosis associated with manufacture of alumina for abrasives or metal production, dyspnea, often beginning quite insidiously, is the most common symptom. Rapid progression, with development of breathlessness with minimal exertion, may occur. Cough productive of frothy mucus sputum was present in most patients. Unlike other pneumoconioses, these patients frequently complained of other symptoms, including substernal discomfort, chest tightness, pleuritic chest pain, weakness, and fatigue.[203,233] Pneumothoraces occur in these patients and may be a recurring problem. Tachypnea, cyanosis, and respiratory distress may be present in advanced disease. Wheezing, rales, and bronchi may be present but are variable findings.

Metal fume fever (see Table 13) associated with inhalation of aluminum fumes (usually by welders or smelters) is characterized by onset of fever, chills, nonproductive cough, myalgias, muscular weakness, nausea, headache, and profuse sweating. Usually the onset is 4 to 6 h after exposure on the first day when a worker starts or returns to work (hence the name "Monday Morning Fever"). The illness usually resolves in 24 h. Some tolerance to the fumes occurs as the work week progresses, but it recurs upon re-exposure after a short period of absence. The diagnosis is made by careful work history. Occasionally, measurements of metal concentrations in the urine, for example zinc or copper, may help if they are increased. There is no specific therapy except removal from exposure or improved ventilation. The pathogenesis is unknown.

### Radiographic Features

The chest radiograph in patients with ''Shaver's disease'' in the early stages reveals a fine lacelike reticular pattern. With time this pattern becomes coarse and more pronounced in the upper-lung zones.[203,251] Mediastinal widening and partial blurring of the hilar shadows can be seen. The diaphragm is often irregular, and tenting is frequently seen. Hyperlucent areas and linear shadows presumably related to subpleural blebs or cysts are felt to be characteristic.[203] Spontaneous pneumothoraces may be seen.

In the recently described cases of potroom workers, welders, and polishers, the radiologic features range from slight, small, irregular opacities to an interstitial infiltrate. There is a trend to increased profusion of the opacities with longer duration of exposure.[249] On the other hand, the majority of cross-sectional studies of smelter workforces have found no evidence of pneumoconiosis on chest X-ray (reviewed in Reference 249).

### Pulmonary Functions

Potroom workers have a lower mean $FEV_1$ and maximal midexpiratory flowrate than control subjects.[248] A restrictive ventilatory defect without evidence of airflow obstruction is most often found in those cases where it has been measured. Arterial hypoxemia at rest, and later with exercise, and impaired diffusing capacity for carbon monoxide are commonly found. In the review of Abramson et al.[249] of cross-sectional studies of smelter workers, they found no documented evidence of significant restrictive ventilatory defects or decreased pulmonary compliance in any of these studies. However, exposures in other settings to alumina have demonstrated pulmonary function abnormalities, that is restrictive spirometry, associated with radiographic abnormalities consistent with interstitial lung disease in a subset of these workers.[249]

### Management and Complications

Prevention is the key in avoiding the development of lung disease from aluminum exposure. In patients with interstitial lung disease, the clinical course may be a progressive one. In ''Shaver's disease,'' onset occurs after a relatively short period of industrial exposure. Rapid and progressive disease occurs and may result in serious disability and death. Some of the symptoms may improve following removal from exposure. The role of immunosuppressive therapy is unknown. It is probably prudent to remove all workers with any radiographic abnormalities from further exposure.

## Uncommon Causes of Metal Dusts Pneumoconiosis

### Cadmium-Induced Lung Disease

Cadmium is a metal recovered as a by-product during the smelting of zinc or lead ores. It is used in electroplating, to make pigments, in plastics, and as one component in cadmium nickel batteries.[261] Cigarette smoking can increase cadmium intake considerably above that obtained through ingestion and inhalation from ambient air.[262]

The major health effects of cadmium are predominantly secondary to renal and pulmonary injury. Acute inhalation of cadmium oxide fumes (for example, from smelting or welding) may cause a delayed toxic fume-type reaction, with upper airway irritation, dyspnea, chest pain, and cough, followed by pulmonary edema. Metal fume fever may also develop. These usually resolve over days, and only rarely is it fatal or does progressive pulmonary fibrosis develop.[262] Chronic cadmium poisoning primarily causes kidney problems. However, chronic cough and sputum production and shortness of breath with exertion can be seen in cadmium-exposed workers. Both chronic bronchitis and emphysema may be found with increased incidence in cadmium workers. The role of cadmium in the production of emphysema is not entirely clear, but it is well documented to cause emphysema (i.e., air space enlargement) and interstitial pulmonary fibrosis in animal models.[263] In addition, workers chronically exposed to high concentrations of airborne cadmium develop a restrictive ventilatory defect and radiographic evidence of mild or moderate interstitial changes.[264] Finally, the role of cadmium as a human lung carcinogen remains to be fully defined.[262] Urinary cadmium may be a useful guide to cadmium exposure but is not used to make the diagnosis of cadmium-related disease.

### Titanium

Titanium is one of the most abundant elements of the earth's crust. It is very resistant to corrosion and is widely used in the construction of parts for commercial airplanes. Titanium dust can be a mild respiratory irritant. Rarely it has been shown to be associated with radiographic features of mild pneumoconiosis. Deposits of titanium dioxide pigments in the pulmonary interstitium associated with cell destruction and slight fibrosis supports the fact that titanium may cause mild pulmonary irritation.[247] A granulomatous pneumonitis has been reported in an aluminum smelter who had a positive LTT to titanium chloride.[265]

### Rare-Earth Metals

Rare-earth metals refers to the lanthanum series with yttrium. These elements are often found together in many ores, such as monazite, xenotine, gasolinite, samarskite, fergusonite, apatite, euxenite, and bastnaesite.[266] All are soft metals with a bright silver luster. These metals have multiple and complex uses that are being actively investigated.[266] The rare earths are relatively nontoxic. Massive acute inhalation may result in an acute pneumonitis and bronchitis.[266]

### Molybdenum

Molybdenum is an abundant mineral obtained from primary mining operation or as a by-product of copper mining. It has been associated with radiographic abnormalities of pneumoconiosis in extremely few cases reported to date.[247] It has not been reported to cause severe physiologic impairment.

### Lead Miner's Pneumoconiosis

Lead miners, like other miners of metal, are at risk for the development of silicosis since they are exposed to dusts with high free silica content.[267] Lead miners are also at risk for systemic lead poisoning. Lead mining results in exposure to lead oxide or lead particles that are inhaled. The deposition rate in the lung varies between 30 and 50%.[267] Lung deposition is due to sedimentation and diffusion by Brownian movement. After deposition in the lung, one third to one half of the lead is absorbed into the blood.

In Masjedi's study of 45 lead miners,[267] he found abnormal lung function with lung restriction in 5 subjects and reticulonodular opacities in 16 subjects. Histopathologic analysis of transbronchoscopic specimens revealed bronchial mucosal edema, squamous metaplasia, mononuclear or neutrophilic infiltration, fibrosis, and deposition of small particles in and outside alveolar macrophages and in the alveolar wall.

Tissue lead levels were found to be extremely high. No characteristic findings of silicosis were found on these small samples, but it is possible that the disease in these lead miners is due to silicosis. Consequently, with the demonstration of increased tissue lead content in the presence of lead exposure and the absence of classic changes of silicosis, one is left to consider that the lung disease results, at least in part, from lead exposure.

### Barium Pneumoconiosis (Baritosis)

Barium is a silver-white metal with numerous uses, from well drilling "mud" to flux in ceramics to magnets in speakers to radiopaque contrast media in radiology departments. Baritosis has been identified in heavily exposed workers following the inhalation of finely divided particles of barium sulfate and barium oxide and lithopone. This results in a benign pneumoconiosis characterized by radiographic changes of small, dense, circumscribed nodules evenly distributed throughout the lung fields without impairment of lung function.[268]

### Iron Oxide and Hematite Pneumoconiosis

Dust and fumes exposure to iron oxide can occur in many settings, including mining, arc welding, metal processing in foundries, and silver polishing. Inhalation of iron oxide fumes or dust, often as an extremely fine particulate produced during the cutting or burning of steel or iron, may cause a benign pneumoconiosis, unless the iron oxide is accompanied by silica or other elements. The radiograph usually reveals linear reticular markings, with a fine, stippling, mottling nodularity.[203] The changes may regress with cessation of exposure.[203] A more severe interstitial process can be seen if the iron oxide inhaled contains sufficient quantities of silica.

### Antimony

Antimony is an intermetal that can form inorganic and organic complexes and that is found in more than 100 minerals.[269] It is used as a flame retardant for plastics, paint, textiles, paper, rubber, and adhesives.[261] During the smelting process, arsenic and sulfur dioxide are liberated. Many of the earlier reports regarding the occupational exposures and subsequent illnesses in these workers likely resulted from the latter exposures than from antimony. Cardiac, gastrointestinal, liver, and reproductive system toxicities are important manifestations of antimony exposure. Pneumoconiosis can occur, but it is not clear if it is directly related to antimony exposure.[261,270] Fine nodular opacities are found, often close to the hilum. Lung function is not significantly impaired, although mild obstructive changes and emphysema are seen.[261,269] Lung cancer deaths were excessive in one study in Britain, but the carcinogenic risk of antimony is not clear.[269]

### Copper

Copper is an abundant and widely used element of the earth's crust. It is frequently complexed with other substances, such as arsenic, cadmium, iron, sulfur, zinc, etc. It is the inhalation of fumes or dust related to smelters, packaging, grinding, cutting, or welding that is the chief hazard. Several respiratory illnesses have resulted from copper exposure:[247] (a) upper respiratory tract irritation, (b) "metal fume fever," and (c) "vineyard sprayer's lung." Vineyard sprayer's lung is primarily of historical interest, since the agent responsible for this disease is no longer available in the U.S. "Bordeaux" mixture contained 1 to 2% aqueous copper sulfate and hydrated lime and was used as a pesticide to prevent mildew in the vineyards. The cases were first reported from Portugal.[271-273] Subclinical, acute, and chronic progressive forms of the disease were described, usually after years of exposure. Symptoms were usually insidious in onset and included progressive weakness, anorexia, weight loss, cough, and dyspnea. The chest roentgenographic pattern included diffuse, micronodular opacities or reticulonodular opacities predominantly in the lower lung zones.[272] In the chronic form of the disease, the chest X-ray revealed symmetric, tumorlike, massive opacities in the upper lobes (similar to the lesions seen in progressive massive fibrosis of coal workers' pneumoconiosis). Lung biopsy revealed a spectrum of changes, characterized by a fibrotic, granulomatous pneumonitis. In addition, copper was found in the lung or mediastinal lymph nodes. Noncaseating granulomas were also seen in the liver.[273] There was a high incidence of lung cancer in these patients.[272]

## REFERENCES

1. **Cherniack, R. M., Colby, T. V., Flint, A., Thurlbeck, W. M., Waldron, J. A., and King, T. E., Jr.,** BAL Cooperative Group Steering Committee, Quantitative assessment of lung pathology in idiopathic pulmonary fibrosis, *Am. Rev. Respir. Dis.,* 144, 892, 1991.

2. **Hyde, D. M., King, T. E., Jr., McDermott, T., Waldron, J. A., Jr., Colby, T. V., Thurlbeck, W. M., Flint, A., Ackerson, L., and Cherniack, R. M.,** Idiopathic pulmonary fibrosis: the quantitative assessment of lung pathology. Comparison of a semi-quantitative versus a morphometric histopathologic scoring system, *Am. Rev. Respir. Dis.,* in press.

3. **Crystal, R. G., Bitterman, P. B., Rennard, S. I., Hance, A. J., and Keogh, B. A.,** Interstitial lung diseases of unknown cause. Disorders characterized by chronic inflammation of the lower respiratory tract, *N. Engl. J. Med.,* 310, 154, 1984.

4. **Crystal, R. G., Fulmer, J. D., Roberts, W. C., Moss, M. L., Line, B. R., and Reynolds, H. Y.,** Idiopathic pulmonary fibrosis: clinical, histologic, radiographic, physiologic, scintigraphic, cytologic and biochemical aspects, *Ann. Intern. Med.,* 85, 769, 1976.

5. **Carrington, C. B., Gaensler, E. A., Coutu, R. E., Fitzgerald, M. X., and Gupta, R. G.,** Natural history and treated course of usual and desquamative interstitial pneumonia, *N. Engl. J. Med.,* 298, 801, 1978.

6. **Colby, T. V. and Carrington, C. B.,** Infiltrative lung disease, in *Pathology of the Lung,* Thurlbeck, W. M., Ed., Thieme Medical Publishers, New York, 1988, 425.

7. **Haslam, P. L., Turton, C. W. G., Lukoszek, A., Salsbury, A. J., Dewar, A., Collins, J. V., and Turner-Warwick, M.,** Bronchoalveolar lavage fluid cell counts in cryptogenic fibrosing alveolitis and their relation to therapy, *Thorax,* 35, 328, 1980.

8. **Watters, L. C., Schwarz, M. I., Cherniack, R. M., Waldron, J. A., Dunn, T. L., Stanford, R. E., and King, T. E., Jr.,** Idiopathic pulmonary fibrosis. Pretreatment bronchoalveolar lavage cellular constituents and their relationships with lung histopathology and clinical response to therapy, *Am. Rev. Respir. Dis.,* 135, 696, 1987.

9. **Liebow, A. A., Steer, A., and Billingsley, J. G.,** Desquamative interstitial pneumonia, *Am. J. Med.,* 39, 369, 1965.

10. **Gaensler, E. A., Goff, A. M., and Prowse, C. M.,** Desquamative interstitial pneumonia, *N. Engl. J. Med.,* 274, 113, 1966.

11. **Epler, G. R., McLoud, T. C., Gaensler, E. A., Mikus, J. P., and Carrington, C. B.,** Normal chest roentgenograms in chronic diffuse infiltrative lung disease, *N. Engl. J. Med.,* 298, 934, 1978.

12. **Livingstone, J. L., Lewis, J. G., Reid, L., and Jefferson, K. E.,** Diffuse interstitial pulmonary fibrosis. A clinical, radiological, and pathological study based on 45 patients, *Q. J. Med.,* 33, 71, 1964.

13. **Woodring, J. H., Barrett, P. A., Rehm, S. R., and Nurenberg, P.,** Acquired tracheomegaly in adults as a complication of diffuse pulmonary fibrosis, *Am. J. Roentgenol.,* 152, 743, 1989.

14. **Muller, N. L., Miller, R. R., Webb, W. R., Evans, K. G., and Ostrow, D. N.,** Fibrosing alveolitis CT-pathologic correlation, *Radiology,* 160, 585, 1986.

15. **Muller, N. L., Guerry-Force, M. L., Staples, C. A., Wright, J. L., Wiggs, B., Coppin, C., Pare, P., and Hogg, J. C.,** Differential diagnosis of bronchiolitis obliterans with organizing pneumonia and usual interstitial pneumonia—clinical, functional, and radiologic findings, *Radiology,* 162, 151, 1987.

16. **Muller, N., Staples, C., Miller, R., Vedal, S., Thurlbeck, W., and Ostrow, D.,** Disease activity in idiopathic pulmonary fibrosis: CT and pathologic correlation, *Radiology,* 165, 731, 1987.

17. **Staples, C., Muller, N., Vedal, S., Abboud, R., Ostrow, D., and Miller, R.,** Usual interstitial pneumonia: correlation of CT with clinical, functional, and radiographic findings, *Radiology,* 162, 377, 1987.

18. **Mathieson, J., Mayo, J., Staples, C., and Muller, N.,** Chronic diffuse infiltrative lung disease: comparison of diagnostic accuracy of CT and chest radiography, *Radiology,* 171, 111, 1989.

19. **Rudd, R. M., Haslam, P. L., and Turner-Warwick, M.,** Cryptogenic fibrosing alveolitis relationships of pulmonary physiology and bronchoalveolar lavage to treatment and prognosis, *Am. Rev. Respir. Dis.,* 124, 1, 1981.

20. **Peterson, M. W., Monick, M., and Hunninghake, G. W.,** Prognostic role of eosinophils in pulmonary fibrosis, *Chest,* 92, 51, 1987.

21. **Haslam, P. L., Hughes, D. A., Dewar, A., and Pantin, C. F. A.,** Lipoprotein macroaggregates in bronchoalveolar lavage fluid from patients with diffuse interstitial lung disease in comparison with idiopathic alveolar lipoproteinosis, *Thorax,* 43, 140, 1988.

22. **Honda, Y., Tsunematsu, K., Suzuki, A., and Akino, T.,** Changes in phospholipids in bronchoalveolar lavage fluid of patients with interstitial lung diseases, *Lung,* 166, 293, 1988.

23. **Robinson, P. C., Watters, L. C., King, T. E., and Mason, R. J.,** Idiopathic pulmonary fibrosis: abnormalities in bronchoalveolar lavage fluid phospholipids, *Am. Rev. Respir. Dis.,* 137, 585, 1988.

24. **McCormack, F. X., King, T. E., Jr., Voelker, D. R., Robinson, P. C., and Mason, R. J.,** Idiopathic pulmonary fibrosis: abnormalities in the bronchoalveolar lavage content of surfactant protein A, *Am. Rev. Respir. Dis.,* 144, 160, 1991.

25. **Englert, M., Yernault, J. C., deCoster, A., and Clumeek, N.,** Diffusing properties and elastic properties in interstitial diseases of the lung, *Prog. Respir. Res.,* 8, 177, 1975.

26. **Fulmer, J. D., Roberts, W. C., von Gal, E. R., and Crystal, R. G.,** Morphologic-physiologic correlates of the severity of fibrosis and degree of cellularity in idiopathic pulmonary fibrosis, *J. Clin. Invest.,* 63, 665, 1979.

27. **Keogh, B. A. and Crystal, R. G.,** Pulmonary function testing in interstitial pulmonary disease. What does it tell us?, *Chest,* 78, 856, 1980.

28. **Watters, L. C.,** Genetic aspects of idiopathic pulmonary fibrosis and hypersensitivity pneumonitis, *Semin. Respir. Med.,* 7, 317, 1986.

29. **Meier-Sydow, J., Rust, M., Kronenberger, H., Thiel, C., Amthor, M., and Riemann, H.,** Long-term follow-up of lung function parameters in patients with idiopathic pulmonary fibrosis treated with prednisone and azathioprine or D-penicillamine, *Prax. Pneumol.,* 33, 680, 1979.

30. **Wright, P. H., Heard, B. E., Steel, S. J., and Turner-Warwick, M.,** Cryptogenic fibrosing alveolitis: assessment by graded trephine lung biopsy histology compared with clinical, radiographic, and physiological features, *Br. J. Dis. Chest,* 75, 61, 1981.

31. **Weese, W. C., Levine, B. W., and Kazemi, H.,** Interstitial lung disease resistant to corticosteroid therapy. Report of three cases treated with azathioprine or cyclophosphamide, *Chest,* 67, 57, 1975.

32. **Meuret, G., Fueter, R., and Gloor, F.,** Early stage of fulminant idiopathic pulmonary fibrosis cured by intense combination therapy using cyclophosphamide, vincristine, and prednisone, *Respiration,* 36, 228, 1978.

33. **Johnson, M. A., Kwan, S., Snell, N. J. C., Nunn, A. J., Darbyshire, J. H., and Turner-Warwick, M.,** Randomized controlled trial comparing prednisolone alone with cyclophosphamide and low dose prednisolone in combination in cryptogenic fibrosing alveolitis, *Thorax,* 44, 280, 1989.

34. **Brown, C. H. and Turner-Warwick, M.,** The treatment of cryptogenic fibrosing alveolitis with immunosuppressant drugs, *Q. J. Med.,* 40, 289, 1971.

35. **Cegla, U. H., Kroidl, R. F., Meier-Sydow, J., Thiel, C., and Czarnecki, G. V.,** Therapy of the idiopathic fibrosis of the lung. Experiences with three therapeutic principles—corticosteroids in combination with azathioprine, D-penicillamine, and para-amino-benzoate, *Pneumonologie,* 152,75, 1975.

36. **Raghu, G., Depaso, W. J., Cain, K., Hammar, S. P., Wetzel, C. E., Dreis, D. F., Hutchinson, J., Pardee, N. E., and Winterbauer, R. H.,** Azathioprine combined with prednisone in the treatment of idiopathic pulmonary fibrosis: A prospective, double-blind randomized, placebo-controlled clinical trial, *Am. Rev. Respir. Dis.,* 144, 291, 1991.

37. **Panos, R. J., Mortenson, R., Niccoli, S. A., and King, T. E., Jr.,** Clinical deterioration in patients with idiopathic pulmonary fibrosis. Causes and assessment, *Am. J. Med.,* 88, 396, 1990.

38. **Colby, T. V. and Myers, J. L.,** The clinical and histologic spectrum of bronchiolitis obliterans including bronchiolitis obliterans organizing pneumonia (BOOP), *Semin. Respir. Med.,* 13, 119, 1992.

39. **Epler, G. R. and Colby, T. V.,** The spectrum of bronchiolitis obliterans, *Chest,* 83, 161, 1983.

40. **Colby, T. V. and Churg, A. C.,** Patterns of pulmonary fibrosis, in 1986 Pathology Annual, Sommers, S. C., Rosen, P. P., and Fechner, R. E., Eds., Appleton-Century-Crofts, Norwalk, CT, 1986, 277.

41. **Troisi, F. M.,** Delayed death caused by gassing in a silo containing green forage, *Br. J. Ind. Med.,* 14, 56, 1957.

42. **Horvath, E. P., Colico, D. G. A., Barbee, R. A., and Dickie, H. A.,** Nitrogen dioxide-induced pulmonary disease, *J. Occup. Med.,* 20, 103, 1978.

43. **Milne, J. E. H.,** Nitrogen dioxide inhalation and bronchiolitis obliterans, *J. Occup. Med.,* 11, 538, 1969.

44. **Yockey, C. C., Eden, B. M., and Byrd, R. B.,** The McConnel missile accident: clinical spectrum of nitrogen dioxide exposure, *JAMA,* 244, 1221, 1980.

45. **Moskowitz, R. L., Lyons, H. A., and Cottle, H. R.,** Silo-filler's disease: clinical, physiologic and pathologic study of patient, *Am. J. Med.,* 36, 457, 1964.

46. **King, T. E., Jr.,** Bronchiolitis obliterans, in *Interstitial Lung Diseases,* Schwarz, M. I., and King, T. E., Jr., Eds., B. C. Decker, Toronto, 1988, 325.

47. **Becklake, M. R., Goldman, H. I., Bosman, A. R., and Freed, C. C.,** The long-term effects of exposure to nitrous fumes, *Am. Rev. Tuberc.,* 76, 398, 1957.

48. **Ramirez, R. J. and Dowell, A. R.,** Silo-filler's disease: nitrogen dioxide-induced lung injury: long-term follow-up and review of the literature, *Ann. Intern. Med.,* 74, 569, 1971.

49. **Jones, G. R., Proudfoot, A. T., and Hall, J. I.,** Pulmonary effects of acute exposure to nitrous fumes, *Thorax,* 28, 61, 1973.

50. **Clutton-Brock, J.,** Two cases of poisoning by contamination of nitrous oxide with higher oxides of nitrogen during anesthesia, *Br. J. Anaesthesiol.,* 39, 388, 1967.

51. **Fleming, G. M., Chester, E. H., and Montenegro, H. D.,** Dysfunction of small airways following pulmonary injury due to nitrogen dioxide, *Chest,* 75, 720, 1979.

52. **Prowse, K.,** Nitrous fume poisoning, *Bull. Eur. Physiopathol. Respir.,* 13, 191, 1977.

53. **Scott, E. U. and Hunt, W. B.,** Silo-filler's disease, *Chest,* 63, 701, 1973.

54. **Rafii, S. and Godwin, M. D.,** Relapse following latent period, *Arch. Pathol.,* 72, 424, 1961.

55. **Darke, C. S. and Warrack, A. J. N.,** Bronchiolitis from nitrous fumes, *Thorax,* 13, 327, 1958.

56. **Lowry, T. and Schuman, L. M.,** "Silo-filler's disease"—a syndrome caused by nitrogen dioxide, *JAMA,* 162, 153, 1956.

57. **Nichols, B. H.,** Clinical effects of inhalation of nitrogen dioxide, *Am. J. Roentgenol. Radium Ther.,* 23, 516, 1930.

58. **Woodford, D. M., Coutu, R. E., and Gaensler, E. A.,** Obstructive lung disease from acute sulfur dioxide exposure, *Respiration,* 38, 238, 1979.

59. **Murphy, D. M. F., Fiarman, R. P., Lapp, N. L., and Morgan, K. C.,** Severe airway disease due to inhalation of fumes from cleansing agents, *Chest,* 69, 372, 1976.

60. **Douglas, W. W., Norman, G., Hepper, G., and Colby, T. V.,** Silo-filler's disease, *Mayo Clin. Proc.,* 64, 291, 1989.

61. **Eichenberger, A., Weber, J., and Hausser, E.,** Pneumopathic des ensileus (Silo-filler's disease), *Schweiz. Med. Wschr.,* 96, 1652, 1966.

62. **Tse, R. L. and Bockman, A. A.,** Nitrogen dioxide toxicity. Report of four cases in firemen, *JAMA,* 212, 1341, 1970.

63. **Rigner, K. G. and Swensson, A.,** Late prognosis of nitrous fume poisoning and follow-up study, *Acta Med. Scand.,* 170, 291, 1961.

64. **Galea, M.,** Fatal sulfur dioxide inhalation, *Can. Med. Assoc. J.,* 91, 345, 1964.

65. **Muller, B.,** Nitrogen dioxide intoxication after a mining accident, *Respiration,* 26, 249, 1969.

66. **Churg, A.,** Small airways disease associated with mineral dust exposure, *Semin. Respir. Med.,* 13, 140, 1992.

67. **Churg, A. and Wright, J. L.,** Small airways disease and mineral dust exposure, *Pathol. Annu.,* 18, 233, 1983.

68. **Churg, A. and Wright, J. L.,** Small airway lesions in patients exposed to non-asbestos mineral dusts, *Hum. Pathol.,* 14, 688, 1983.

69. **Wright, J. L. and Churg, A.,** Morphology of small-airway lesions in patients with asbestos exposure, *Hum. Pathol.,* 15, 68, 1984.

70. **Churg, A., Wright, J. L., Wiggs, B., Pare, P. D., and Lazar, N.,** Small airways disease and mineral dust exposure, *Am. Rev. Respir. Dis.,* 131, 139, 1985.

71. **Wright, J. L. and Churg, A.,** Severe diffuse small airways abnormalities in long term chrysotile asbestos miners, *Br. J. Ind. Med.,* 42, 556, 1985.

72. **Wright, J. L., Tron, V., Wiggs, B., and Churg, A.,** Cigarette smoke potentiates asbestos-induced airflow abnormalities, *Exp. Lung Res.,* 14, 537, 1988.

73. **Begin, R., Boileau, R., and Peloquin, S.,** Asbestos exposure, cigarette smoking, and airflow limitation in long-term Canadian chrysotile miners and millers, *Am. J. Ind. Med.,* 11, 55, 1987.

74. **Becklake, M. R.,** Chronic airflow limitation: its relationship to work in dusty occupations, *Chest,* 88, 608, 1985.

75. **Becklake, M. R.,** Occupational exposures: evidence for a causal association with chronic obstructive pulmonary disease, *Am. Rev. Respir. Dis.,* 140, S85, 1989.

76. **Kennedy, S. M., Wright, J. L., Mullen, J. B., Pare, P. D., and Hogg, J. C.,** Pulmonary function and peripheral airway disease in patients with mineral dust or fume exposure, *Am. Rev. Respir. Dis.,* 132, 1294, 1985.

77. **Tyler, W. S., Julian, M. D., and Hyde, D. M.,** Respiratory bronchiolitis following exposures to photochemical air pollutants, *Semin. Respir. Med.,* 13, 94, 1992.

78. **Graham, D. E. and Koren, H. S.,** Biomarkers of inflammation of ozone-exposed humans. Comparison of the nasal and bronchoalveolar lavage, *Am. Rev. Respir. Dis.,* 142, 152, 1990.

79. **Paulo, M. and Gong, H. J.,** Respiratory effects of ozone: whom to protect, when, and how?, *J. Respir. Dis.,* 12, 482, 1991.

80. **Koren, H. S., Devlin, R. B., Graham, D. E., Mann, R., McGee, M. P., Horstman, D. H., Kozumbo, W. J., Becker, S., House, D. E., McDonnell, W. F., and Bromberg, P. A.,** Ozone-induced inflammation in the lower airways of human subjects, *Am. Rev. Respir. Dis.,* 139, 407, 1989.

81. **Jones, R. N., Turner-Warwick, M., Ziskind, M., and Weill, H.,** High prevalence of antinuclear antibodies in sandblasters' silicosis, *Am. Rev. Respir. Dis.,* 113, 393, 1976.

82. **Seaton, A.,** Silicosis, in *Occupational Lung Diseases,* 2nd ed., Morgan, W. K. C., and Seaton, A., Eds., W.B. Saunders, Philadelphia, 1984, 251.

83. **Silicosis and Silicate Disease Committee,** Diseases associated with exposure to silica and nonfibrous silicate minerals, *Arch. Pathol. Lab. Med.,* 112, 673, 1988.

84. **Hauglustaine, D., Van Damme, B., Daenens, P., and Michielsen, P.,** Silicon nephropathy: a possible occupational hazard, *Nephron,* 26, 219, 1980.

85. **Arnalich, F., Lahoz, C., Picazo, M. L., Monereo, A., Arribas, J. R., Martinez Ara, J., and Vasquez, J. J.,** Polyarteritis nodosa and necrotizing glomerulonephritis associated with long-standing silicosis, *Nephron,* 51, 544, 1989.

86. **Michel, R. D. and Morris, J. F.,** Acute silicosis, *Arch. Intern. Med.,* 113, 850, 1964.

87. **Buechner, H. A. and Ansari, A.,** Acute silico-proteinosis, *Dis. Chest,* 55, 274, 1969.

88. **Suratt, P. M., Winn, W. C., Brody, A. R., Bolton, W. K., and Giles, R. D.,** Acute silicosis in tombstone sandblasters, *Am. Rev. Respir. Dis.,* 115, 521, 1977.

89. **Begin, R., Masse, S., Sebastien, P., Martel, M., Bosse, J., Dubois, F., Geoffroy, M., and Labbe, J.,** Sustained efficacy of aluminum to reduce quartz toxicity in the lung, *Exp. Lung Res.,* 13, 205, 1987.

90. **Begin, R., Ostiguy, G., Fillion, R., and Colman, N.,** Computed tomography scan in the early detection of silicosis, *Am. Rev. Respir. Dis.,* 144, 697, 1991.

91. **Bergin, C. J., Mueller, N. L., Vedal, S., and Chan-Yeung, M.,** CT in silicosis: correlation with plain films and pulmonary function tests, *Am. J. Roentgenol.,* 146, 477, 1986.

92. **Teculescu, D. B., Stanescu, D. C., and Pilat, L.,** Pulmonary mechanics in silicosis, *Arch. Environ. Health,* 14, 461, 1967.

93. **Musk, A. W., Peters, J. M., Wegman, D. H., and Fine, L. J.,** Pulmonary function in granite dust exposure: a four-year follow-up, *Am. Rev. Respir. Dis.,* 115, 769, 1977.

94. **Graham, W. G. B., O'Grady, R. V., and Dubuc, B.,** Pulmonary function loss in Vermont granite workers, *Am. Rev. Respir. Dis.,* 123, 25, 1981.

95. **Ng, T. P., Tsin, T. W., O'Kelly, F. J., and Chan, S. L.,** A survey of the respiratory health of silica-exposed gemstone workers in Hong Kong, *Am. Rev. Respir. Dis.,* 135, 1249, 1987.

96. **Doll, N. J., Stankus, R. P., Hughes, J., Weill, H., Ramesh, C. G., Rodriguez, M., Jones, R. N., Alspaugh, M. A., and Salvaggio, J. E.,** Immune complexes and autoantibodies in silicosis, *J. Allergy Clin. Immunol.,* 68, 281, 1981.

97. **Youinou, P., Ferec, C., Cledes, J., et al.,** Immunological effect of silica dust analyzed by monoclonal antibodies, *J. Clin. Lab. Immunol.,* 16, 207, 1985.

98. **Christman, J., Emerson, R. J., Graham, W. G. B., and Davis, G. S.,** Mineral dust and cell recovery from the bronchoalveolar lavage of healthy Vermont granite workers, *Am. Rev. Respir. Dis.,* 132, 393, 1985.

99. **Calhoun, W. J., Christman, J. W., Ershler, W. B., Graham, W. G., and Davis, G. S.,** Raised immunoglobulin concentrations in bronchoalveolar lavage fluid of healthy granite workers, *Thorax,* 41, 266, 1986.

100. **Koskinen, H., Tiilikainen, A., and Nordman, H.,** Increased prevalence of HLA-Aw19 and of the phenogroup Aw19, B18 in advanced silicosis, *Chest,* 83, 849, 1983.

101. **Honda, H., Hirayama, K., Kikuchi, I., et al.,** HLA and silicosis in Japan, *N. Engl. J. Med.,* 319, 1610, 1988.

102. **Kreiss, K., Danilovs, J. A., and Newman, L. S.,** Histocompatibility antigens in a population based silicosis series, *Br. J. Ind. Med.,* 46, 364, 1989.

103. **Snider, D.,** The relationship between tuberculosis and silicosis, *Am. Rev. Respir. Dis.,* 118, 455, 1978.

104. **Hong Kong Chest Service/Tuberculosis Research Centre, M. M. R. C.,** A controlled clinical comparison of 6 and 8 months of antituberculosis chemotherapy in the treatment of patients with silicotuberculosis in Hong Kong, *Am. Rev. Respir. Dis.,* 143, 262, 1991.

105. **Owens, G. R. and Medsger, T. A.,** Systemic sclerosis secondary to occupational exposure, *Am. J. Med.,* 85, 114, 1988.

106. **Sluis-Cremer, G. K., Hessel, P. A., Hnizado, E., et al.,** Silica, silicosis and progressive systemic sclerosis, *Br. J. Ind. Med.,* 42, 838, 1985.

107. **Cowie, R. L.,** Silica dust-exposed mine workers with scleroderma (systemic sclerosis), *Chest,* 92, 260, 1987.

108. **Rodnan, G. P., Benedek, T. G., Medsger, T. A., et al.,** The association of progressive systemic sclerosis (scleroderma) with coal miner pneumoconiosis and other forms of silicosis, *Ann. Intern. Med.,* 66, 323, 1967.

109. **Zschunke, E., Ziegler, V., and Haustein, U. F.,** Occupationally induced connective tissue disorders, in *Occupational Skin Disease,* Adam, R. A., Ed., Grune and Stratton, Orlando, FL, 1989.

110. **Haustein, U. F., Ziegler, V., Herrmann, K., Mehlhorn, J., and Schmidt, C.,** Silica-induced scleroderma, *J. Am. Acad. Dermatol.,* 22, 444, 1990.

111. **Bailey, W. C., Brown, M., Buechner, H. A., Weill, H., Ichinose, H., and Ziskind, M.,** Silicomycobacterial disease in sand blasters, *Am. Rev. Respir. Dis.,* 110, 115, 1974.

112. **Sluis-Cremer, G. K., Hessel, P. A., Hnizado, E., and Churchill, A. R.,** Relationship between silicosis and rheumatoid arthritis, *Thorax,* 41, 596, 1986.

113. **Finkelstein, M., Kusiak, R., and Suranyi, G.,** Mortality among miners receiving workmen's compensation for silicosis in Ontario 1940–1975, *J. Occup. Med.,* 24, 1982.

114. **Finkelstein, M. M., Liss, G. M., Krammer, F., and Kusiak, R. A.,** Mortality among Surface-Industry Workers Receiving Workers' Compensation Awards for Silicosis in Ontario 1940–1984. A Preliminary Report, Ontario Workers' Compensation Board, Toronto, 1986.

115. **Kurppa, K., Gudbergsson, H., Hannunkari, I., et al.,** Lung cancer among silicotics in Finland, in *Silica, Silicosis and Cancer. Controversy in Occupational Medicine,* Goldsmith, D. F., Winn, D. M., and Shy, C. M., Eds., Praeger, New York, 1986.

116. **Westerholm, P., Ahlmark, A., Maasing, R., and Segelberg, I.,** Silicosis and lung cancer—a cohort study, in *Silica, Silicosis, and Cancer. Controversy in Occupational Medicine,* Goldsmith, D. F., Winn, D. M., and Shy, C. M., Eds., Praeger, New York, 1986, 327–333.

117. **Costello, J. and Graham, W. G.,** Vermont granite workers' mortality study, *Am. J. Ind. Med.,* 13, 483, 1988.

118. **International Agency for Research on Cancer,** *Monographs on the Evaluation of the Carcinogenic Risk of Chemicals to Humans. Silica and Some Silicates,* IARC, Lyon, 1987, 42.

119. **Pairon, J. C., Brochard, P., Jaurand, M. C., and Bignon, J.,** Silica and lung cancer: a controversial issue, *Eur. Respir. J.,* 4, 730, 1991.

120. **Sharma, S. K., Pande, J. N., and Verma, K.,** Effect of prednisone treatment in chronic silicosis, *Am. Rev. Respir. Dis.,* 143, 814, 1991.

121. **Mossman, B. T. and Gee, J. B. L.,** Medical progress: asbestos-related diseases, *N. Engl. J. Med.,* 320, 1721, 1989.

122. **Mossman, B. T., Bignon, J., Corn, M., Seaton, A., and Gee, J. B. L.,** Asbestos: scientific developments and implications for public policy, *Science,* 247, 294, 1990.

123. **Murphy, R. L., Becklake, M. R., Gaensler, E. A., Gee, B. L., Goldman, A. M., Lewinsohn, H. C., Mitchell, R. S., Utell, M. J., and Weill, H.,** The diagnosis of nonmalignant diseases related to asbestos, *Am. Rev. Respir. Dis.,* 134, 363, 1986.

124. **Cordier, S. P. L., Brochard, P., Bignon, J., Ameille, J., and Proteau, J.,** Epidemiologic investigation of respiratory effects related to environmental exposure to asbestos inside insulated buildings, *Arch. Environ. Health,* 42, 303, 1987.

125. **Oliver, L. C., Sprince, N. L., and Greene, R.,** Asbestos-related disease in public school custodians, *Am. J. Ind. Med.,* 19, 303, 1991.

126. **Manino, D., Etzel, R., and Cobb, N.,** Deaths with asbestosis in the United States, 1968 through 1988, *Am. Rev. Respir. Dis.,* 145, 334, 1992.

127. **Chesson, J., Hatfield, J., Schultz, B., Dutrow, E., and Blake, J.,** Airborne asbestos in public buildings, *Environ. Res.,* 51, 100, 1990.

128. **Sawyer, R. N. and Spooner, C. M.,** Sprayed asbestos-containing material in buildings: a guidance document, *Environ. Prot. Agency (U.S.) Publ.,* 450, 2, 1978.

129. **Craighead, J. E., Abraham, J. L., Churg, A., Green, F. H. Y., Kleinerman, J., Pratt, P. C., Seemayer, T. A., Vallyathan, V., and Weill, H.,** The pathology of asbestos-associated diseases of the lungs and pleural cavities: diagnostic criteria and proposed grading schema, *Arch. Pathol. Lab. Med.,* 106, 540, 1982.

130. **Hourihane, D. O. and McCaughey, W. T. E.,** Pathological aspects of asbestosis, *Postgrad. Med. J.,* 42, 613, 1966.

131. **Brody, A. R., Hill, L. H., Adkins, B., and O'Connor, R. W.,** Chrysotile asbestos inhalation in rats: deposition pattern and reaction of alveolar epithelium and pulmonary macrophages, *Am. Rev. Respir. Dis.,* 123, 670, 1981.

132. **Pinkerton, K. E., Pratt, P. C., Brody, A. R., and Crapo, J. D.,** Fiber localization and its relationship to lung reaction in rats after chronic inhalation of chrysotile asbestos, *Am. J. Pathol.,* 117, 484, 1984.

133. **Chang, L. Y., Overby, L. H., Brody, A. R., and Crapo, J. D.,** Progressive lung cell reactions and extracellular matrix production after a brief exposure to asbestos, *Am. J. Pathol.,* 131, 156, 1988.

134. **Lemaire, I.,** Characterization of the bronchoalveolar cellular respnse in experimental asbestosis. Different reactions depending on the fibrogenic potential, *Am. Rev. Respir. Dis.,* 131, 144, 1985.

135. **Arden, M. G. and Adamson, I. Y. R.,** Collagen synthesis and degradation during the development of asbestos-induced pulmonary fibrosis, *Exp. Lung Res.,* 18, 9, 1992.

136. **Rom, W. N., Travis, W. D., and Brody, A. R.,** Cellular and molecular basis of the asbestos-related diseases, *Am. Rev. Respir. Dis.,* 143, 408, 1991.

137. **Kamp, D. W., Graceffa, P., Pryor, W. A., and Weitzman, S. A.,** The role of free radicals in asbestos-induced diseases, *Free Rad. Biol. Med.,* 12, 293, 1992.

138. **Xaubet, A., Rodriguez-Roisin, R., Bombe, J. A., Marin, A., Roca, J., and Agusti-Vidal, C. A.,** Correlation of bronchoalveolar lavage and clinical and functional findings in asbestosis, *Am. Rev. Respir. Dis.,* 133, 848, 1986.

139. **Garcia, J. G. N., Griffith, D. E., Cohen, A. B., and Callahan, K. S.,** Alveolar macrophage from patients with asbestos exposure release increased levels of leukotriene B4, *Am. Rev. Respir. Dis.,* 139, 1494, 1989.

140. **Schwartz, D. A., Galvin, J. R., Merchant, R. K., Dayton, C. S., Bermeister, L. F., Merchant, J. A., and Hunninghake, G. W.,** Influence of cigarette smoking on bronchoalveolar lavage cellularity in asbestos-induced lung disease, *Am. Rev. Respir. Dis.,* 145, 400, 1992.

141. **deShazo, R. D., Morgan, J., Bozelka, B., et al.,** Natural killer cell activity in asbestos workers: interactive effects of smoking and asbestos exposure, *Chest,* 94, 482, 1988.

142. **Robinson, B. W., Rose, A. H., Hayer, A., et al.,** Increased pulmonary gamma interferon production in asbestosis, *Am. Rev. Respir. Dis.,* 138, 278, 1988.

143. **Delclos, G. L., Flitcraft, D. G., Brousseau, K. P., Windsor, N. T., Nelson, D. L., Wilson, R. K., and Lawrence, E. C.,** Bronchoalveolar lavage analysis gallium-67 lung scanning and soluble interleukin-2 receptor levels in asbestos exposure, *Environ. Res.,* 48, 164, 1989.

144. **Gellert, A. R., Macey, M. G., Uthayakumar, S., et al.,** Lymphocyte subpopulations in bronchoalveolar lavage fluids of asbestos workers, *Am. Rev. Respir. Dis.,* 132, 824, 1985.

145. **Yoneda, T. H. K., Narita, N., et al.,** NK cell activity in asbestosis, *Eur. J. Respir. Dis.,* 68, 64, 1986.

146. **Roggli, V. L.,** Human disease consequences of fiber exposures: a review of human lung pathology and fiber burden data, *Environ. Health Perspect.,* 88, 295, 1990.

147. **De Vuyst, P., Jedwab, J., Dumortier, P., Vandermorten, G., Vande Wever, R., and Yernault, J. C.,** Asbestos bodies in bronchoalveolar lavage, *Am. Rev. Respir. Dis.,* 126, 972, 1982.

148. **Sebastien, P., Armstrong, B., Monchaux, G., and Bignon, J.,** Asbestos bodies in bronchoalveolar lavage fluid and in the lung parenchyma, *Am. Rev. Respir. Dis.,* 137, 75, 1988.

149. **Schwartz, D. A., Galvin, J. R., Bermeister, L. F., Merchant, R. K., Dayton, C. S., Merchant, J. A., and Hunninghake, G. W.,** The clinical utility and reliability of asbestos bodies in bronchoalveolar fluid, *Am. Rev. Respir. Dis.,* 144, 684, 1991.

150. **Fraser, R. G., Pare, J. A. P., Pare, P. D., Fraser, R. S., and Genereux, G. P.,** Pleuropulmonary Disease Caused by Inhalation of Inorganic Dust (Pneumoconiosis) in *Diagnosis of Diseases of the Chest, Vol. III,* Fraser, R. G., Rosen, P. P., Fraser, R. S., and Genereux, G. P., Eds., 1990, 2346.

151. **Freundlich, I. M. and Greening, R. R.,** Asbestos and associated medical problems, *Radiology,* 89, 224, 1967.

152. **Mintzer, R. A. and Cugell, D. W.,** The association of asbestos-induced pleural disease and rounded atelectasis, *Chest,* 81, 457, 1982.

153. **Hillerdal, G.,** Rounded atelectasis—clinical experience with 74 patients, *Chest,* 95, 836, 1989.

154. **Kipen, H. M., Lilis, R., Suzuki, Y., Valciukas, J. A., and Selikoff, I. J.,** Pulmonary fibrosis in asbestos insulation workers with lung cancer: a radiological and histopathological evaluation, *Br. J. Ind. Med.,* 44, 96, 1987.

155. **Aberle, D. R., Gamsu, G., and Ray, C. S.,** High-resolution CT of benign asbestos-related diseases: clinical and radiographic correlation, *AJR,* 151, 883, 1988.

156. **Staples, C. A., Gamsu, G., Ray, C. S., and Webb, W. R.,** High resolution computed tomography and lung function in asbestos-exposed workers with normal chest radiographs, *Am. Rev. Respir. Dis.,* 139, 1502, 1989.

157. **Muller, N. L. and Miller, R. R.,** Computed tomography of chronic diffuse infiltrative lung disease, part 1, *Am. Rev. Respir. Dis.,* 142, 1206, 1990.

158. **Aberle, D. R., Gamsu, G., Ray, C. S., and Feuerstein, I. M.,** Asbestos-related pleural and parenchymal fibrosis: detection with high-resolution CT, *Radiology,* 166, 729, 1988.

159. **Lynch, D., Gamsu, G., and Aberle, D.,** Conventional and high resolution CT in the diagnosis of asbestos-related diseases, *Radiographics,* 9, 523, 1989.

160. **Muller, N. L. and Miller, R. R.,** Computed tomography of chronic diffuse infiltrative lung disease, part 2, *Am. Rev. Respir. Dis.,* 142, 1440, 1990.

161. **Fournier-Massey, G. and Becklake, M. R.,** Pulmonary function profiles in Quebec asbestos workers, *Buss Physiopathol. Respir.,* 11, 429, 1975.

162. **Turner-Warwick, M. and Haslam, P.,** Circulating rheumatoid and antinuclear factors in asbestos workers, *Br. Med. J.,* 3, 492, 1970.

163. **Heffner, J. E. and Repine, J. E.,** Pulmonary strategies of antioxidant defenses, *Am. Rev. Respir. Dis.,* 140, 531, 1989.

164. **Begin, R., Cantin, A., and Masse, S.,** Effects of cyclophosphamide treatment in experimental asbestosis, *Exp. Lung. Res.,* 14, 823, 1988.

165. **Selikoff, I. J.,** Historical developments and perspectives in inorganic fiber toxicity in man, *Environ. Health Perspect.,* 88, 269, 1990.

166. **Becklake, M. R.,** Asbestos and other fiber-related diseases of the lungs and pleura, *Chest,* 100, 248, 1991.

167. **Doll, R.,** Mortality from lung cancer in asbestos workers, *Br. J. Ind. Med.,* 12, 81, 1955.

168. **Davis, J. M. G. and Cowie, H. A.,** The relationship between fibrosis and cancer in experimental animals exposed to asbestos and other fibers, *Environ. Health Perspect.,* 88, 305, 1990.

169. **Hughes, J. M. and Weill, H.,** Asbestosis as a precursor of asbestos-related lung cancer: results of a prospective mortality study, *Br. J. Ind. Med.,* 48, 229, 1991.

170. **Warnock, M. L. and Isenberg, W.,** Asbestos burden and the pathology of lung cancer: results of a prospective mortality study, *Chest,* 89, 20, 1986.

171. **Turner-Warwick, M., Lebowitz, M., Burrows, B., and Johnson, A.,** Cryptogenic fibrosing alveolitis and lung cancer, *Thorax,* 35, 496, 1980.

172. **Hammond, E. C., Selikoff, I. J., and Seidman, H.,** Asbestos exposure, cigarette smoking and death rates, *Ann. N.Y. Acad. Sci.,* 330, 473, 1979.

173. **Tron, V., Wright, J. L., Wiggs, B., and Churg, A.,** Cigarette smoke makes airway and early parenchymal asbestos-injured lung disease worse in the guinea pig, *Am. Rev. Respir. Dis.,* 136, 271, 1987.

174. **Gamble, J. F., Fellner, W., and Dimeo, M. J.,** An epidemiologic study of a group of talc workers, *Am. Rev. Respir. Dis.,* 119, 741, 1979.

175. **Davis, G. S. and Calhoun, W. J.,** Occupational and environmental causes of interstitial lung disease, in *Interstitial Lung Diseases,* Schwarz, M. I. and King, T. E., Jr., Eds., B.C. Decker, Toronto, 1988, 63.

176. **Vallyathan, N. V. and Craighead, J. E.,** Pulmonary pathology in workers exposed to nonasbestiform talc, *Hum. Pathol.,* 12, 28, 1981.

177. **Marschke, G., Haber, L., and Feinberg, M.,** Pulmonary talc embolization, *Chest,* 68, 824, 1975.

178. **Arnett, E. N., Battle, W. E., and Roberts, W. C.,** Intravenous injection of talc-containing drugs intended for oral use. A cause of pulmonary granulomatosis and pulmonary hypertension, *Am. J. Med.,* 60, 711, 1976.

179. **Pare, J. P., Cote, G., and Fraser, R.,** Long-term follow-up of drug abusers with intravenous talcosis, *Am. Rev. Respir. Dis.,* 139, 233, 1989.

180. **Schmidt, R. A., Glenny, R. W., Godwin, J. D., Hampson, N. B., Cantino, M. E., and Reichenbach, D. D.,** Panlobular emphysema in young intravenous ritalin abusers, *Am. Rev. Respir. Dis.,* 143, 649, 1991.

181. **Fisher, E., Watkins, G., Lam, N. V., Tsuda, H., Hermann, C., Johal, J., and Liu, H.,** Objective pathological diagnosis of coal workers' pneumoconiosis, *JAMA,* 245, 1829, 1981.

182. **Davis, J. M. G., Chapman, J., Collings, P., Douglas, A. N., Fernie, J., Lamb, D., and Ruckley, V. A.,** Variations in the histological patterns of the lesions of coal workers' pneumoconiosis in Britain and their relationship to lung dust content, *Am. Rev. Respir. Dis.,* 128, 118, 1983.

183. **Doll, N. J., Stankus, R. P., and Barkman, H. W.,** Immunopathogenesis of asbestosis, silicosis, and coal workers' pneumoconiosis, *Clin. Chest Med.,* 4, 3, 1983.

184. **Lippmann, M., Eckert, H. L., Hahon, N., and Morgan, W. K. C.,** Circulating antinuclear and rheumatoid factors in coal miners, *Ann. Intern. Med.,* 79, 807, 1973.

185. **Heise, E. R., Mentnech, S. S., Olenchock, S. A., Kutz, S. A., Morgan, W. K. C., Merchant, J. A., and Major, P. C.,** HLA-A1 and coalworkers' pneumoconiosis, *Am. Rev. Respir. Dis.,* 119, 903, 1979.

186. **Wallaert, B., Rossi, G. A., and Sibille, Y.,** Clinical guidelines and indications for bronchoalveolar lavage (BAL): collagen-vascular diseases, *Eur. Respir. J.,* 3, 942, 1990.

187. **Borm, P. J. A., Palmen, N., Engelen, J. M., and Buurman, W. A.,** Spontaneous and stimulated release of tumor necrosis factor-alpha (TNF) from bold monocytes of miners with coal workers' pneumoconiosis, *Am. Rev. Respir. Dis.,* 138, 1589, 1988.

188. **Janssen, Y. M. W., Engelen, J. J. M., Giancola, M. S., Low, R. B., Vacek, P., and Borm, P. J. A.,** Serum type III procollagen N-terminal peptide in coal miners, *Exp. Lung Res.,* 18, 1, 1992.

189. **Musk, A. W., Cotes, J. E., Bevan, C., and Campbell, M. J.,** Relationship between type of simple coalworker's pneumoconiosis and lung function. A nine-year follow-up study of subjects with small rounded opacities, *Br. J. Ind. Med.,* 38, 313, 1981.

190. **Nemery, B., Brasseur, L., Veriter, C., and Frans, A.,** Impairment of ventilatory function and pulmonary gas exchange in non-smoking coalminers, *Lancet,* ii, 1427, 1987.

191. **Elmes, P. C.,** Inorganic dusts, in *Hunter's Diseases of Occupations,* Raffle, P. A. B., Lee, W. R., McCallum, R. I., and Murray, R., Eds., Little, Brown, Boston, 1987, 634.

192. **Davies, D. and Cotton, R.,** Mica pneumoconiosis, *Br. J. Ind. Med.,* 40, 22, 1983.

193. **Sepulveda, M. J., Vallyathan, V., Attfield, M. D., Piacitelli, L., and Tucker, J. H.,** Pneumoconiosis and lung function in a group of kaolin workers, *Am. Rev. Respir. Dis.,* 127, 231, 1983.

194. **Oldham, P. D.,** Pneumoconiosis in Cornish china clay workers, *Br. J. Ind. Med.,* 40, 131, 1983.

195. **Kennedy, T., Rawlings, W. J., Baser, M., and Tockman, M.,** Pneumoconiosis in Georgia kaolin workers, *Am. Rev. Respir. Dis.,* 127, 215, 1983.

196. **Lapenas, D., Gale, P., Kennedy, T., Rawlings, W., and Dietrich, P.,** Kaolin pneumoconiosis. Radiologic, pathologic and mineralogic findings, *Am. Rev. Respir. Dis.,* 130, 282, 1984.

197. **Wagner, J. C., Pooley, F. D., Gibbs, A., Lyons, J., Sheers, G., and Moncrieff, C. B.,** Inhalation of china stone and china clay dusts: relationship between the mineralogy of dust retained in the lungs and pathological changes, *Thorax,* 41, 190, 1985.

198. **Oakes, D., Douglas, R., Knight, K., Wusteman, M., and McDonald, J. C.,** Respiratory effects of prolonged exposure to gypsum dust, *Am. Occup. Hyg.,* 26, 833, 1982.

199. **Vocaturo, G., Colombo, F., Zanoni, M., Rodi, F., Sabbioni, R., and Pietra, R.,** Human exposure to heavy metals. Rare earth pneumoconiosis in occupational workers, *Chest,* 83, 780, 1983.

200. **Enterline, P. E., Marsh, G. M., and Esmen, N. A.,** Respiratory disease among workers exposed to man-made mineral fibers, *Am. Rev. Respir. Dis.,* 128, 1, 1983.

201. **Konzen, J. L.,** Observations on fiberglass in relation to health, in *Occupational Medicine: Principles and Practical Applications,* 2nd ed., Zenz, C., Ed., Year Book Medical Publ., Chicago, 1988, 1067.

202. **Gross, P., et al.,** Lungs of workers exposed to fiber glass: a study of their pathologic changes and their dust content, *Arch. Environ. Health,* 23, 67, 1971.

203. **Levy, S. A.,** An overview of occupational pulmonary disorders, in *Occupational Medicine: Principles and Practical Applications,* Zenz, C., Ed., Year Book Medical Publ., Chicago, 1988, 201.

204. **Durand, P., Bégin, R., Samson, L., Cantin, A., Massé, S., Dufresne, A., Perreault, G., and Laflamme, J.,** Silicon carbide pneumoconiosis: a radiographic assessment, *Am. J. Ind. Med.,* 20, 37, 1991.

205. **Begin, R., Boctor, M., Bergeron, D., Cantin, A., Berthiaume, Y., Peloquin, S., Bisson, G., and Lamoureux, G.,** Radiographic assessment of pleuropulmonary disease in asbestos workers: posteroanterior, four view films, and computed tomograms of the thorax, *Br. J. Ind. Med.,* 41, 373, 1984.

206. **Peters, J. M., Smith, T. J., Bernstein, L., Wright, W. W., and Hammond, S. K.,** Pulmonary effects of exposures in silicon carbide manufacturing, *Br. J. Ind. Med.,* 41, 109, 1984.

207. **Smith, T. J., Hammond, S. K., Laidlaw, F., and Fine, S.,** Respiratory exposures associated with silicon carbide production: estimation of cumulative exposures for an epidemiological study, *Br. J. Ind. Med.,* 41, 100, 1984.

208. **Gauthier, J. J., Ghezzo, H., and Martin, R. R.,** Pneumoconiosis following carborundum (silicon carbide) exposure, *Am. Rev. Respir. Dis.,* 131, A191, 1985.

209. **Osterman, J. W., Greaves, I. A., Smith, T. J., Hammond, S. K., Robins, J. M., and Theriault, G.,** Work related decrement in pulmonary function in silicon carbide production workers, *Br. J. Ind. Med.,* 46, 708, 1989.

210. **Funahashi, A., Schlueter, D., Pintar, K., Siegesmund, K., Mandel, G. S., and Mandel, N. S.,** Pneumoconiosis in workers exposed to silicon carbide, *Am. Rev. Respir. Dis.,* 129, 635, 1984.

211. **De Vuyst, P., Vande Weyer, R., De Coster, A., Marchandise, F. X., Dumortier, P., Ketelbant, P., Jedwab, J., and Yernault, J. C.,** Dental technician's pneumoconiosis, *Am. Rev. Respir. Dis.,* 133, 316, 1986.

212. **Hayashi, H. and Kajita, A.,** Silicon carbide in lung tissue of a worker in the abrasive industry, *Am. J. Ind. Med.,* 14, 145, 1988.

213. **Osterman, J. W., Greaves, I. A., Smith, T. J., Hammond, S. K., Robins, J. M., and Theriault, G.,** Respiratory symptoms associated with low level sulphur dioxide exposure in silicon carbide production workers, *Br. J. Ind. Med.,* 46, 629, 1989.

214. **Cantin, A., Allard, C., and Begin, R.,** Increased alveolar plasminogen activator in early asbestosis, *Am. Rev. Respir. Dis.,* 139, 604, 1989.

215. **Kriebel, D., Brain, J. D., Sprince, N. L., and Kazemi, H.,** The pulmonary toxicity of beryllium, *Am. Rev. Respir. Dis.,* 137, 464, 1988.

216. **Cullen, M. R., Cherniack, M. G., and Kominsky, J. R.,** Chronic beryllium disease in the United States, *Semin. Respir. Med.,* 7, 203, 1986.

217. **Freiman, D. G. and Hardy, H. L.,** Beryllium disease: the relation of pulmonary pathology to clinical course and prognosis based on a study of 130 cases from the U.S. Beryllium Case Registry, *Hum. Pathol.,* 1, 25, 1970.

218. **Reeves, A. L.,** The immunotoxicity of beryllium, in *Immunotoxicity,* Gibson, G. G., Hubbard, R., and Parke, D. V., Eds., Academic Press, London, 1983, 261.

219. **Newman, L. S., Kreiss, K., King, T. E., Jr., Seay, S., and Campbell, P. A.,** Pathologic and immunologic alterations in early stages of beryllium disease: re-examination of disease definition and natural history, *Am. Rev. Respir. Dis.,* 139, 1479, 1989.

220. **Sprince, N. L., Kanarek, D. J., Weber, A. L., Chamberlin, R. I., and Kazemi, H.,** Reversible respiratory disease in beryllium workers, *Am. Rev. Respir. Dis.,* 117, 1011, 1978.

221. **Cullen, M. R., Kominsky, J. R., Rossman, M. D., Cherniack, M. G., Rankin, J. A., Balmes, R. R., Kern, J. A., Daniele, R. P., Palmer, L., Naegel, G. P., McManus, K., and Cruz, R.,** Chronic beryillium disease in a precious metal refinery, *Am. Rev. Respir. Dis.,* 135, 201, 1987.

222. **Rossman, M. D., Kern, J. A., Elias, J. A., Cullen, M. R., Epstein, P. E., Preuss, O. P., Markham, T. N., and Daniele, R. P.,** Proliferative response of bronchoalveolar lymphocytes to beryllium, *Ann. Intern. Med.,* 108, 687, 1988.

223. **Saltini, C., Winestock, K., Kirby, M., Pinkston, P., and Crystal, R. G.,** Maintenance of alveolitis in patients with chronic beryllium disease by beryllium-specific helper T cells, *N. Engl. J. Med.,* 320, 1103, 1989.

224. **Maceira, J. M., Fukuyama, K., W. L., E.,** Appearance of T-cell subpopulations during the time course of berylllium-induced granulomas, *Arch. Environ. Health,* 38, 302, 1983.

225. **Aronchick, J. M., Rossman, M. D., and Miller, W. T.,** Chronic beryllium disease: diagnosis, radiographic findings, and correlation with pulmonary function tests, *Radiology,* 163, 677, 1987.

226. **Rose, C. S. and Newman, L. S.,** Hypersensitivity pneumonitis and chronic beryllium disease, in *Interstitial Lung Diseases,* 2nd ed., Schwarz, M. I. and King, T. E., Jr., Eds., Mosby Year Book, Philadelphia, 1993, 231.

227. **Kreiss, K., Newman, L., Mroz, M., and Campbell, P.,** Screening blood test identifies subclinical beryllium disease, *Occup. Med.,* 31, 603, 1989.

228. **Mroz, M. M., Kreiss, K., Lezotte, D. C., Campbell, P. A., and Newman, L. S.,** Reexamination of the blood lymphocyte transformation test in the diagnosis of chronic beryllium disease, *J. Allergy Clin. Immunol.,* 88, 54, 1991.

229. **Sjögren, I., Hillerdal, G., Andersson, A., and Zetterström, O.,** Hard metal lung disease: importance of cobalt in coolants, *Thorax,* 35, 653, 1980.

230. **Demedts, M., Gheysens, B., Nagels, J., et al.,** Cobalt lung in diamond polishers, *Am. Rev. Respir. Dis.,* 130, 130, 1984.

231. **Coates, E. O., Jr. and Watson, J. H. L.,** Diffuse interstitial lung disease in tungsten carbide workers, *Ann. Intern. Med.,* 75, 709, 1971.

232. **Sluis-Cremer, G. K., Thomas, R. G., and Solomon, A.,** A report of 4 cases, *S. Afr. Med. J.,* 71, 598, 1987.

233. **Davis, G. S. and Calhoun, W. J.,** Occupational and environmental causes of interstitial lung disease, in *Interstitial Lung Diseases,* 2nd ed., Schwarz, M. I. and King, T. E., Jr., Eds., Mosby Year Book, Philadelphia, 1993, 179.

234. **Balmes, J. R.,** Respiratory effects of hard metal dust exposure, in *Occupational Medicine - State of the Art Reviews, Vol 2,* Rosenstock, L., Ed., 1987, 327.

235. **Austenfeld, J. L. and Colby, T. V.,** Recognizing lung diseases induced by hard metal exposure, *J. Respir. Dis.,* 10, 65, 1989.

236. **Cugell, D. W., Morgan, W. K., Perkins, D. G., and Rubin, A.,** The respiratory effects of cobalt, *Arch. Intern. Med.,* 150, 177, 1990.

237. **Liebow, A. A.,** Defunction and classification of interstitial pneumonias in human pathology, *Prog. Respir. Res.,* 8, 1, 1975.

238. **Abraham, J. L. and Spragg, R. G.,** Documentation of environmental exposure using open lung biopsy, transbronchial biopsy, and bronchoalveolar lavage in giant cell interstitial pneumonia (GIP), *Am. Rev. Respir. Dis.,* 119, 197, 1979.

239. **Abraham, J. L.,** Lung pathology in 21 cases of giant cell interstitial pneumonia (GIP) suggest GIP is pathognomonic in cobalt (hard metal) disease, *Chest,* 91, 312, 1987.

240. **Ohori, N. P., Sciurbu, F. C., Owens, G. R., et al.,** Giant cell interstitial pneumonia and hard metal pneumoconiosis. A clinicopathologic study of four cases and review of the literature, *Am. J. Surg. Pathol.,* 13, 581, 1989.

241. **Davison, A. G., Haslam, P. L., Corrin, B., et al.,** Interstitial lung disease and asthma in hard metal workers: bronchoalveolar lavage, ultrastructural and analytic finding, and results of bronchial provocation tests, *Thorax,* 38, 119, 1983.

242. **Shirakawa, T., Kusaka, Y., Fujimura, N., Goto, S., Kato, M., Heki, S., and Morimoto, K.,** Occupational asthma from cobalt sensitivity in workers exposed to hard metal dust, *Chest,* 95, 29, 1989.

243. **Kusaka, Y., Yokoyama, K., Sera, Y., Yamamoto, S., Sone, S., Kyono, H., Shirakawa, T., and Goto, S.,** Respiratory diseases in hard metal workers: an occupational hygiene study in a factory, *Br. J. Ind. Med.,* 43, 474, 1986.

244. **Meyer-Bisch, C., Pham, Q. I., Mur, J. M., Massin, N., Moulin, J. J., Teculescu, D., Carton, B., Pierre, F., and Baruthio, F.,** Respiratory hazards in hard metal workers: a cross sectional study, *Br. J. Ind. Med.,* 46, 302, 1989.

245. **Kusaka, Y., Nakano, Y., and Shirakawa, T.,** Lymphocyte transformation with cobalt in hard metal asthma, *Ind. Health,* 27, 155, 1989.

246. **Nemery, B., Nagels, J., Verbexen, E., Dinsdale, D., and Demedts, M.,** Rapidly fatal progression of cobalt lung in a diamond polisher, *Am. Rev. Respir. Dis.,* 141, 1373, 1990.

247. **Rowe, D. M., Solomayer, J. A., and Zenz, C.,** Other metals, in *Occupational Medicine: Principles and Practical Applications,* 2nd ed., Zenz, C., Ed., Year Book Medical Publ., Chicago, 1988, 639.

248. **Chan-Yeung, M., Wong, R., Maclean, L., Tan, F., Schulzer, M., Enarson, D., Martin, A., Dennis, R., and Grzybowski, S.,** Epidemiologic health study of workers in an aluminum smelter in British Columbia, *Am. Rev. Respir. Dis.,* 127, 465, 1983.

249. **Abramson, M. J., Wlodarczyk, J. H., Saunders, N. A., and Hensley, M. J.,** Does aluminum smelting cause lung disease?, *Am. Rev. Respir. Dis.,* 139, 1042, 1989.

250. **Jederlinic, P. J., Abraham, J. L., Churg, A., Himmelstein, J. S., Epler, G. R., and Gaensler, E. A.,** Pulmonary fibrosis in aluminum oxide workers. Investigation of nine workers, with pathologic examination and microanalysis in three of them, *Am. Rev. Respir. Dis.,* 142, 1179, 1990.

251. **Gilks, B. and Churg, A.,** Aluminum-induced pulmonary fibrosis: do fibers play a role?, *Am. Rev. Respir. Dis.,* 136, 176, 1987.

252. **Chen, W., Monnat, J. R., Chen, M., and Mottet, N. K.,** Aluminum induced pulmonary granulomatosis, *Hum. Pathol.,* 9, 705, 1978.

253. **De Vuyst, P., Du Mortier, P., Rickaert, F., Van der Weyer, R., Lenclud, C., and Yernault, J. C.,** Occupational lung fibrosis in an aluminum polisher, *Eur. J. Respir. Dis.,* 68, 131, 1986.

254. **Miller, R. R., Churg, A. M., Hutcheon, M., and Lam, S.,** Pulmonary alveolar proteinosis and aluminum dust exposure, *Am. Rev. Respir. Dis.,* 130, 312, 1984.

255. **Herbert, A., Sterling, G., Abraham, J., and Corrin, B.,** Desquamative interstitial pneumonia in an aluminum welder, *Hum. Pathol.,* 13, 694, 1982.

256. **Musk, A. W., Greville, H. W., and Tribe, A. E.,** Pulmonary disease from occupational exposure to an artificial aluminum silicate for cat litter, *Br. J. Ind. Med.,* 37, 367, 1980.

257. **Vallyathan, V., Bergeron, W. N., Robichaux, P. A., and Craighead, J. E.,** Pulmonary fibrosis in an aluminum arc welder, *Chest,* 81, 372, 1982.

258. **Eklund, A., Arns, R., Blaschke, E., Hed, J., Hjertquist, S.-O., Larsson, K., Lögren, H., Nyström, J., Sköld, C., and Tronling, G.,** Characteristics of alveolar cells and soluble components in bronchoalveolar lavage fluid from non-smoking aluminium potroom workers, *Br. J. Ind. Med.,* 46, 782, 1989.

259. **Kennedy, M. C. S.,** Aluminum powder inhalations in the treatment of silicosis of pottery workers and pneumoconiosis of coal miners, *Br. J. Ind. Med.,* 13, 85, 1956.

260. **Dubois, F., Begin, R., Cantin, A., Masse, S., Martel, M., Bilodeau, G., Dufresne, A., Perreault, G., and Sebastien, P.,** Aluminum inhalation reduces silicosis in a sheep model, *Am. Rev. Respir. Dis.,* 137, 1172, 1988.

261. **Waldron, H. A.,** Metals cadmium nickel batteries, in *Hunter's Diseases of Occupations,* Raffle, P. A. B., Lee, W. R., McCallum, R. I., and Murray, R., Eds., Little, Brown, Boston, 1987, 239.

262. **Oberdörster, G.,** Pulmonary toxicity and carcinogenicity of cadmium, *J. Am. Coll. Toxicol.,* 8, 1251, 1989.

263. **Snider, G. L., Lucey, E. C., Fris, B., Jung-Legg, Y., Stone, P., and Franzblau, C.,** Cadmium-chloride-induced air-space enlargement with interstitial pulmonary fibrosis is not associated with destruction of lung elastin, *Am. Rev. Respir. Dis.,* 137, 918, 1988.

264. **Smith, T. J., Petty, T. L., Reading, J. C., and Lakshiminarayan, S.,** Pulmonary effects of chronic exposure to airborne cadmium, *Am. Rev. Respir. Dis.,* 114, 161, 1976.

265. **Reline, S., et al.,** Granulomatous disease associated with pulmonary deposition of titanium, *Br. J. Ind. Med.,* 10, 652, 1986.

266. **Knight, A. L.,** The rare earths, in *The Rare Earths,* 2nd ed., Zenz, C., Ed., Year Book Medical Publ., Chicago, 1988, 609.

267. **Masjedi, M. R., Estineh, N., Bahadori, M., Alavi, M., and Sprince, N. L.,** Pulmonary complications in lead miners, *Chest,* 96, 18, 1989.

268. **Doig, A.,** Baritosis: a benign pneumoconiosis, *Thorax,* 31, 1, 1976.

269. **Dickerson, O. B. and Smith, T. H. F.,** Antimony, arsenic, and other compounds, in *Occupational Medicine: Principles and Practical Applications,* 2nd ed., Zenz, C., Ed., Year Book Medical Publ., Chicago, 1988, 509.

270. **Cooper, D. A., Pendergrass, E. P., Vorwald, A. J., et al.,** Pneumoconiosis in workers in an antimony industry, *Am. J. Roentgenogr.,* 103, 495, 1968.

271. **Pimentel, J. C. and Marques, F.,** 'Vineyard sprayer's lung': a new occupational disease, *Thorax,* 24, 678, 1969.

272. **Villar, T. G.,** Vineyard sprayer's lung. Clinical aspects, *Am. Rev. Respir. Dis.,* 110, 545, 1974.

273. **Pimentel, J. C. and Menezes, P.,** Liver granulomas containing copper in vineyard sprayer's lung, *Am. Rev. Respir. Dis.,* 111, 189, 1975.

# Chapter 10

# Conditions with an Uncertain Relationship to Air Pollution: Sick Building Syndrome, Multiple Chemical Sensitivities, and Chronic Fatigue Syndrome

*Philip Witorsch, M.D. and Sorell L. Schwartz, Ph.D.*

## CONTENTS

## INTRODUCTION

Sick building syndrome (SBS), multiple chemical sensitivities (MCS), and chronic fatigue syndrome (CFS) are conditions that have been said to be associated with exposure to air pollutants, although the validity and significance of any such associations are highly questionable and remain to be substantiated by credible scientific studies.[1] Thus, while this entire area continues to be scientifically and medically quite problematic and uncertain, it has achieved a level of social and financial significance, primarily as a result of the fact that claims related to such situations have become major issues in toxic tort and workers' compensation litigation, legislative matters relating to the workplace and the environment, and issues relative to definitions of disability.[1-3]

Although considerable attention has been devoted to these conditions over the past 15 to 20 years, including international and national conferences, symposia published in the scientific, engineering, and medical literature, committee and commission reports, and assessments by a number of professional organizations, there, nevertheless, continues to exist considerable uncertainty and misunderstanding relative to definitions, characterization, diagnostic criteria, and scientific and clinical aspects of these conditions.[1,3-28]

## SICK BUILDING SYNDROME

"Building-associated illness" has been defined as "the summary term for health problems in buildings" and includes the entities of building-related illness (BRI) and sick building syndrome (SBS) (Table 1).[24] BRI is generally used to refer to "recognized diseases with a defined pathophysiology that can be attributed to a building,"[24] i.e., "with a clear etiology such as hypersensitivity diseases, infections, and complaints related to specific contaminants."[29] BRI is made up of specific clinical syndromes that are clearly causally related to specific and generally readily

Table 1  **Building-Associated Illness**

I. Building-related illness (BRI)
    1. Allergic or irritative rhinitis
    2. Acute viral, mycoplasmal, chlamydial, etc., URIs
    3. *Legionella pneumophilia* infections (Legionnaire's disease, Pontiac fever)
    4. Hypersensitivity pneumonitis and humidifier fever
    5. Occupational and environmental asthma
    6. Irritative reaction to fibrous glass
    7. Reaction to mineral fibers and/or dusts
    8. Carbon monoxide intoxication
    9. Reactions to carbonless carbon paper
    10. Tuberculosis
    11. Varicella, measles, smallpox

II. Sick building syndrome (SBS)

identifiable indoor air pollutants or contaminants. These may include, among others, the following:

- building-related allergic or irritative rhinitis
- acute upper respiratory infections probably caused by viruses and/or related microorganisms
- Legionnaire's disease and Pontiac fever, caused by the bacterium, *Legionella pneumophilia*
- hypersensitivity pneumonitis and humidifier fever due to immunological reactions to microbial contaminants
- occupational/environmental asthma triggered by exposure to a specific building-related allergen or exacerbation of pre-existing asthma related to an allergic or irritant reaction to building air contaminants, such as microbial agents or chemicals
- irritative reactions to fibrous glass
- reactions to mineral fibers and dusts
- reactions to detergent residues
- carbon monoxide intoxication
- reactions to components of carbonless copy paper
- reactions to ozone
- building-associated airborne outbreaks of tuberculosis, varicella, measles, and smallpox (Table 1).[24,25,27,29-37]

   In contrast, SBS refers to a variable constellation of nonspecific complaints, including mucous membrane irritation, behavioral/"central nervous system" symptoms, dry, irritated, pruritic skin, chest tightness, and allergic-like symptoms, as well as fatigue and headache, which appear to have a temporal association with being in a particular building. Typical symptoms include eye irritation, nose and throat irritation, lethargy, and headache. In addition, some investigators have also reported complaints of hoarseness, skin irritation, reduced memory, reduced power of concentration, dizziness, nausea, and altered perception of odors and taste (Table 2). Patients complaining of this syndrome usually have no abnormal physical findings or clinical laboratory abnormalities that can be associated with the condition. In the majority of instances, the symptoms usually dissipate very shortly after leaving the building, and there is no scientifically acceptable evidence that this syndrome is associated with any long-term physical effects or residue.[1,8,24,25,27,29,38-43]

Table 2　**SBS: Typical Complaints**

1. Mucous membrane irritation (nose, throat, eyes)
2. CNS-like, including behavioral symptoms
3. Dry, irritated, pruritic skin
4. Chest tightness
5. Fatigue
6. Headache
7. Lethargy
8. Hoarseness
9. Memory loss
10. Difficulty with concentration
11. Dizziness
12. Nausea
13. Altered perception of odors and taste

Table 3　**SBS: Suggested Causes**

1. Carbon dioxide
2. Formaldehyde
3. Volatile organic compounds (VOCs)
4. Reduced fresh air circulation and "excessive" recirculation of indoor air
5. 4-Phenylcyclohexene (4-PCH)
6. Textiles, upholstery, curtains, drapes
7. Paper and plastic products
8. Respirable particulates
9. High or low temperatures
10. High or low humidity
11. Bioaerosols, including microorganisms and microbial products (e.g., endotoxins, mycotoxins)
12. Environmental tobacco smoke (ETS)
13. Psychogenic factors, including mass psychogenic illness
14. Negative ions

The specific cause of SBS is generally unclear. For the most part, investigations of outbreaks or episodes of SBS generally have not shown elevations of any individual pollutant or contaminant to levels that would be considered likely to cause disease or even discomfort. Although the disorder has, at various times, been attributed to a variety of indoor environmental pollutants, including $CO_2$, formaldehyde, and volatile organic compounds (VOCs), there does not appear to be any consistent relationship with a particular chemical or physical agent.[1,8,19,25,38,39,41,43–52] Similarly, while some investigators have reported an association with low fresh air ventilation rates and excessive recirculation of indoor air, this has not been a consistent finding in other studies.[8,25,29,33,38,39,52,53] Other substances or conditions that have been said to be causally related, but for which there is as yet no firm scientific evidence, include carpeting, 4-phenylcyclohexene (4-PCH), textiles, upholstery, curtains, paper, respirable particulates, temperature, humidity, microorganisms, and microbial products and bioaerosols, including endotoxins and mycotoxins.[8,25,38,39,41,43–61] It has also been suggested that psychogenic factors may play a role in at least some instances of SBS, and that at least some outbreaks in certain office buildings may represent examples of mass psychogenic illness (Table 3).[1,8,19,25,38,39,46,62–67]

Although it has been suggested that modification of the building's heating, ventilation, and air conditioning (HVAC) system, including increased levels of ventilation and/or cleaning of the ventilation system and ducts, will prevent or eliminate further episodes of this syndrome among building occupants,[25] this has not been the case universally.[29,47] Measurements of levels of various pollutants in instances of SBS have generally failed to reveal any individual agent to be elevated to an unsafe level.[1,8,25,38] It has been proposed that SBS does not represent a single entity but, rather, a nonspecific reaction to a variable number of environmental factors, including psychogenic factors, with the particular causative factor varying from building to building.[25,27]

Various studies directed at determining the cause of SBS have produced variable and often conflicting results. Finnegan and co-workers in the U.K. reported a markedly high prevalence of SBS complaints in buildings with mechanical ventilation, with or without air recirculation, compared to buildings with natural ventilation. They also noted a higher prevalence in buildings with humidification of the air, compared to buildings without such humidification or with natural ventilation. They found no significant difference in levels of CO, $O_3$, formaldehyde, positive and negative ions, temperature, relative humidity, or air velocity between a building with natural and one with mechanical ventilation.[48] In a subsequent study, some of the same investigators found no significant difference in symptom prevalence related to particular type of ventilating or air conditioning system. They also noted a higher prevalence of symptoms in women than in men, and in clerical and secretarial workers than in managers and professionals.[42]

Investigators in Denmark reported a significant association in office workers between symptoms of SBS and female gender, history of hay fever, history of migraine, smoking, air quality problems at home, job category, handling carbonless copy paper, photocopying more than 25 sheets per day, VDT work for more than 1 h per day, lack of variation in work, dissatisfaction with supervisor, dissatisfaction with colleagues, job satisfaction inhibited by work quantity and little influence, and high work speed. They also noted an association between dust levels, area of fleecy materials, surface of open shelves in the work space, and air temperature. They found no association with type of ventilation system.[50,51] While the results of this Danish Town Hall study would tend to suggest that dust and microbial agents play a role in SBS, other investigators have reported results that are not consistent with such a causation.[52]

Some investigators have reported data said to suggest a role of VOCs in causing SBS.[57–59] On the other hand, other investigators have reported results not consistent with such a hypothesis.[8,38,53] There have been similarly conflicting results reported with regard to a suggestion made of negative air ions as a cause of SBS.[54,55] Other possible factors that have been suggested include low relative humidity and psychogenic factors, but these suggestions are also currently controversial.[43,45,60,61,67]

Thus, at the present time, the cause of SBS, if there is a single cause, remains to be elucidated. As noted above, it has been suggested that the causes of SBS may be variable, with different factors operating in different situations, and even that in some instances more than one factor may be operative.[25,27] While this is certainly an interesting hypothesis, it remains to be scientifically validated.

In summary, SBS appears to be a self-limited condition, without any long-term or permanent sequelae, that is probably multifactorial in etiology, with different individual factors or combinations of factors often operative in different situations. In some instances, psychogenic factors, triggered by such things as job dissatisfaction, may play a major role. In other situations, inadequate ventilation with buildup of ordinary air pollutants may be important, while in yet other circumstances, contamination of the HVAC system with one or another microbial or chemical or physical material may be at fault. Clearly, considerable work needs to be done in this area, particularly with regard to epidemiology and elucidation of potential causative agents, including psychogenic factors.

## MULTIPLE CHEMICAL SENSITIVITIES

Multiple chemical sensitivities (MCS) is a term that has been applied to individuals who complain of recurrent episodes of symptoms, which in the majority of cases are nonspecific and which appear to relate to a number of organ systems. Typical symptoms include inability to concentrate, feeling ''spaced-out,'' lack of energy, easy fatiguability, heightened perception of odors, mental confusion, and depression. In some cases, other symptoms are reported as well, including gastrointestinal tract symptoms, eye, ear, nose, and throat complaints, respiratory problems, cardiovascular irregularities, skin disorders, genitourinary symptoms, and muscle, joint, and bone pain. Many of the complaints reported by these individuals are allergy-like and include such symptoms as rhinitis, headache, irritability, insomnia, palpitations, other cardiovascular symptoms, and food and chemical intolerance (Table 4). The condition is characteristically a polysymptomatic one and at any given time may involve any organ or many organ systems in a patient concurrently. The episodes of symptoms are attributed by the patients (and their physicians, a number of whom are clinical ecologists) to chemical exposure encountered in a variety of settings, including chemicals in indoor air.[1,2,9,14–17,28,68–74]

Individuals who claim to have MCS often relate the onset of their problem to a specific acute or chronic chemical exposure, following which, symptoms are said to be precipitated by exposure to a wide variety of environments and chemically unrelated compounds, which are claimed to be present in the environment at very low (and at times virtually unmeasurable) concentrations. The chemical exposures said to initiate the condition, as well as those said to trigger subsequent episodes of symptoms, involve concentrations far below those established to cause adverse or noxious effects on any generally accepted physical or toxicological (including immunological) basis and, in many instances, at concentrations that are virtually unmeasurable.[1,3,5,9,11,14–17,22,28,69–71,74–77] Most commonly, the offending agent is an airborne substance, and frequently its presence can be recognized by an odor or, less commonly, by the presence of some type of eye or skin sensation.[18,77–84] The substances that have been claimed to trigger episodes of MCS symptoms include, among others, the following: volatile organic compounds (VOCs), tobacco smoke, formaldehyde, perfumes and colognes, deodorants, vehicle exhaust, diesel fumes, cleaning products and detergents, pesticides, petroleum products and byproducts, and fungal and other biological products and agents (Table 5).[1,3–5,9,14–17,22,28,70,71,73,84–87]

It is characteristic of the condition that, on physical examination, these individuals are found to have no objective findings that are consistent with the MCS complaints. There are also no demonstrable corresponding pathologic findings or laboratory abnormalities using generally accepted laboratory methods. When such patients are found to have objective physical, laboratory, or pathological findings, these are virtually invariably attributable to some other condition that the patient can be demonstrated to have. Such other conditions are generally much more probable causes of any objective findings and corresponding symptoms than the alleged MCS.[1,3,5,9,11,14–16,18,20,22,28,70,71,77,78] A past medical history of atopic allergies or of emotional stressors, or both, is quite common, in our experience (Table 6). There are reports of patients with MCS who also have vocal cord dysfunction (VCD) syndrome, a condition characterized by an abnormality in vocal cord function that can be demonstrated to occur upon provocation with various stimuli. In such cases, MCS-provocative stimuli may also be provocative stimuli for VCD. It has been suggested that psychogenic factors may be important in the genesis of VCD.[89]

Table 4 **MCS: Typical Symptoms**

1. Inability to concentrate
2. Feeling "spaced-out"
3. Lack of energy
4. Easy fatiguability
5. Heightened perception of odors
6. Mental confusion
7. Depression
8. Gastrointestinal symptoms
9. Eye, ear, nose, throat irritation
10. Respiratory difficulty
11. Cardiovascular irregularities, palpitations
12. Skin disorders
13. Genitourinary problems
14. Muscle, bone, and joint pain
15. Rhinitis
16. Headaches
17. Irritability
18. Insomnia
19. Chest pain
20. Food and chemical intolerance

Table 5 **Substances Claimed to Trigger MCS**

1. Volatile organic compounds (VOCs)
2. Environmental tobacco smoke (ETS)
3. Formaldehyde
4. Perfumes and colognes
5. Deodorants
6. Vehicle exhaust
7. Diesel fumes
8. Cleaning products
9. Detergents
10. Pesticides
11. Petroleum products and byproducts
12. *Candida albicans*
13. Other biological products and agents

The concept of MCS was introduced by a group of alternative medicine practitioners known as clinical ecologists, who suggested that MCS was a distinct disease entity. A number of other names have been used for MCS, primarily by clinical ecologists, and include environmental illness, chemical hypersensitivity syndrome, cerebral allergies, chemically induced immune dysregulation, 20th century disease, total allergy syndrome, ecologic illness, and chemical AIDS.[1,3,4,9,16,68,69,74,75,85] As noted above, evaluations of individuals with MCS complaints have not revealed consistent pathologic lesions or specific laboratory abnormalities on conventional tests, and there is no specific diagnostic test for this condition, nor, for that matter, even a clear definition of the condition.[16,28]

The clinical ecologists have put forth a number of theories that purport to explain the basis and mechanisms of MCS. These theories are based on the premise that synthetic chemicals and other environmental substances can produce adverse effects at levels of exposure far below what can be explained by accepted mechanisms of

Table 6  **MCS: Clinical Findings**

1. No relevant findings on physical examination
2. No relevant laboratory abnormalities
3. No demonstrable pathologic findings
4. Past history of atopic allergy common
5. Past history of emotional stressors common

disease, including toxicologic and immunologic mechanisms. It has been further suggested that these adverse effects include damage to and "dysregulation" of the immune system, so that the sufferer becomes "hypersensitive" to many common substances at extremely low levels, this "hypersensitivity" involving mechanisms other than conventional allergic (immunologic) reactions. It has been proposed that this "hypersensitivity" causes MCS sufferers to react to levels of substances in the environment that are innocuous to normal people.[68,69,72,74,75,85,90] This etiologic and pathogenetic construct, which assumes mechanisms and concepts that are inconsistent with known biological mechanisms and, thus, are biologically implausible, also remains to be validated by studies using acceptable scientific methodology.[1-5,16]

Clinical ecologists have also proposed certain diagnostic tests for MCS, including so-called sublingual or intradermal provocation-neutralization tests. As a matter of fact, such tests have been proposed for both diagnostic and therapeutic purposes. While there have been a number of uncontrolled studies using such techniques, which have resulted in a large body of anecdotal information, to date there are no well-controlled studies that demonstrate that such techniques have either diagnostic or therapeutic value.[1-6,10,13,16,18,20,21,23,68,69,74,75,91]

A subset of MCS is the so-called Candida hypersensitivity syndrome. In this alleged entity, it is proposed that patients develop MCS as a consequence of the overgrowth of and systemic infection with *Candida albicans*, which is said to produce a toxin that impairs immune function in susceptible individuals.[85] The validity of this proposed mechanism has never been supported by other than anecdotal reports, and there are very serious questions about its plausibility. In fact, recent studies cast considerable doubt on this proposed mechanism.[1,7,16,28,92]

It is not surprising that, when critically reviewed by a number of scientific groups, considerable doubt and skepticism have been raised regarding the concept of MCS as a valid diagnostic condition or disease entity. The concept of MCS as a disease entity is made problematic by the number, variety, and nonspecificity of symptoms, the lack of objective physical signs, laboratory abnormalities, or demonstrable pathologic findings, and the overlapping of many of the nonspecific symptoms with those of recognized disorders, including psychiatric disorders that have been demonstrated to preexist in a number of the patients. Other problems include the absence of credible scientific support for the proposed immune imbalance mechanism or any of the other postulated mechanisms, the absence of data from well-controlled, scientifically acceptable clinical studies to support the proposed diagnostic and therapeutic techniques, and the highly questionable biologic plausibility of the pathogenetic concepts that have been put forth.[1-6,10,13,16,18,20-23,28,70] While a number of articles and books have been written in support of the concept of MCS, these have been found to be primarily testimonial and lacking in scientifically acceptable data.[1-7,68,69,72,74,75,85,90]

On the other hand, a number of investigators have reported their evaluations of series of patients claiming to have MCS in the peer-reviewed scientific medical literature.[18,21,22,70,88,93] As a group, the results of these studies permit some generalizations regarding some characteristics of patients alleged to have MCS. Many of these patients have been seen in the context of litigation or compensation proceedings. Women predominate. Most of the patients are generally well-educated, and professional credentials in science, law, or a health- or environment-related field are not unusual, although a diversity of occupations are represented. Many of these individuals attribute the onset of their MCS to a work-related exposure, a not unexpected finding in view of the fact that many of them are plaintiffs in legal proceedings, where they are attempting to obtain compensation for alleged work-related injury. Terr, in a study of 90 patients with alleged work-related MCS, reported that examination of prior medical records revealed that two thirds of these individuals had been treated for the same symptoms for many years before their reported occupational exposure.[22] Typically, these patients report heterogeneous histories with varied, nonspecific symptoms that involve many organ systems. As noted earlier, there are no consistent abnormal physical findings, laboratory abnormalities, or pathologic findings. Both the initiating exposure and subsequent exacerbating exposures are attributed to a quite varied and large number of environmentally encountered chemicals, as also noted earlier.

It is typical for such patients to consult many physicians over a period of years, without obtaining a satisfactory (to them) explanation of their symptoms. When they ultimately come under the care of a clinical ecologist who makes a diagnosis of MCS, they often find this satisfactory, even though there is little in the way of improvement of symptoms once the diagnosis is made. When seen by physicians other than clinical ecologists, psychiatric diagnoses, including somatoform disorders, anxiety disorders, and depression, are the most common alternatives to MCS that are diagnosed.[1,3,21,22,28,70,93–98] In fact, it is the relatively high prevalence of underlying and associated psychiatric illness in such patients, together with a lack of any credible evidence supporting a role for chemical sensitivity or immunologic dysfunction, that has raised the question of a causal role for psychogenic mechanisms in this condition. This suggestion appears to be supported by a number of published investigations addressing this area. Black et al. reported a significantly higher prevalence of mood disorders, anxiety disorders, and somatoform disorders in their series of MCS patients, compared to a group of "normal" controls.[88] Simon and co-workers reported their evaluation of 37 aerospace workers who developed mucosal irritation related to exposure to phenol-formaldehyde in their workplace, which resolved upon cessation of exposure. However, 13 of the workers reported subsequent development of symptoms that were diagnosed as MCS by a clinical ecologist. When these 13 individuals were compared with the 24 co-workers who only had transient symptoms, the MCS patients were found to have a significantly higher prevalence of pre-existing psychiatric disorders, including anxiety, depression, and somatization, as well as a higher prevalence of medically unexplained symptoms than the 24 "controls."[97]

Thus, the findings reported in the peer-reviewed, scientific medical literature support the proposition that psychogenic factors play a significant if not pivotal role in the genesis of symptoms among many individuals said to have MCS. Some investigators have reported findings that, in at least some of these individuals, suggest that psychogenic mechanisms involving conditioning related to perceived odors may play a role. That is, an initial experience involving an irritating concentration of a chemical associated with an odor results in conditioning of the individual to the odor, so that on subsequent occasions, the person experiences discomfort associated with perception of the odor, even when the concentrations of the chemical are well below levels that could plausibly cause the symptoms. This is said to be then followed by generalization of this reaction to odorants other than the initial chemical, a phenomenon referred to as "behavioral sensitization to odorants with stimulus generalization."

While an interesting possibility and consistent with a number of clinical observations, such a mechanism remains to be confirmed by controlled scientific evaluation.[81-83,99]

It has also been suggested that individuals who develop MCS have a preexisting, underlying personality or emotional disorder that makes them more likely to accept an unsubstantiated environmental explanation for their symptoms. It has been pointed out that even though clinical ecology methods of treatment are almost always ineffective, patients said to have MCS choose to undertake the often extreme lifestyle changes recommended as part of such treatment, rather than to accept a psychiatric diagnosis or psychiatric treatment.[28] It has also been proposed that, in at least some MCS patients, the condition has features that are more characteristic of a belief system or a subculture than a psychological or physical disease.[100,101]

It is apparent that many of the symptoms reported in patients with SBS, as noted earlier, such as chest tightness, headache, cough, confusion, eye and mucous membrane irritation, etc., are similar if not identical to many symptoms reported by individuals said to have MCS. In addition to having such nonspecific symptoms in common, the two entities also have in common the absence of an identified specific causative agent, lack of evidence of a clear, biologically plausible etiologic and pathogenetic mechanism, lack of agreement over the definition and inclusion and exclusion criteria, and a plausible role of psychogenic factors and mechanisms. An important difference would appear to be the fact that in SBS, the symptoms relate to a specific building, dissipate rapidly upon leaving that site, and do not recur except upon reentering that building, while MCS involves episodes that recur over a period of months or years in a variety of different settings. It should be noted, however, that this difference is somewhat blurred by the fact that MCS has also been alleged to be a sequela or consequence of SBS.[87,90]

## CHRONIC FATIGUE SYNDROME

Another alleged syndrome that shares many features with MCS is the so-called chronic fatigue syndrome (CFS). Patients said to have CFS complain of severe weakness and exhaustion over many months or even years, which may be associated with difficulty concentrating, memory loss, confusion, depression, insomnia, flu-like symptoms, sore throat, headache, fever, and muscle and/or joint pain (Table 7). In most patients with CFS, there are no characteristic physical findings, no diagnostic laboratory tests, and no demonstrable pathological findings, a constellation of negative findings reminiscent of that seen in patients with MCS (Table 8).[1,102-106]

It had been suggested that patients with CFS had a chronic infection with Epstein-Barr (EB) virus, but careful clinical studies, including prospective clinical and virologic studies, as well as a placebo-controlled trial of acyclovir (which inhibits EB virus replication both *in vitro* and *in vivo*), have failed to confirm such a relationship.[1,104,105,107-111] One recent study reported data suggesting a possible association with human T-lymphotrophic virus type II (HTLV-II), but this remains to be confirmed by other investigators.[112] Other viral infections have been proposed, as well as immunologic mechanisms, but these remain to be confirmed by appropriate further investigation.[1,104,105,113] A high prevalence of atopic allergy has been reported, representing another point of similarity with MCS (Tables 6 and 8).[114] It is notable that a majority of the patients with CFS in most series appear to have a psychiatric disorder, yet another similarity with MCS (Tables 6 and 8).[1,102,106,115] Whether CFS represents a variant of MCS, a single disorder, or a heterogeneous group of patients whose symptoms may reflect a multiplicity of somatic and/or psychogenic causes continues to be a matter of current controversy.[1,104,111]

Table 7  **CFS: Typical Complaints**

1. Severe weakness and exhaustion
2. Difficulty concentrating
3. Memory loss
4. Confusion
5. Depression
6. Insomnia
7. Flu-like symptoms
8. Sore throat
9. Headaches
10. Fever
11. Muscle and joint pain

Table 8  **CFS: Clinical Findings**

1. No characteristic physical findings
2. No definitive laboratory abnormalities
   a. No evidence of any relationship with EB virus infection
3. No demonstrable pathological findings
4. High prevalence of atopic allergy
5. Psychiatric disorders common

# REFERENCES

1. **Council on Scientific Affairs, American Medical Association,** Council report: clinical ecology, *JAMA,* 268, 3465, 1992.
2. **Committee on Occupational Medicine, American College of Occupational Medicine,** Environmental illness, *J. Occup. Med.,* 32, 211, 1990.
3. **American College of Physicians,** American College of Physicians position statement: clinical ecology, *Ann. Intern. Med.* 111, 168, 1989.
4. **American Academy of Allergy and Immunology,** American Academy of Allergy and Immunology position statement: clinical ecology, *Allergy Clin. Immunol.,* 78, 269, 1986.
5. **Task Force on Clinical Ecology, California Medical Association Scientific Board,** Clinical ecology: a clinical appraisal, *W. J. Med.,* 144, 239, 1986.
6. **Council on Scientific Affairs, American Medical Association,** In vitro diagnostic testing and immune therapy for allergy: report I, part II of the Allergy Panel, *JAMA,* 258, 1505, 1987.
7. **American Academy of Allergy and Immunology,** American Academy of Allergy and Immunology position statement: candidiasis hypersensitivity syndrome, *J. Allergy Clin. Immunol.,* 78, 271, 1986.
8. **Bardana, E. J., Jr.,** Building-related illness, in *Occupational Asthma,* Bardana, E. J., Jr., Montanaro, A., and O'Hollaren, M. T., Eds., Hanley and Belfus, Philadelphia, 1991, 1.
9. **Cullen, M. R.,** The worker with multiple chemical sensitivities: an overview, *Occup. Med.,* 2, 655, 1989.
10. **Terr, A. I.,** Multiple chemical sensitivities immunologic critique: clinical ecology theories and practice, *Occup. Med.,* 2, 683, 1987.
11. **Davies, J. W. and Wilkins, K.,** Eds., Proceedings of the environmental sensitivities workshop, *Chronic Dis. Can.,* January, 1991.

12.  **Committee on Advances in Assessing Human Exposure to Airborne Pollut-
     ants, Board on Environmental Studies and Toxicology, Commission on Geo-
     sciences, Environment, and Resources, National Research Council,** Human
     Exposure Assessment for Airborne Pollutants: Advances and Opportunities, Na-
     tional Academy of Sciences, Washington, D.C., 1991.

13.  **American Academy of Allergy,** American Academy of Allergy position state-
     ment: controversial techniques, *J. Allergy. Clin. Immunol.,* 67, 333, 1981.

14.  **Bascom, R.,** Chemical Hypersensitivity Syndrome Study: Options for Actions, A
     Literature Review and Needs Assessment, State of Maryland: Department of the
     Environment, 1989.

15.  **Committee on Environmental Hypersensitivities,** Report of the Ad Hoc Com-
     mittee on Environmental Hypersensitivity Disorders, Ministry of Health, Toronto,
     Ontario, 1985.

16.  **Kahn, E. and Letz, G.,** Clinical ecology: environmental medicine or unsubstan-
     tiated theory, *Ann. Intern. Med.,* 111, 104, 1989.

17.  **Board on Environmental Studies and Human Toxicology, Commission on Life
     Sciences, National Research Council,** Multiple Chemical Sensitivities: Ad-
     dendum to Biologic Markers in Immunotoxicology, National Academy Press,
     Washington, D.C., 1992.

18.  **Patterson, R., Harris, K. E., Stopford, W., Vander Heiden, G., Grammer,
     L. C., and Bunn, W.,** Irritant symptoms and immunologic responses to multiple
     chemicals: importance of clinical and immunologic correlations, *Int. Arch. Allergy
     Appl. Immunol.,* 85, 467, 1988.

19.  **Ryan, S. M. and Morrow, L. A.,** Dysfunctional buildings or dysfunctional
     people: an examination of the sick-building syndrome and allied disorders, *J. Con-
     sult. Clin. Psychol.,* 60, 220, 1992.

20.  **Salvaggio, J. E.,** Clinical and immunologic approach to patients with alleged en-
     vironmental injury, *Ann. Allergy,* 66, 493, 1991.

21.  **Staudenmayer, H. and Selner, J. C.,** Neuropsychophysiology during relaxation
     in generalized universal "allergic" reactivity to the environment: a comparison
     study, *J. Psychosom. Res.,* 34, 259, 1990.

22.  **Terr, A. I.,** Clinical ecology in the workplace, *J. Occup. Med.,* 31, 257, 1989.

23.  **VanArsdel, B. P. and Larson, E. B.,** Diagnostic tests for patients with suspected
     allergic disease: utility and limitations, *Ann. Intern. Med.,* 110, 304, 1989.

24.  **Cone, J. E. and Hodgson, M. J.,** Preface, in Problem buildings: building-asso-
     ciated illness and the sick building syndrome, *Occup. Med.,* 4, x, 1989.

25.  **Marbury, M. C. and Woods, J. E., Jr.,** Building-related illnesses, in *Indoor Air
     Pollution,* Samet, J. M. and Spengler, J. D., Eds., The Johns Hopkins University
     Press, Baltimore, MD, 1991, 306.

26.  **National Research Council, Committee on Indoor Air Quality,** Policies and
     Procedures for Control of Indoor Air Quality, National Academy Press, Wash-
     ington, D.C., 1987.

27.  **Hodgson, M. J.,** Clinical diagnosis in management of building-related illness and
     the sick-building syndrome, *Occup. Med.,* 4, 593, 1989.

28.  **Terr, A. I.,** Multiple chemical sensitivities syndrome, *Immunology and Allergy
     Clinics of North America,* 12, 897, 1992.

29.  **Kreiss, K.,** The epidemiology of building-related complaints and illness, *Occup.
     Med.,* 4, 575, 1989.

30.  **Brundage, J. F., Scott, R. N., Leadnar, W. M., et al.,** Building-associated risk
     of febrile acute respiratory disease in army trainees, *JAMA,* 259, 2108, 1988.

31.  **Burge, H. A.,** Indoor air and infectious disease, *Occup. Med.,* 4, 713, 1989.

32.  **Fink, J. N.,** Hypersensitivity pneumonitis, *Clin. Chest Med.,* 13, 303, 1992.

33. **Kreiss, K. and Hodgen, M. J.,** Building-associated epidemics, in *Indoor Air Quality,* Walsh, P. J., Dudney, C. S., and Copenhaper, E. D., Eds., CRC Press, Boca Raton, FL, 1984, 87.

34. **Schneider, T.,** Man-made mineral fibers and other fibers in the air and in settled dust, *Environ. Int.,* 12, 61, 1986.

35. **Kreiss, K., Gonzalez, M. G., Conright, K. L., and Scheere, A. R.,** Respiratory irritation due to carpet shampoo: two outbreaks, *Environ. Int.,* 8, 337, 1982.

36. **Flachsbart, P. G. and Ott, W. R.,** A rapid method for surveying CO concentrations in high-rise buildings, *Environ. Int.,* 12, 255, 1986.

37. **LaMarte, F. P., Merchant, J. A., and Casale, T. B.,** Acute systemic reactions to carbonless carbon paper associated with histamine release, *JAMA,* 260, 242, 1988.

38. **Bardana, E. J., Jr., Montanaro, A., and O'Hollaren, M. T.,** Building-related illness, *Clin. Rev. Allergy,* 6, 61, 1988.

39. **Berglund, B., Bjorksten, B., Erikkson, B., Lindvall, T., Moghissi, A. A., Nygren, A., Sevenstrand, I., Snaar, J., Stolwijk, J. A. J., and Sundell, J.,** Stockholm statement on building-related allergic and other adverse effects on human health and comfort, *Environ. Int.,* 18, 115, 1992.

40. **Donnell, H. D., Bagby, J. R., Harmon, R. G., Crellin, J. R., Chasiki, C., Gright, M. F., van Tuinen, M., and Metzger, R. W.,** Report of an illness outbreak at the Harry S. Truman state office building, *Am. J. Epidemiol.,* 129, 550, 1989.

41. **Molhave, L.,** The sick buildings: a subpopulation among the problem buildings, in *Indoor Air '87. Proceedings of the Fourth International Conference on Indoor Air Quality and Climate,* Vol. 2, Seifert, B., Esdorn, H., Fischer, M., et al., Eds., Institute for Water, Soil and Air Hygiene, Berlin, 1987, 469.

42. **Burge, S. et al.,** Sick-building syndrome: a study of 4373 office workers, *Ann. Occup. Hyg.,* 31, 493, 1987.

43. **Waller, R. A., Atkins, W. S., and Partners, U. K.,** Case study of a sick building, in *Proceedings of the Third International Conference on Indoor Air Quality and Climate,* Vol. 3, Bergund, B., Linvall, T., and Sundell, J., Eds., Swedish Council for Building Research, Stockholm, 1984, 349.

44. **Hodgson, M. J., Frohliger, J., Permar, E., et al.,** Symptoms and micro-environmental measures in non-problem buildings, *J. Occup. Med.,* 33, 527, 1991.

45. **Robertson, A. S., Burge, P. S., Hedge, A., Simms, J., Gill, F. S., Finnegan, M., Pickering, C. A. C., and Dolton, G.,** Comparison of health problems related to work and environmental measurements in two office buildings with different ventilation systems, *Br. Med. J.,* 291, 373, 1985.

46. **Skov, P., Valbjen, O., and Pedersen, P. B.,** Influence of personal characteristics, job-related factors and psychosocial factors on the sick-building syndrome, *Scand. J. Work Environ. Health,* 15, 286, 1989.

47. **Salisbury, S. A.,** A typically frustrating building investigation, *Ann. ACGIH,* 10, 129, 1984.

48. **Finnegan, M. J., Pickering, C. A., and Burge, P. S.,** The sick-building syndrome: prevalence studies, *Br. Med. J.,* 289, 1573, 1984.

49. **Goldstein, B. V.,** Predicting the risk of indoor air pollutants, *Toxicol. Ind. Health,* 7, 195, 1991.

50. **Skove, P. and Valbjorn, O.,** The sick-building syndrome in the office environment: the Danish townhall study, in *Indoor Air '87. Proceedings of the Fourth International Conference on Indoor Air Quality and Climate,* Vol. 2, Seifert, B., Esdorn, H., Fischer, M., et al., Eds., Institute for Water, Soil and Air Hygiene, Berlin, 1987, 239.

51. **Valbjorn, O. and Skove, P.,** Influence of indoor climate on the sick-building syndrome prevalence, in *Indoor Air '87. Proceedings of the Fourth International Conference on Indoor Air Quality and Climate,* Vol. 2. Seifert, B., Esdorn, H., Fischer, M., et al., Eds., Institute for Water, Soil and Air Hygiene, Berlin, 1987, 593.

52. **Harrison, J., et al.,** The sick-building syndrome: further prevalence studies and investigations of possible causes, in *Indoor Air '87. Proceedings of the Fourth International Conference on Indoor Air Quality and Climate,* Vol. 2, Seifert, B., Esdorn, H., Fischer, M., et al., Eds., Institute for Water, Soil and Air Hygiene, Berlin, 1987, 487.

53. **Menzies, R., Tamblyn, R., Farant, J., Hanley, J., Nunes, F., and Tamblyn, R.,** The effect of varying levels of outdoor-air supply on the symptoms of sick building syndrome, *N. Engl. J. Med.,* 328, 821, 1993.

54. **Hawkins, L. H. and Morris, L.,** Air ions and the sick-building syndrome, in *Proceedings of the Third International Conference on Indoor Air Quality and Climate,* Vol. 3, Bergund, B., Linvall, T., and Sundell, J., Eds., Swedish Council for Building Research, Stockholm, 1984, 197.

55. **Finnegan, M. J., et al.,** Effect of negative ion generators in a sick building, *Br. Med. J.,* 295, 1195, 1987.

56. **Morey, P. R.,** Case presentations: problems caused by moisture spaces of office buildings, *Ann. ACGIH,* 10, 121, 1984.

57. **Molhave, L., Bach, B., and Petersen, O. F.,** Human reactions during controlled exposures to low concentrations of organic gases and vapors known as normal indoor air pollutants, in *Proceedings of the Third International Conference on Indoor Air Quality and Climate,* Vol. 3, Bergund, B., Linvall, T., and Sundell, J., Eds., Swedish Council for Building Research, Stockholm, 1984, 431.

58. **Bach, B., Molhave, L., and Petersen, O. F.,** Human reactions during controlled exposures to low concentrations of organic gases and vapors known as normal indoor air pollutants: performance tests, in *Proceedings of the Third International Conference on Indoor Air Quality and Climate,* Vol. 3, Bergund, B., Linvall, T., and Sundell, J., Eds., Swedish Council for Building Research, Stockholm, 1984, 397.

59. **Kjaergaard, S., Molhave, L., and Petersen, O. F.,** Human reactions to indoor air pollution: n-becane, in *Indoor Air '87. Proceedings of the Fourth International Conference on Indoor Air Quality and Climate,* Vol. 2, Seifert, B., Esdorn, H., Fischer, M., et al., Eds., Institute for Water, Soil and Air Hygiene, Berlin, 1987, 97.

60. **Arundel, A. V., et al.,** Indirect health effects of relative humidity in indoor environment, *Environ. Health Perspect.,* 65, 351, 1986.

61. **Breysse, P. A.,** The office environment: how dangerous?, in *Proceedings of the Third International Conference on Indoor Air Quality and Climate,* Vol. 3, Bergund, B., Linvall, T., and Sundell, J., Eds., Swedish Council for Building Research, Stockholm, 1984, 315.

62. **Bauer, R. M., Greve, K. W., Besch, E. L., Schramke, C. J., Crouch, J., Hicks, A., Ware, M R., and Lyles, W. B.,** The role of psychological factors in the report of building-related symptoms in sick-building syndrome, *J. Consult. Clin. Psychol.,* 60, 213, 1992.

63. **Boxer, P. A.,** Indoor air quality: a psychosocial perspective, *J. Occup. Med.,* 32, 425, 1990.

64. **Boxer, P. A., Singal, M., and Hartle, R. W.,** An epidemic of psychogenic illness in an electronics plant, *J. Occup. Med.,* 26, 381, 1984.

65. **Colligan, M. J.,** The psychological effects of indoor air pollution, *Bull. N.Y. Acad. Med.,* 57, 1014, 1981.

66. **Evans, V. G. W., Colone, S. D., and Shearer, D. F.,** Psychological reactions to air pollution, *Environ. Res.,* 45, 1, 1988.

67. **Morris, L. and Hawkins, L.,** The role of stress in the sick-building syndrome, in *Indoor Air '87. Proceedings of the Fourth International Conference on Indoor Air Quality and Climate,* Vol. 2, Seifert, B., Esdorn, H., Fischer, M., et al., Eds., Institute for Water, Soil and Air Hygiene, Berlin, 1987, 566.

68. **Ashford, N. A. and Miller, C. S.,** *Chemical Exposure: Low Levels and High Stakes,* Van Nostrand Reinhold, New York, 1991.

69. **Bell, I. R.,** Clinical ecology, in *A New Medical Approach to Environmental Illness,* Common Knowledge Press, Bolinas, CA, 1982.

70. **Terr, A. I.,** Environmental illness: a clinical review of 50 cases, *Arch. Intern. Med.,* 146, 145, 1986.

71. **Cullen, M. R., Pace, P. E., and Redlich, C. A.,** The experience of the Yale occupational and environmental medicine clinics with multiple chemical sensitivities, 1986–1991, *Toxicol. Ind. Health,* 8, 15, 1992.

72. **Levin, A. S. and Byers, V. S.,** Environmental illness: a disorder of immune regulation, *Occup. Med.,* 2, 669, 1987.

73. **Ross, G. H.,** History and clinical presentation of the chemically sensitive patient, *Toxicol. Ind. Health,* 8, 21, 1992.

74. **Rea, N. J., Bell, J. R., Suits, C. W., and Smiliey, R. E.,** Food and chemical susceptibility after environmental chemical overexposure: case history, *Ann Allergy,* 41, 101, 1978.

75. **Randolph, T. G. and Moss, R. W.,** *An Alternative Approach to Allergies: The New Field of Clinical Ecology Unravels the Environmental Causes of Mental and Physical Illness,* Lippincott & Cromwell, New York, 1980.

76. **Kilburn, K. H.,** How should we think about chemically reactive patients?, *Arch. Environ. Health,* 48, 4, 1993.

77. **Fiedler, N., Maccia, C., and Kipen, H.,** Evaluation of chemically sensitive patients, *J. Occup. Med.,* 34, 529, 1992.

78. **Cone, J. E. and Shusterman, D.,** Health effects of indoor odorants, *Environ. Health Perspect.,* 85, 53, 1991.

79. **Doty, R. L., Deems, D. A., Frye, R. E., and Shapiro, A.,** Olfactory sensitivity, nasal resistance, and autonomic function in patients with multiple chemical sensitivities, *Arch. Otolaryngol. Head Neck Surg.,* 114, 1422, 1988.

80. **Schiffman, S. S. and Nagle, H. T.,** Effect of environmental pollutants on taste and smell, *Otolaryngol. Head Neck Surg.,* 106, 693, 1992.

81. **Shusterman, D.,** Critical review: the health significance of environmental odor pollution, *Arch. Environ. Health,* 47, 76, 1992.

82. **Shusterman, D., Balmes, J., and Cone, J.,** Behavioral sensitization to irritants/odorants after acute over-exposures, *J. Occup. Med.,* 30, 565, 1988.

83. **Schusterman, D., Lipscomb, J., Neurta, R., and Satin, K.,** Symptom prevalence and odor-worry interaction near hazardous waste sites, *Environ. Health Perspect.,* 94, 25, 1991.

84. **Winneke, G. and Kastka, J.,** Comparison of odor-annoyance data from different industrial sources: problems and implications, *Dev. Toxicol. Environ. Sci.,* 15, 129, 1987.

85. **Crook, W. G.,** *The Yeast Connection,* 3rd ed., Professional Books, Jackson, TN, 1989.

86. **Simon, G. E.,** Epidemic multiple chemical sensitivity in an industrial setting, *Toxicol. Ind. Health,* 8, 41, 1992.

87. **Welch, L. S. and Sokas, R.,** Development of multiple chemical sensitivity after an outbreak of sick-building syndrome, *Toxicol. Ind. Health,* 8, 47, 1992.

88. **Black, D. W., Rathe, A., and Goldstein, R. B.,** Environmental illness: a controlled study of 26 subjects with "20th century disease," *JAMA,* 264, 3166, 1990.

89. **Selner, J. C., Staudenmayer, H., Koepke, J. W., Harvey, R., and Christopher, K.,** Vocal cord dysfunction: the importance of psychologic factors and provocation challenge testing, *J. Allergy Clin. Immunol.,* 79, 726, 1987.

90. **Rogers, S. A.,** Diagnosing the tight building syndrome or diagnosing chemical hypersensitivity, *Environ. Int.,* 15, 75, 1989.

91. **Jewitt, D. L., Fine, G., and Greenburg, M. H.,** A double-blind study of symptom provocation to determine food sensitivities, *N. Engl. J. Med.,* 323, 429, 1989.

92. **Dismukes, W. E., Wade, J. S., Lee, J. Y., Dockery, B. K., and Hain, J. D.,** A randomized double-blind trial of nystatin therapy for the candidiasis hypersensitivity syndrome, *N. Engl. J. Med.,* 323, 1717, 1990.

93. **Stewart, D. E. and Raskin, J.,** Psychiatric assessment of patients with "20th century disease" ("total allergy syndrome"), *Can. Med. Assoc. J.,* 133, 1001, 1985.

94. **Brodsky, C. M.,** Psychological factors contributing to somatoform diseases attributed to the workplace: the case of intoxication, *J. Occup. Med.,* 25, 459, 1983.

95. **Rosenberg, S. J., Freedman, M. R., Schmailing, K. B., and Rose, C.,** Personality styles of patients asserting environmental illness, *J. Occup. Med.,* 32, 678, 1990.

96. **Schottenfeld, R. S.,** Workers with multiple chemical sensitivities: a psychiatric approach to diagnosis and treatment, *Occup. Med.,* 2, 739, 1987.

97. **Simon, G. E., Katon, W. J., and Sparks, P. J.,** Allergic to life: psychological factors in environmental illness, *Am. J. Psychiat.,* 147, 901, 1990.

98. **Simon, G. E., Daniell, W., Stockbridge, H., Claypoole, K., and Rosenstock, L.,** Immunologic, psychological, and neuropsychological factors in multiple chemical sensitivity: a controlled study, *Ann. Intern. Med.,* 119, 97, 1993.

99. **Bolla-Wilson, K., Wilson, R. J., and Bleecker, M. L.,** Conditioning of physical symptoms after neurotoxic exposure, *J. Occup. Med.,* 30, 684, 1988.

100. **Brodsky, C. M.,** Multiple chemical sensitivity and other "environmental illness:" a psychiatrist's view, *Occup. Med.,* 2, 695, 1987.

101. **Brodsky, C. M.,** Allergic to everything: a medical subculture, *Psychosomatics,* 24, 731, 1983.

102. **Kroenke, K.,** Chronic fatigue syndrome: is it real?, *Postgrad. Med.,* 89, 44, 1991.

103. **Holmes, G. P., Kaplin, J. E., Gantz, N. M., et al.,** Chronic fatigue syndrome: a working case definition, *Ann. Intern. Med.,* 108, 387, 1988.

104. **Swartz, M. N.,** The chronic fatigue syndrome: one entity or many?, *N. Engl. J. Med.,* 319, 1726, 1988.

105. **Gold, D., Bowden, R., Sixbey, J., et al.,** Chronic fatigue: a prospective clinical and virologic study, *JAMA,* 264, 48, 1990.

106. **Shafran, S. V.,** The chronic fatigue syndrome, *Am. J. Med.,* 90, 730, 1991.

107. **Hellinger, W. C., Smith, T. F., Van Scoy, R. E., Spitzer, P. G., Forgacs, P., and Edson, R. S.,** Chronic fatigue syndrome and the diagnostic utility of antibody to Epstein-Barr virus early antigen, *JAMA,* 260, 971, 1988.

108. **Straus, S. E., Dale, J. K., Tobi, M., Lawley, T., Preble, O., Blaese, R. M., Hallahan, C., and Henle, W.,** Acyclovir treatment of the chronic fatigue syndrome: lack of efficacy in a placebo-controlled trial, *N. Engl. J. Med.,* 319, 1692, 1988.

109. **Holmes, G. P., Kaplan, J. E., Stewart, J. A., Hunt, B., Pinsky, P. F., and Schonberger, L. B.,** A cluster of patients with a chronic mononucleosis-like syndrome: is Epstein-Barr virus the cause?, *JAMA,* 257, 2297, 1987.

110. **Buchwald, D., Sullivan, J. L., and Komaroff, A. L.,** Frequency of "chronic active Epstein-Barr virus infection" in a general medical practice, *JAMA,* 257, 2303, 1987.

111. **Straus, S. E.,** EB or not EB—that is the question, *JAMA,* 257, 2335, 1987.

112. **DeFreitas, E., Hilliard, B., Cheney, P. R., et al.,** Retroviral sequences relative to human T-lymphotrophic virus type II in patients with chronic fatigue immune dysfunction syndrome, *Proc. Natl. Acad. Sci. U.S.A.,* 88, 2922, 1991.

113. **Landay, A. L., Jessop, C., Lennette, E. T., and Levy, J. A.,** Chronic fatigue syndrome: clinical condition associated with immune activation, *Lancet,* 338, 707, 1991.

114. **Straus, S. E., Dale, J. K., Wright, R., and Metcalfe, D. D.,** Allergy and the chronic fatigue syndrome, *J. Allergy Clin. Immunol.,* 81, 791, 1988.

115. **Krusei, M. J. P., Dale, J., and Straus, S. E.,** Psychiatric diagnoses in patients who have chronic fatigue syndrome, *J. Clin. Psychiatr.,* 50, 53, 1989.

# INDEX

## A

Absorption, 42
Accumulation, 21
Acebutolol, 218
Acrolein, 31
  as carcinogen, 41
  functional changes caused by, 35
  in tobacco smoke, 12
Acrylonitrile, 41, 71
Actinomycetes, 108
Activated charcoal, 221
Activated oxygen species, see Oxygen radicals
Adaptation, 21
Addictive responses, 21
Additivity, definition, 42
Adduct formation, 137, see also DNA adducts
Adenocarcinomas, 50, 51, 62
Adenopathy, in beryllium disease, 258
Adrenal function studies, 159
Adult respiratory distress syndrome, 222
Aerodynamic diameter, 28, 42
Aerosols, 3, 20, 209
  biological, 13–14
  carcinogens, 40
  definition, 42
  inhalation injury, 218–223
Airborne substances, classification of, 20–21
Air conditioners, 208, 209
Air flow resistance, see Pulmonary function tests
Air spaces, definition, 42
Air volume flow rate, 5
Airway conductance, definition, 42
Airway resistance, definition, 42
Airways, definition, 42
Albumin, 33
Aldehydes, 35
Algorithms, causality, 141–142, 145–151
Alkylating agents, 67, 72
Allergen, definition, 42
Allergic alveolitis, see Hypersensitivity
      pneumonitis
Allergic rhinitis, 37, 196
Allergic sensitization, 32
Allergist, 164
Allergy, immune mechanisms, see
      Immunological mechanisms
Alumina, 251

Aluminum/aluminosis, 208
  and bronchiolitis, 221, 223
  bronchoalveolar lavage findings, 33
  diagnosis of occupational lung disease, 256
  metal fume fever, 266
  radiographic findings, 211
Aluminum chloride, 264
Aluminum oxide, 221, 223
Aluminum pneumoconiosis, 264–267
  cellular patterns, bronchoalveolar lavage, 212
  histopathology, 265
  management and complications, 267
  pulmonary function tests, 267
  radiographic findings, 267
  signs and symptoms, 266–267
  work environment, 264–265
Aluminum potroom workers, 212
Aluminum silicates, 208
Alveolar-arterial oxygen tension, see also Blood
      gases
  abnormal pattern in cardiopulmonary exercise
      testing, 182–183
  impairment evaluation, 178
Alveolar duct fibrosis, 99
Alveolar ducts, 21–22
Alveolar epithelial carcinomas, 41
Alveolar macrophage, see also Macrophages
  bronchoalveolar lavage findings, 33
  definition, 42
Alveolitis, see Hypersensitivity pneumonitis
Ambient air, 4, 7–9
  air quality standards, 1
  meteorological considerations, 3
  modeling, 6
  pollutant properties, sources, effects, 7–9
American Cancer Society, 100
American Society of Heating, Refrigeration,
      and Air-Conditioning Engineers, 4
American Thoracic Society impairment
      definitions, 168, 169
Ames assay, 79, 80
Amine precursor uptake and decarboxylation
      (APUD) cells, 50
4–Aminobiphenyl (4ABP), 64
Amiodarone, 218
Ammonia, 2, 31, 32, 221, 223
Amosite, 13, 113
Amphiboles, 73, 113, 232, 235